# The Cinderella.2 Manual

Jürgen Richter-Gebert • Ulrich H. Kortenkamp

# The Cinderella.2 Manual

Working with
The Interactive Geometry Software

 Springer

Jürgen Richter-Gebert
TU München
Zentrum Mathematik (M10)
LS Geometrie
Boltzmannstr. 3
85748 Garching
Germany
richter@ma.tum.de

Prof. Ulrich H. Kortenkamp
Pädagogische Hochschule
Karlsruhe
Bismarckstr. 10
76133 Karlsruhe
Germany
kortenkamp@cinderella.de

ISBN 978-3-662-50082-8     e-ISBN 978-3-540-34926-6
DOI 10.1007/978-3-540-34926-6
Springer Heidelberg Dordrecht London New York

Mathematics Subject Classification (2010): 97U50, 97U70, 97N80, 97G10, 97U60, 97U80, 00A66, 00A79, 00A20, 51P05, 51N15

Printed on acid-free paper

Springer is part of Springer Science+Business Media (www.springer.com)

# Preface to Cinderella 2.6

*The best thing about the future is that it comes only one day at a time.*

*Abraham Lincoln*

Another five years passed. Five years full of work, projects, ideas, collaboration, teaching, research,.... During this time Cinderella got involved in many different projects. We took most of them as opportunities to enhance the capabilities of Cinderella in several (sometimes unexpected) directions. Some of these features are still experimental, but many of them are included in the current release of Cinderella.2 and in this documentation.

A course on the interrelation of mathematics and music, taught in 2008 in collaboration with the Deutsches Museum in Munich, was the starting point for a variety of ways to create audio output with Cinderella. The internet portal www.mathe-vital.de, initiated in 2007, was another seed for many new developments. Among them support for TeX-like typesetting of formulas, advanced operators for linear algebra and calculus, and advanced drawing of functions. An exhibition together with the Deutsche Museum triggered the development of new ways to deal with images and their transformations. The growing desire to handle 3D objects was the reason to create a plugin structure.

There are still many half official features that are not covered in full detail or at all by this documentation. Among these are support for multitouch devices, graph algorithms, background jobs, and many more. We are quite sure that many of these features will be made officially available later as well.

In a sense between Cinderella 2.0 and the current release 2.6 the project grew up and by now is in a state in which it could be called a device for mathematical visualization that goes far beyond geometry. We hope that you will enjoy creating such in Cinderella, or just *doing maths*, as much as we and our students do.

Another main feature of version 2.6 is the documentation you are just reading. For the first time since a couple of years the documentation again covers the complete official functionality that is available by the current version of Cinderella. The printed version of the manual has almost 500 pages, and you may wonder why we need them to describe the functionality of just one program. There are many answers to this and we will here mention at two of them: The first one is the obvious one – we added so much to Cinderella. The second answer is related to you, dear reader: We tried to produce a manual that is as readable and as accessible as possible. We included about 600 screenshots, many code examples for the scripting language, step-by-step tutorials and also some technical and scientific background information. Although Cinderella is a very mighty tool, many tasks can be performed in amazingly simple manners. After mastering the basics you can browse through this manual and just pick the aspects that are most interesting for you: geometry, physics, mathematical programming, music, function plotting, fractals, etc.

While we prepared the final version of the documentation for Version 2.6 again many people were very helpful. In particular we here want to thank Elena Kohler and Stefan Kranich for their extensive help in the final stages of the manuscript, and Stephan Berndts for his software to translate our documentation Wiki into LaTeX.

A great and very special thanks also goes to Ralph Möllers, the owner of Terzio Verlag. He gave us the permission to use the beautiful pictures of *Ritter Rost* for explaining the image operations of Cinderella. The copyright of these Images is held by Terzio Verlag and we are really proud to be allowed to use them.

Cinderella is a constant work-in-progress. We are sure that there are still many features that might be a valuable completion of the software. Our documentation might at some places be misleading or even wrong, there may be several bugs or inaccuracies. We are listening: Please just visit our website at http://cinderella.de for updates and errata. If you want to report a bug you can easily do so by sending a mail to bugs@cinderella.de. And if you just want to talk to us: you can contact us at authors@cinderella.de.

*September 2011*
*Jürgen Richter-Gebert, Munich*
*Ulrich Kortenkamp, Karlsruhe*

## Preface to Cinderella 2.0

> *I may not always be perfect, but I am always me!*
>
> *Found on a T-shirt*

Seven and a half years have passed since the first release of Cinderella in 1998, and the project went through several metamorphoses that we did not foresee ourselves. Now, we feel that enough new aspects and features were added to justify a new major release: *Cinderella.2*.

In a sense programming Cinderella turned out to be a kind of "never ending story" and there is always one more feature that should be added or another one that could be improved (or, not infrequently, debugged). So, the current version may not be perfect in all aspects, but it has so many substantially new features that it would be a pity not to release it and see what people will do with it.

So, what do you have to expect from the new version? First of all, the most obvious change is that Cinderella is no longer "only" a geometry program (nevertheless the geometry part has been improved significantly). The present release consists of three major parts: *Cinderella* – the geometry engine, *CindyScript* – a functional programming language and *CindyLab* – a physics simulation engine. At first sight, these three parts can be used almost independently from each other. However, Cinderella (p. 67), CindyScript (p. 219) and CindyLab (p. 167) are designed to work hand in hand and to take as much advantage from each other as possible – if you look at our new logo, this logo symbolizes the three parts interacting with each other in various ways. Although it would be an exciting story, we will not explain here how we ended up with this final design, because this would fill too many pages. In short, it was a long process, driven by demands and requests of our users, our own desire for cool software scenarios, several conferences on scientific visualization and multimedia and last but not least several days and nights in which we were following fruitful paths (or sometimes dead ends).

In the geometry part of *Cinderella* there are many substantial improvements. Transformations and Transformation groups have been added, there are many more tools for constructing conics, it is even possible to construct fractal objects. Also, the direct construction of regular polygons has been made possible. Transformations and transformation groups turned out to be a great help to make more advanced constructions and we encourage the users to really take advantage of those concepts. One of the mostly requested features were "macros". Cinderella.2 now comes with a copy/paste/redefine concept that facilitates the re-usage of already available construction parts. It is also possible to encapsulate parts of a construction into a toolbar button that can be reused in other constructions as well.

CindyLab provides an environment in which the points of Cinderella constructions may become masses, and in which segments may act like springs or other force-generating objects. Although CindyLab only provides a particle/mass/force simulation paradigm, it is a quite powerful tool. We already had a lot of fun experimenting with solar systems with several suns, strange mechanical devices and simulated billiard tables.

Finally, CindyScript is a functional programming language that was designed with applications in geometry but also other parts of mathematics in mind. CindyScript grew out of the desire to have a kind of function plotter available in Cinderella. Well, as things sometimes happen in computer science one starts with a simple solution for a demand and observes that by similar techniques much more could be achieved. Roughly the history was like: We want to calculate formulas - why don't we use this for manipulating the position of points? - why don't we add control structures? - why don't we add high level matrix operations? - ... and list operations ... and recursion structures ... and even more powerful function plotting? This is how we finally ended up with a full-featured, mathematically oriented functional, real-time, high level programming language. We are quite sure that, so far, we do not even imagine what is possible by fully exploiting the advantages of a dynamic geometry environment in connection to a programming language. At least we are very surprised how our students who already work with CindyScript use it! We encourage all readers to build really cool and unexpected applications with this tool.

Still there are several bells and whistles that do not fall under the above three program parts. For instance Cinderella.2 supports the recognition of hand sketches that makes it possible to use Cinderella with a pen tablet an interactive whiteboard or a PDA.

There are several features that did not make their way to the final release. The decision whether a feature entered the release was mainly made by stability considerations. We will add many of these features one by one over the next few months whenever we think that they work reasonably well. Among the things that will come are: native support for geometric bases, a recording tool (*CINErella*) for geometric tutorial films, a hardware simulator, and many more. So we recommend updating frequently.

It is almost impossible to mention all the people that have been helpful in finishing Cinderella.2 by comments, user feedbacks, beta-testing, etc. Still first of all we want to give a great "thank you" and a big excuse to our families. Finishing Cinderella took a lot of our spare time and our families very often missed us as fathers and husbands. Uncountable many weekends and nights were sacrificed for finishing yet another feature or chasing yet another bug. We both hope that in the future there will be more time for all the other things that are also important in live.

We also want to thank Dirk Materlik who got involved into the team during his diploma thesis about sketch recognition and later in the Matheon Visage project, and besides his work there helped to resolve many critical design issues (the Inspector (p. 153) wouldn't be what it is without him). Also a great thank you to to Gunter Gemmel who in an overnight-hack gave us the gift of an implementation of the PSLQ algorithm (p. 355). Many people have contributed by actively using and commenting several beta and pre-beta versions. Here are some of them, in no special order (and a big excuse to all those who we forgot in this list): Hermann Vogel, Gunter Gemmel, Martin von Gagern, Peter Lebmeir, Vanessa Krummeck, Thorsten Orendt, Andreas Fest, Carola Dietrich, Wolf Dieter Heker, David Bakin, Christof Boeckler, Gerhard Bischoff, Alexander Elkins, Dan Beaton, Camille Wormser, Franz Klement.

Finally there are two people who definitely deserve a special mention. One of them is David Kramer, our copy editor from Harvard. He carefully read every single line we wrote for the documentation and helped us as non-native speakers to end up with an at least linguistically understandable documentation. Thank you David! The second person is, in a sense, almost a member of the Cinderella team. It is Martin Peters, our responsible editor at Springer Verlag. He always had an open ear for our new problems, always understood that we need even more time to end-up with a release version, was always helpful in finding solutions to publishing issues and was extremely active in making several important contacts. And, the most important thing, always gave us the freedom to make our own decisions while trusting that this is the best way to end up with the best possible outcome. Martin, thanks a lot for your patience and confidence!

*May 2006*
*Jürgen Richter-Gebert, Munich*
*Ulrich Kortenkamp, Schwäbisch Gmünd*

## Preface to Cinderella 1.2

*Cinderella* is a program for doing geometry on a computer. In its present form it is the product of a sequence of three projects carried out between 1993 and 1998. It is based on various mathematical theories ranging from the great discoveries of the geometers of the nineteenth century to newly developed methods that find their first applications in this program.

The idea for the first of these projects was born in 1992 during a combinatorics conference at the Mittag-Leffler Institute in Sweden, when Henry Crapo and Jürgen Richter-Gebert were taking a trip on a boat called *Cinderella*. At that time, Jürgen Richter-Gebert had developed symbolic methods for automatic theorem proving in geometry [25], and both of them dreamed of computer software that would allow one to input geometric configurations with just a few mouse clicks and then ask the computer about properties of these configurations.

Henry and Jürgen started the project on a NeXT platform, which at that time was famous for its marvelous software architecture. *Cinderella* became the working title for the project, and this title turned out to be unremovable.

A few weeks of development produced the first working prototype. The program was based on principles from projective geometry and invariant theory. It was able to find *readable* algebraic proofs for many theorems of projective geometry about points, lines, and conics [5].

However, as a platform NeXT gradually declined in popularity, and with it the initial enthusiasm for *Cinderella*. After the summer of 1995 almost no further progress was made. At a conference on computational geometry at Mt. Holyoke College, in South Hadley, Massachusetts, it was almost impossible to give a software demonstration due to the vanishing of NeXT computers and their operating system.

In August 1996, right after that Mt. Holyoke conference, we (Ulli Kortenkamp and Jürgen Richter-Gebert, at that time working at the Technical University of Berlin in the group of Günter M. Ziegler) decided to start a new project, based entirely on the platform-independent language Java. At that time, the language Java was relatively new, and at first both of us were very skeptical about using an interpreted (presumably slow) language as the basis for a program that requires a large amount of computation in real time. But we tried anyway.

The goal of this second project was to have the old functionality that was available in the NeXT version, substantially extended by features of Euclidean and non-Euclidean geometry. We also wanted functionality for geometric loci. Moreover, since Java is designed to be "Internet-aware," the new program should be able to run inside a web browser. In particular, we wanted to be able to create student exercises for the web. The theorem-proving facilities of the program should be used to automatically check the correctness of the student's solution.

Conferences, competitions, and their deadlines are often driving forces for rapid development. A first working version was presented at the "CGAL startup meeting" in September 1996 at the Technical University (ETH) Zurich. A second version won the "Multimedia Innovation Award" at the Multimedia Transfer of the ASK Karlsruhe in January 1997.

During 1997, Jürgen Richter-Gebert became an assistant professor at the ETH Zurich. This change forced another break in the development. Ulli Kortenkamp moved to Zurich in September.

At the same time, we began negotiating for the publication of *Cinderella*. Originally, we planned to polish and finish the second project. However, things turned out differently.

The second version, like other computer programs for geometry, suffered from seemingly unavoidable mathematical inconsistencies. These inconsistencies came from ambiguities in operations like "Take the intersection of a circle and a line." There may be two, one, or no intersections depending on the position of the circle and the line. While dragging a construction, the program has to decide which intersection point to choose. This seemingly innocuous ambiguity may lead to terrible inconsistencies in the behavior of a construction. It may happen that while you move a point only a little bit, whole parts of the construction flip over.

At the beginning of 1998 it turned out that this problem of jumping elements was indeed solvable. However, it was clear that implementing the theory would not be an easy job. Every configuration had to be embedded in a complex vector space. Results of analytic function theory had to be used to avoid "singular situations." If we wanted to use those new insights, we had to rewrite the mathematical kernel of the program from scratch. The program had to perform approximately twenty to one hundred times as many computations as previously, a challenge for us and for Java.

We decided to do this and ended up with the third project, whose outcome you see here. In a period of unbelievably intensive work (that stretched our patience and that of our families to the extreme) we rewrote the whole program again, tuning the program to higher performance at every opportunity.

It turned out to be a good idea to undertake this effort. The benefits of the newly developed theory were much greater than we had originally thought. Based on the new methods we were able to do reliable randomized theorem checking. This proved to be much more useful than the old symbolic methods. It was also possible to generate complete geometric loci by generic methods, which is a novelty to the best of our knowledge.

The present program is a mixture of old geometry from the nineteenth century, complex analysis, our new methods, and modern software technology. We hope you will enjoy it as much as we do.

*Jürgen Richter-Gebert, Ulli Kortenkamp*

*Zurich, December 1998*

# Contents

# Part I
# General Concepts

# Chapter 1

# Introduction

## 1.1 What Is Cinderella?

Why did we write *Cinderella*? Aren't there enough programs that are suitable for doing mathematics, and in particular for producing mathematical graphics? Indeed, there are many programs, but *Cinderella* is special in many respects.

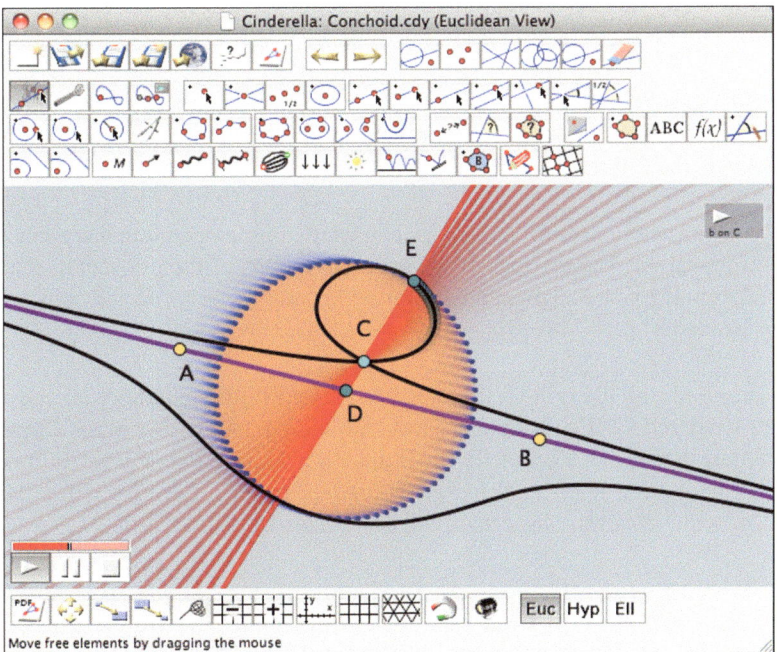

*A screenshot of Cinderella performing an animation*

We want to point out the major features of this software. *Cinderella ...*

- *... is a mouse-driven interactive geometry program:* With a few mouse clicks you can construct simple or complex geometric configurations. No programming or keyboard input is necessary. After you have completed a construction you can pick a base element with the mouse and drag it around, while the entire construction follows your moves consistently. This enables you to explore the dynamic behavior of a drawing.

- ... *has built-in automatic proving facilities:* While you construct your configuration *Cinderella* reports any nontrivial facts that occur.

- ... *allows simultaneous manipulation and construction in different views:* You can view and manipulate the same configuration in the usual Euclidean plane, on a sphere, and even in Poincaré's hyperbolic disk.

- ... *has "native support" for non-Euclidean geometries:* In *Cinderella* you can easily switch between Euclidean, hyperbolic, and elliptic geometry. Depending on the context, your actions are always interpreted correctly.

- ... *has advanced facilities for geometric loci:* The unique mathematical methods of *Cinderella* guarantee that complete real branches of the loci and not only parts of them are drawn.

- ... *is "Internet-aware":* The entire program is written in Java. Each construction can be exported immediately to an interactive web page. Even student exercises and animations can be created in this way.

- ... *produces high-quality printouts:* You can generate camera-ready PostScript or PDF files of your drawings. This vector output is superior to screenshot pictures and uses the full resolution of the printer.

- ...*is based on mathematical theory:* The whole implementation has a mathematical foundation. The theories of the great geometers of the nineteenth century, as well as many new insights, make *Cinderella* a highly reliable and consistent tool for geometry.

In addition to this list (which was already present in the handbook of the first release), Cinderella.2 is equipped with many new features that make Cinderella.2 much more than "just" a geometry program. For a detailed list of new features please refer to the section *What Is New in Cinderella.2* (p. 20). Here we will only give a brief overview of several aspects of the new release. So *Cinderella.2...*

- ... *comes with powerful modes for geometric transformations:* You can construct several kinds of geometric transformations starting with simple translations or reflections up to projective or Möbius transformations. These transformation modes are extremely useful for simplifying constructions.

- ... *allows the construction of fractals:* Combining several transformations, it is possible in *Cinderella.2* to construct so-called "iterated function systems." These are objects that are self-similar with respect to several transformations. These objects have fascinating geometric properties, and many well-known fractals are included in this class of objects.

- ... *is freely programmable:* One of the main new features in Cinderella.2 is the existence of a full-featured high-level programming language. The language CindyScript was designed to work seamlessly with interactive drawing environments. It is a functional language that allows powerful high-level programming. By adding only a few lines of code one can achieve significant control over the behavior of a construction.

- ... *has built-in simulation facilities:* Cinderella.2 comes with a special simulation engine CindyLab that can be used to construct physical simulations. CindyLab is based on a mass-particle/forces paradigm. One can simply draw an experiment and start the simulation. In particular, the combination of physics simulation with geometry or with the programming environment opens the possibility of surprising insights and amazing interactive simulations.

- ... *supports audio output:* Cinderella.2 has advanced functions for audio output. Via the build in MIDI interface of your computer it provides access to the generation of melodies and sound. By this one can on the one hand accompany mathematical demonstrations with sound effects. On the other hand one can experiment with the structure of sound itself on a very fundamental level.

- ... *provides advanced formula rendering:* Cinderella.2 has a built-in formula renderer that supports the TeX formula description language. By this, complex mathematical formulas can be inserted in mathematical visualizations. Formulas may even change dynamically with the drawing.

- ... *supports image rendering and transformations:* As a new feature starting from Cinderella.2.6 we also support using pixel graphic images within CindyScript. Images may be transformed and deformed in various ways. It is also possible to create custom images with self created content.

- ... *supports pen-driven devices:* A special interface for recognition of hand-drawn sketches makes Cinderella usable with purely pen-driven devices (such as electronic whiteboards and Tablet PCs). A hand-drawn sketch will be automatically recognized and converted into an interactive construction.

Thus Cinderella.2 consists of three major program parts: the geometry program, the scripting language, and the simulation engine. The following chapters will highlight several aspects of these three parts of the program.

### *1.1.1 A Dynamic Geometry Program*

Here we will briefly sketch how to use Cinderella.2 as a plain geometry program (without taking advantage of the scripting facilities and the simulation engine). In particular, we will emphasize several application areas in which it is advantageous to use *Cinderella*.

### 1.1.1.1 Sample Applications

The application areas of *Cinderella* reach from pure Euclidean (and non-Euclidean) geometry via physics (optics, for example) to computational kinematics and computer-aided design (CAD). The following sample applications offer potential scenarios in which you can benefit by using *Cinderella*.

### Exact Drawings

Suppose that you are writing a scientific publication for which you need one or more figures. When the drawings are a bit more complicated, it is usually almost impossible to produce a perfect one on the first attempt.

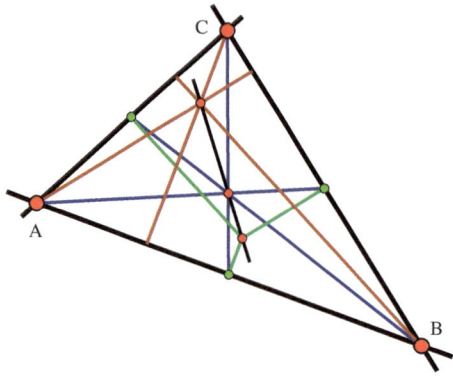

*The Euler line of a triangle*

Either you will have many elements in almost the same place, or some parts of the figure will not fit onto the page.

With *Cinderella* you start by making a computer sketch of a construction. This sketch might not look like the final drawing you expected. However, it provides all the relations that are essential for the construction. After you have made a sketch, you select the base elements and move them around until the picture satisfies all your aesthetic requirements. During the "move phase" you will always have a valid instance of your geometric construction. Finally, you use *Cinderella's* inspector to adjust the color and size of each geometric element.

### Geometric Calculator

Sometimes it happens that you want to get a feeling for some geometric situation. Either you have read something interesting in a geometry book or you have a new idea yourself.

You do the construction with *Cinderella* and start to play with it. Through geometric exploration you gain new insights, and often hidden properties of the construction are revealed. *Cinderella's* mathematically consistent implementation ensures that no strange effects occur that do not come genuinely from the configuration.

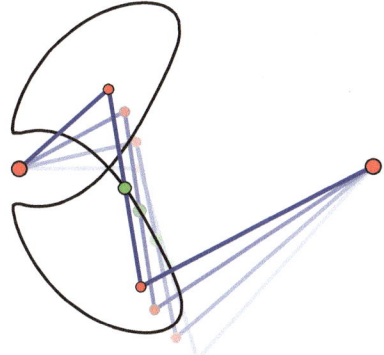

*Dynamics of a three-bar linkage*

When you want to communicate your research to other colleagues, you can create an interactive web page and make it available on the Internet. Then your colleagues will have instant access to the configuration and can interact with it locally in their Java-enabled web browsers.

**Interactive Worksheets and Student Exercises**

Another interesting application is the generation of interactive worksheets or student exercises. With Cinderella it is easy to export any construction to the internet. By this you can create an infinite variety of interactive mathematical worksheets. By suitable programming techniques it is even possible to create interactive student exercises. Imagine that you want to teach students how to construct the circumcenter of a triangle using only straightedge and compass. First you do the construction yourself. Using this sample construction you can create an exercise that guides a user step by step through the construction task. The required programming techniques are presented in the section Interactive Exercises (p. 433).

The students can solve the exercises on their own computers and arrive at a solution completely by themselves or by following hints you have provided. No matter what construction a student may have used to solve the problem, *Cinderella's* integrated automatic theorem-checking facilities can decide whether it is correct. The student's creativity for finding a solution is not restricted by the program.

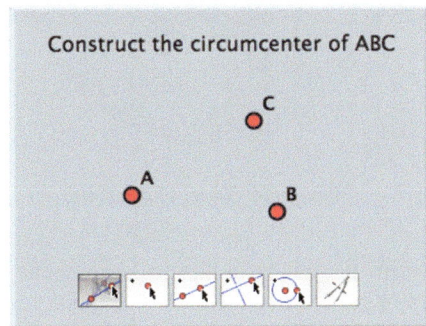

*An exercise for students*

### 1.1.1.2 Design and Features

Several major design goals guided us in the development of *Cinderella*. We want to mention the three most important ones to give you an impression of the overall architecture of the program.

**General Approaches**

*Cinderella* is designed to cover a wide range of geometric disciplines. The program provides "native support" for *Euclidean geometry*, *hyperbolic geometry*, *elliptic geometry*, and *projective geometry*.

This means that you do not have to simulate hyperbolic geometry by making complicated Euclidean constructions. You can use the "hyperbolic mode" of *Cinderella* and the constructions will behave like elements of the hyperbolic plane.

*Cinderella* achieves this by implementing very general mathematical approaches to geometry that form a common background for all of the areas above. Much of the mathematics behind *Cinderella* makes use of the great, and unfortunately almost forgotten, geometric achievements of the geometers of the nineteenth century. We would like to mention just a few of them: *Monge* and *Poncelet*, who "invented" projective geometry; *Plücker*, *Grassmann*, *Cayley*, and *Möbius*, who developed a beautiful algebraic language to deal with projective geometry; *Gauss*, *Bolyai*, and *Lobachevsky*, who "discovered" what is today called hyperbolic geometry; and finally, *Klein*, *Cayley*, and *Poincaré*, who managed to obtain a unified description of Euclidean and non-Euclidean geometries in terms of projective geometry and complex numbers. For an excellently written and exciting introduction to the historical development of geometry in the nineteenth century we recommend the book of Yaglom [34]. In addition, the historical book by Felix Klein [11] is a very interesting introduction into this topic.

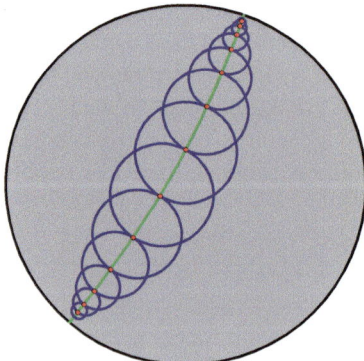

*Hyperbolic circles of equal size*

*Projective geometry* forms the background for the incidence geometry part of *Cinderella*, and *Cayley-Klein geometries* form the backbone for the metric part of *Cinderella*.

**Mathematical Consistency**

To put it metaphorically, "The geometric constructions done with *Cinderella* should behave as if they lived in a reasonable geometric universe. In this universe no unnatural things should happen."

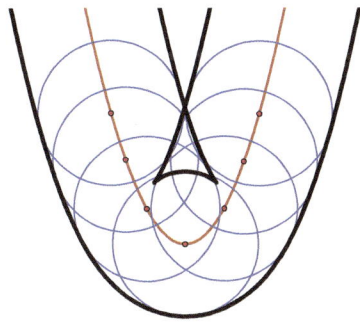

*Offset curve of a parabola, a challenge for most CAD systems*

Other systems for interactive geometry suffer from mathematical inconsistencies. You do a construction, drag around the base elements, and suddenly one part of the construction jumps from one place to another. This behavior is unfortunately common even in software systems for parametric CAD.

*Cinderella* completely resolves this problem by using a new theory. It makes use of features from *complex analysis* and merges them with the "old geometry" mentioned above.

Based on this theory, it was possible to equip *Cinderella* with an *automatic theorem checker* that governs most of the internal decisions *Cinderella* makes. This theorem checker is also used for automatic feedback operations in student exercises. Another benefit of this approach is that you have a generic tool to construct correct and complete geometric loci, which are real branches of algebraic curves.

**Modular Design**

*Cinderella* is designed to be as modular as possible. This architecture makes *Cinderella* well prepared for further extensions in many directions.

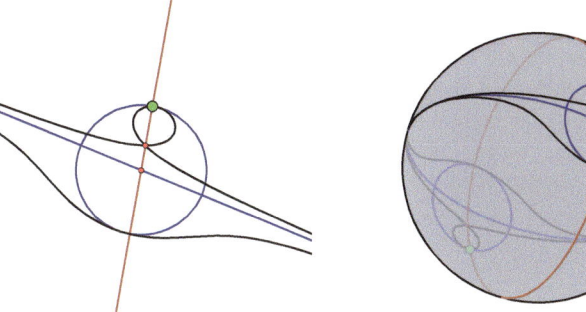

*Conchoidal curve in a Euclidean...*          *...and in a spherical view*

Even the present release benefits from the modular approach. For instance, it is possible to view the same geometric constructions simultaneously in many geometric contexts. A construction in hyperbolic geometry can be simultaneously displayed, and manipulated, in the "Poincaré model" of hyperbolic geometry and in the "Beltrami-Klein model." The simultaneous use of different views makes it possible to gain a deeper understanding of a configuration. For example, the "behavior at infinity" of a configuration becomes immediately visible in a spherical view.

**Transformations and Transformation Groups**

Cinderella provides powerful support for all kinds of geometric transformations. This often makes it possible to encapsulate high-level geometric operations into a very simple construction principles. Transformations are used in several contexts. Cinderella.2 supports simple reflections, translations, and rotations, as well as projective transformations and Möbius transformations.

Projective transformations are a very useful tool for constructing perspectively correct drawings of three-dimensional scenes. Furthermore, Cinderella provides facilities for constructing transformation groups generated by several transformations.

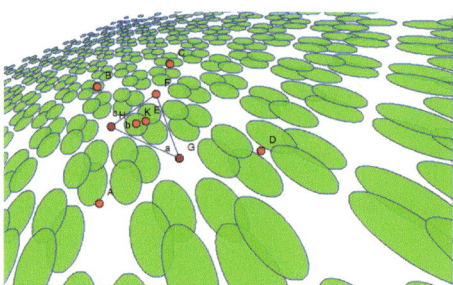

*Perspective view of an ornament*

For example, the picture above was created as the perspective image of the application of a transformation group to a circle. The transformation group was generated by $60°$ rotations around the corners of a regular triangle.

**Fractals**

Cinderella also provides a simple approach for the construction of fractal structures. The fractals constructible by Cinderella are so-called iterated function systems. They are constructed using a collection of geometric transformations. The iterated function system is a cloud of points that is self-similar with respect to the defining transformations. Using the interactive features of Cinderella, the program provides a unique experimental environment for the exploration of spaces of fractal objects.

*A simple fractal*

## 1.1.2 An Environment for Physics Simulation

Compared to previous releases, *Cinderella.2* offers a completely new suite of applications. It provides an engine for physical mass-particle/force simulations. With this engine it is possible to draw a physical experiment and execute it by pressing a play button. It is possible to combine physics elements freely with geometric elements. A detailed description can be found in the CindyLab documentation. Here we will give only a brief overview.

### 1.1.2.1 Sample Applications

CindyLab is an open environment for physics simulation. It offers several elementary objects with which it is possible to create simple as well as very complex physical scenarios.

**Virtual Physical Workbench**

The philosophy of CindyLab is very similar to the philosophy of the geometric part of Cinderella. There one has several construction modes to set up a geometric configuration as well as a move mode in which one can explore the behavior of the construction. In CindyLab one has several tools to set up a physical experiment. By clicking a play button one can run the experiment. It is also possible to interact with the running experiment by moving free points of the construction.

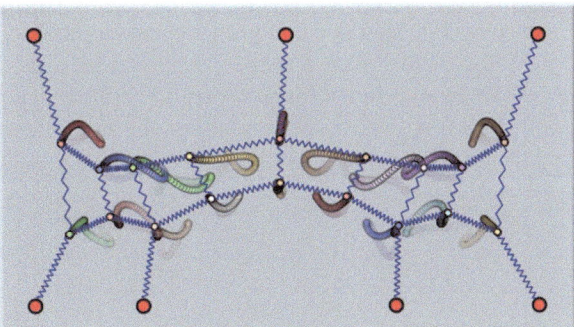

*An experiment with several springs*

CindyLab is a very useful environment for free experimentation with scenarios. Often one discovers surprising effects that stimulate other experiments. Since there are no predefined experiments, CindyLab is indeed a virtual construction kit.

### Explaining

For educational purposes, CindyLab is very well suited for creating well-defined experiments that exemplify well-known physical effects. The simulation engine is reasonably exact numerically, so that it provides reliable results for many interesting situations. In particular, it is possible to display and modify parameters of simulation objects via the CindyScript programming language. Since one can also easily export CindyLab constructions to HTML-pages, one can easily create interactive physics tutorials.

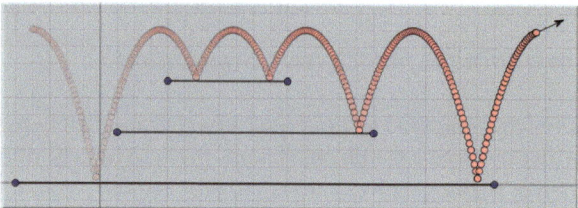

*Demonstration of the conservation of energy*

The range of applications reaches from completely predefined experiments that run by clicking the play button via experiments in which the student has to adjust parameters to explore some situation to open scenarios in which the student has actually to arrange the objects to get the desired interaction.

### 1.1.2.2 Design and Features

### Exact Integrators

The simulation engine of CindyLab is based on a mass-particle/force model. Each moving point is modeled as a pointlike particle, and the interactions are modeled as forces between the particles (or between particle and environment). The forces influence the acceleration of the particles.

The numerical simulation engine of CindyLab is based on an *explicit Runge–Kutta integrator*. There are many possible choices for such an integrator. We chose an integrator that represents a reasonable compromise among the desiderata of numerical reliability, flexibility, and speed. The specific integrator used in CindyLab is a Dormand-Prince-45 integrator. In the Inspector (p. 153) (the tool used to control the parameters of Cinderella) it is possible to readjust the numerical accuracy of the integrator. Thus it is also possible to model sensitive scenarios numerically.

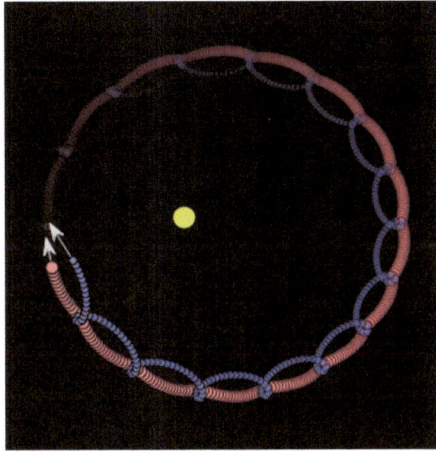

*A sun/planet/moon system*

## Interaction with Geometry

CindyLab is designed to interact seamlessly with the geometry part of Cinderella and with the language CindyScript. This opens several possibilities. On the one hand, one can easily enhance the visual appearance of simulations by drawing geometric decorations. It is also possible to make geometric analyses of simulated scenarios by adding a few geometric constructions. For instance, the picture below demonstrates the analysis of a planet orbiting a sun. It reveals a hidden property of the velocity vector in this situation, namely, that the velocity vector traces out a circular path. The geometric analysis was done using a simple translation that maps the velocity vector to a fixed point.

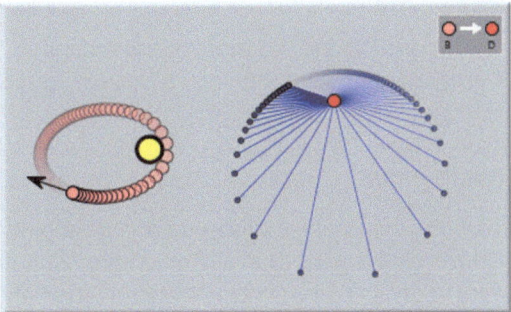

*Analysis of the velocity vector in a sun/planet system*

## Interaction with Scripting

In a similar way, CindyLab can also interact with the language CindyScript. This allows all physical parameters of a simulation to be read, and many of them can be influenced directly. This offers the possibility of a detailed numerical analysis of an experiment. CindyScript also provides several operators that are designed especially for collaboration with CindyLab. In particular, it is possible to perform curve plotting of physical parameters directly or to draw the flux of a force field.

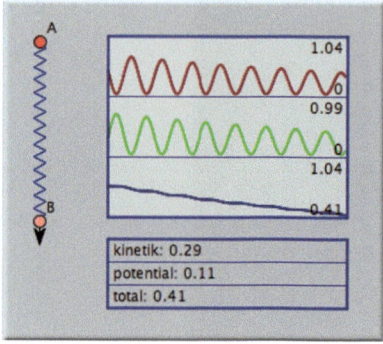
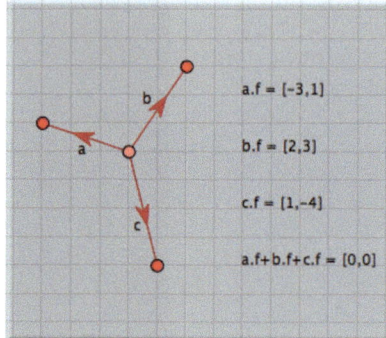

*A harmonic pendulum*                    *Forces in equilibrium*

The interaction of physics and scripting makes possible a wide variety of applications. In particular, it is possible to simulate robotic constructions in which a physical robot is simulated by CindyLab and controlled by CindyScript.

### 1.1.3 A Programming Environment

The CindyScript programming language is a powerful tool for adding all kinds of enhancements to Cinderella constructions. It is designed in a way such that the interplay with the geometric part of Cinderella and the simulations of CindyLab is as smooth as possible. Its functional design allows for high-level programming that enables rapid prototyping of interactive scenarios. The CindyScript language takes advantage of the fact that it runs in an environment in which on the one hand, facilities are present for graphical output and mouse-driven input (the geometric part), and on the other hand, physical simulations can be "outsourced" to the CindyLab simulation engine. As a consequence, the user can focus on programming the core problem. Very often, only a few lines of code are needed to achieve the desired behavior. This is in strong contrast to common programming languages, where a substantial part of programming goes into the creation of a graphical mouse-driven user interface.

#### 1.1.3.1 Sample Applications

The possibilities of CindyScript are endless and restricted only by the imagination of the user (and perhaps disk space), as in any other programming language. Nevertheless, we here want to present a few core application areas of the language.

#### Enhanced Drawing Output

One of the core facilities of CindyScript is that one can programmatically produce graphical output in a geometric view. When using a dynamic geometry system, one very often encounters situations in which it would be desirable to extend a geometric construction by more and more elements to such an extent that it becomes unreasonable to do this constructively. The programming facilities of CindyScript can be used to create these elements automatically.

CindyScript offers access to geometric transformations, such that it is very easy to create functionally very complicated geometric drawings with only a few lines of code.

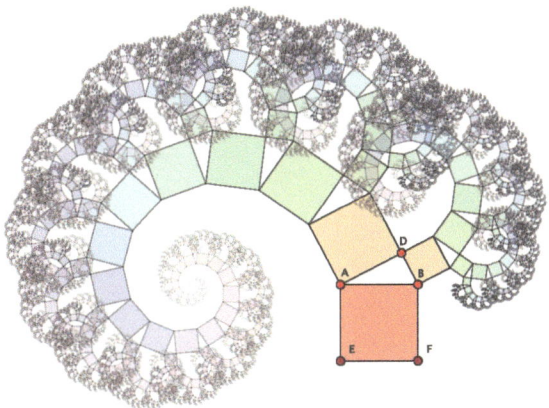

*A scripted "Pythagoras tree"*

**Programmatic Drawing**

There is a special class of applications that cannot be covered at all by a classical approach to dynamic geometry. Constructions in dynamic geometry are inherently like unbranched programs, in which all calculations are always performed. Thus it is inherently difficult to include logical decisions or algorithmic behavior in a dynamic geometry program. Usually, these problems are resolved by conditionally controlling the visibility behavior of geometric elements by intersection properties (so-called *Boolean points*). *Cinderella.2* also offers this functionality, but at the same time, CindyScript provides a much better and more elegant way of dealing with this problem.

CindyScript is a full-featured high-level programming language in which it is easy to implement arbitrary algorithmic behavior. Thus even complicated algorithms can be included on a general level in a dynamic geometry environment. Graphical output can be easily included in such algorithms.

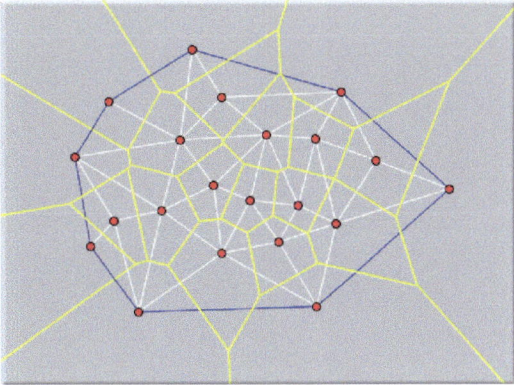

*The Delaunay and Voronoi diagrams of a point set*

**Analysis of Mathematical Functions**

In particular, CindyScript offers advanced routines for function plotting. Thus it is possible to draw and analyze functions dealing with derivatives, extrema, and so on. Functions can be entered directly in a drawing, and results can be shown instantly. Making use of this together with the HTML export features of Cinderella, it is easy to generate interactive worksheets that allow for a wide variety of modes of analysis of functions. Functions can also be defined by geometric dependencies of the geometric constructions, which facilitates combining input and analytic parts in one example.

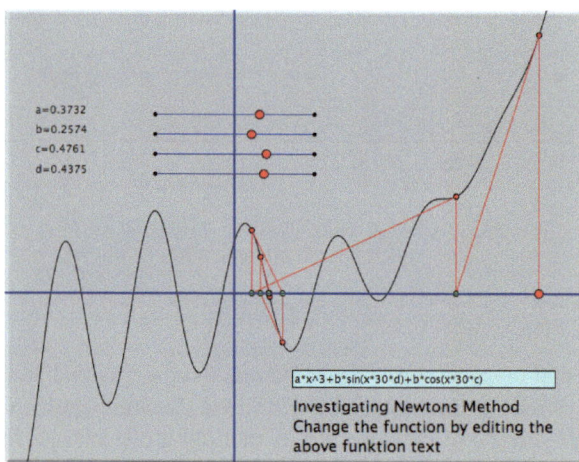

*Newton iteration to find a root of a function*

**Controlling the Behavior of Constructions**

A particularly interesting scenario arises when CindyScript controls the positions of free elements. The movements controlled by CindyScript usually have priority over those performed by the user. Thus you can influence the effect of user input. The following example illustrates this in a very simple scenario. The picture below demonstrates Pick's theorem, which allows one to calculate the area of a triangle by calculating the integer points in the polygon and the integer points on the boundary.

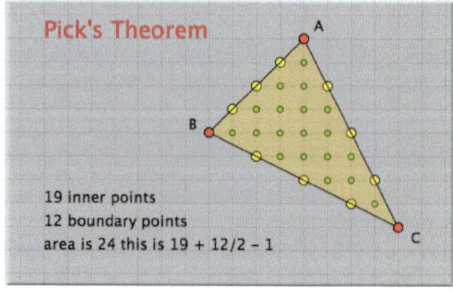

*A demonstration of Pick's theorem*

Pick's theorem applies only to polygons whose vertex coordinates are pairs of integers, a restriction that is usually not supported by a dynamic geometry program. However, by adding the following three lines of code one can alter the behavior of the three free points:

```
A.xy=round(A);
B.xy=round(B);
C.xy=round(C);
```

Although the points are freely movable, this code forces each vertex to snap to the closest integral point. All the text and all the green and yellow points in this example are also generated with CindyScript.

### 1.1.3.2 Design and Features

CindyScript was designed to suit the needs of programming mathematical problems in an environment in which the input parameters change dynamically due to user interaction. On the one hand, this was achieved by providing a language that has both powerful data types (real numbers, complex numbers, lists, vectors, matrices, etc.) and powerful operations that can work on these data. On the other hand, it was achieved by taking into account the real-time requirements for interactive user input already when the language was designed.

### High-Level Functional Programming

CindyScript is a functional language. This means that all operations are expressed as the application of functions to data. Programming functionally may take some time to get used to, but it offers a very high expressiveness, such that it is possible to describe complex situations with just a few lines of code. For instance, the following three lines of code calculate and print a list of the prime numbers smaller than 100:

```
divisors(x):=select(1..x,mod(x,#)==0);
primes(n):=select(1..n,length(divisors(#))==2);
print(primes(100));
```

The first line defines a function that returns a list of all divisors of a number x. The second line provides a function that returns a list of all numbers smaller than n that have exactly two divisors. These are exactly the prime numbers smaller than n. The output of the algorithm is the following line:

```
[2,3,5,7,11,13,17,19,23,29,31,37,41,43,47,53,59,61,67,71,73,79,83,89,97]
```

Using these high-level programming facilities together with graphics input and output offers possibilities that are not available in most other programming environments. The picture below shows an iteratively defined plant-like structure that was programmed in CindyScript. The code for generation of this plant is shown next to it. It essentially uses only elementary drawing functions, advanced geometric transformations, and functional programming to create the entire drawing.

```
list(x):=(
  gsave();
  repeat(length(x),turtle(x_#));
  grestore();
);

turtle(x):=(
  (if(x=="F",foreward));
  (if(x=="+",left));
  (if(x=="-",right));
  (if(x=="[",open));
  (if(x=="]",close));
);

foreward:=(draw((0,0),(l,0));translate((l,0)));
left:=rotate(angle);
right:=rotate(-angle);
open:=gsave();
close:=grestore();

l=0.2;
angle=A.x/4;

n=4;
s="F";
repeat(n,s=replace(s,"F","F[+F]F[-F]F"));
rotate(pi/2);
list(s)
```

*Growing a plant in a view*

### Real-Time Calculations

CindyScript code can be evaluated for even the tiniest move of a geometric configuration. In this way, it is possible to achieve real-time interaction with geometric
drawings. A user selects a point, moves it, and then can directly observe the changing results of the calculations. Although CindyScript is an interpreted language, it is
still fast enough to create a fluent drawing impression for most standard situations.
The possibility of dynamically running programs that immediately react to user input opens a wide variety of applications and interactive scenarios. Here the applications range from explanations at an elementary- and high-school level through
university courses up to investigations in sophisticated mathematical research. The
example below shows a snapshot of an illustration of linear regression using the
least squares method. The picture is dynamically updated when points are moved or
added.

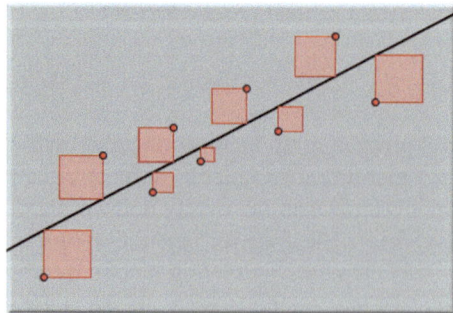

*Linear regression by Gaussian approximation*

**Exact Timing**

CindyScript provides various possibilities for synchronizing programs both with real time and with the time of a physics simulation. Thus one can easily create simulations that involve timing behavior. This can be very useful in programming interactive games or edutainment scenarios in which, for instance, the task of solving an exercise is bound to time restrictions. The ability to synchronize CindyScript with the timing of physics simulations can be used in two directions. Thus one can use this technique to synchronize a physics simulation with real time. Alternatively, it is possible to couple programming time and computation time, making it possible to slow down or speed up a simulation synchronously with a computation.

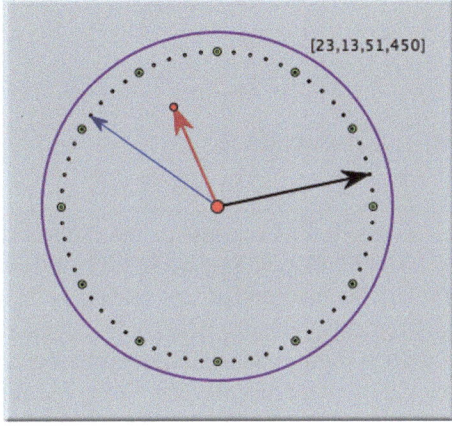

*A real-time clock*

**Advanced Plotting**

One line of CindyScript code is sufficient for simple function plotting. For example, `plot(sin(x))` will immediately invoke the graph of the sine function to be plotted in the view. However, Cinderella also provides several possibilities and enhancements that allow for the direct display of maxima, minima, and inflection points. With these features it is easy to make a visually meaningful analysis of functions. It is also possible to generate two-dimensional function plots that use color values for different function values. It is thus possible to create images of two-dimensional data quite easily. The picture below shows the density distribution of the function `sin(dist(x-A)*dist(x-B))` for two points $A$ and $B$.

*Cassinian ovals as waves*

## 1.2  What Is New in Cinderella.2

This chapter provides a very brief overview of the differences between the previously published versions of Cinderella (versions 1.0 to 1.4) and the current release *Cinderella.2*. The official version name of this release is "2.6". This release offers many substantial improvements, enhancements, and extensions to the old versions. Measured in bytes of written source code, the Cinderella.2 release has a size of about 7 MB. This is approximately five times the size of the old releases. The reason for this significant growth is the fact that Cinderella.2 contains several areas that open completely new possibilities: While Cinderella 1.4 was solely a geometry program, the new release, Cinderella.2, offers (among other things) programming facilities, full support of transformations, many enhanced graphical features, and a complete physics simulation engine.

The following list of items is a rough overview of the major changes. The items appear in no particular order. For details on the topics the reader is referred to the corresponding chapters of the reference part of the manual. For the reader already familiar with Cinderella, this chapter may be a useful guide for exploring the new features.

### *1.2.1  Enhanced Graphics Primitives*

#### Circular Arcs and Angle Marks

*Cinderella.2* offers circular arcs (p. 93) as fully supported geometric elements. Arcs can be defined by clicking three points. The first and third points define its endpoints. The arc is supported by the unique circle that passes through all three points. Like a circle, an arc may be intersected with other geometric elements. Free points can also be bound to an arc. Furthermore, a new graphics primitive has been introduced: a little arc for marking an angle.

**Active Segments**

While in Cinderella 1.2, segments were of a purely graphical nature and could not be used for constructions, in *Cinderella.2* segments (p. 84) are fully featured geometric objects. They behave very similarly to lines. A free point bound to a segment cannot be moved away from the segment. It is also possible to define loci whose movers move only along a segment.

**Polygon Modes**

Several modes have been added that facilitate the construction of regular polygons (p. 112). This is particularly important for the construction of regular symmetric patterns based on the symmetry of polygons. Symmetric polygons are very useful in combination with the transformation modes (p. 122).

**Conic Modes**

Several new modes have been added for the construction of conic sections (p. 110). In particular, it is possible to construct ellipses and hyperbolas by the position of the foci and a perimeter point.

**Triangular Mesh**

In addition to the quadrangular mesh, it is also possible to use a regular triangular mesh for snapping points in Euclidean view (p. 114).

**Background Images**

It is possible to load a bitmap as a background image. In particular, this enables the analysis of geometric situations that come from exterior sources by overlaying pictures with dynamic drawings.

**PDF and Bitmap Export**

To include Cinderella pictures in other documents, an export to PDF and several image formats are provided. The PDF export also provides the possibility of exact scaling.

## 1.2.2 Enhanced Animations, Traces and Loci

**New Animation Concept**

Compared to the old concept of animations, the new animation (p. 107) model is much more flexible. While an animation is running, it is still possible to move the base elements of the construction. Also, several animations can be running and controlled at the same time. In particular, it is possible to control the individual speed of each of the animations. The new animations work seamlessly together with the physics simulations that can be constructed using CindyLab (p. 167).

**Traces**

All geometric elements can leave traces on the screen. Traces become dimmer with age, creating a nice visual effect. This is a very useful feature for improving the dynamic appearance in printouts. Traces may also be used to study the movement of dependent elements. Sometimes, traces are more flexible than loci for this purpose.

**Literal Loci**

Loci are usually drawn as solid lines. In *Cinderella.2* it is also possible to generate a locus in a way that resembles a trace of the geometric object. For this, a sequence of copies of the traced element is produced on the screen. This feature is particularly valuable in studying traces of objects that cannot be represented by a solid line (such as traces of circles and polygons).

## 1.2.3 Transformation Geometry

**Transformations**

Transformations (p. 122) are one of the major improvements in *Cinderella.2*. One can define reflections, rotations, translations, similarities, affine transformations, projective transformations, and even Möbius transformations. Transformations are very powerful construction tools in many situations. Arbitrary geometric element can be mapped by transformations.

**Transformation Groups**

Several transformations can be grouped together. A transformation group (p. 145) now allows a geometric element to be mapped iteratively by all the members of the transformation group. With this tool it is easy to generate ornamental structures even in advanced settings such as hyperbolic geometry.

**Iterated Function Systems**

Like a transformation group, an iterated function system (p. 141) comprises several transformations. However, in this situation a cloud of points is generated that is self-similar with respect of all the translations involved. Usually, an iterated function system is a fractal object that showcases a very interesting and rich structure.

**Bases (experimental feature)**

It is also possible to perform all drawing operations with respect to a basis that has been chosen in advance. The basis can be either a translation basis, a similarity basis, an affine basis, or a projective basis. Using bases, it is very easy to construct perspectively correct pictures of geometric situations.

## 1.2.4 Enhanced Construction Facilities

**Copy and Paste**

It is possible to copy and paste parts of a geometric construction. Together with the redefine (p. 415) operation, this forms a powerful concept for creating large constructions.

**Tool from Selection**

With the tool from selection mode, one can create individual construction tools that contain their own constructions as macro operations.

**Redefine**

The redefine (p. 415) feature is very helpful for changing the dependency structure of a construction. With this tool one can convert a dependent point to a free point and vice versa. Especially in combination with copy/paste, this is a very useful concept.

### *1.2.5 Fundamentally New Program Parts*

**The Inspector**

In *Cinderella.2* the inspector became the central controlling unit for all properties of a configuration and its objects. In the inspector (p. 153) one can control and access the appearance properties, the tracing and physical properties, and the definition properties of any object.

**CindyScript**

As already mentioned in the introduction, CindyScript is one of the most important extensions in Cinderella. CindyScript is a fully featured functional programming language that is designed to work seamlessly together with the other parts of Cinderella. Using CindyScript, it is possible to control the behavior of a dynamic construction. It is also possible to add arbitrary drawing information to a picture.

**CindyLab**

The second major enhancement is CindyLab. This is an engine for physics simulations based on a point-mass/forces simulation model. In CindyLab one can simply draw a physical experiment and immediately simulate it. In particular, the interplay of physics and geometry can be studied with Cinderella.

**Pen-Driven Devices**

*Cinderella.2* offers an input mode that is designed to support pen-driven devices such as tablet PCs, electronic whiteboards, and PDAs. In this mode Cinderella analyzes hand-drawn strokes and translates them (if possible) into geometric constructions.

## 1.3 Technical Background

*This section from the original 1999 manual is reproduced here for historic reasons.*

Normally you would, and should, not care about the programming language and other technical details of software you just want to use. Nevertheless, we want to tell you about the computer science background of *Cinderella*.

*Cinderella* was written in Java, the platform independent programming language developed by Sun Microsystems. This means that the software can be run on any computer, irrespective of its operating system, provided that there is a thing called the "Java Virtual Machine" (JVM) for this system. These Java Virtual Machines are available from Sun Microsystems for Windows 95/98/NT and Solaris, and there exist ports for Linux, OS/2, MacOS, BeOS, AIX, HP-UX and many more. In fact, you probably already have a JVM installed on your computer, since Netscape Navigator and Microsoft Internet Explorer have a built-in JVM. This, in turn, means that you can run Java programs inside a web browser. These programs are known as "Applets".

We do not want to explain Java in full detail here, instead we recommend the official Java home page at http://www.javasoft.com as a starting point for further reading. However, we do want to tell you about some of the consequences of choosing Java for *Cinderella*.

It is a fact that, although Microsoft Windows is the dominating operating system today, many mathematics departments have a variety of Unix workstations. Even within the same working group you can find a mixture of different systems. Java enables everybody, regardless of the choice of platform, to use *Cinderella* in the same way. It is even possible to install and use the *same code on all of your computers*. For us, we could use our favorite operating system (Linux) for development, and at the same time we were sure to reach the largest possible audience.

Second, the fact that you can run Java software inside web browsers has been used for the web export functionality of *Cinderella*. This means that you can publish constructions easily, spice up your personal home page with animations, or assign construction homework to your students. Our license agreement gives you great freedom in redistributing the necessary parts of *Cinderella*, but please obey the few restrictions that come with it.

Java is an *interpreted* language, as opposed to *compiled* languages like C or C++, which are the usual languages used for most software. Interpreted languages have some technical advantages, but they suffer from an additional translation step which slows the program down. Java (or the virtual machine) has been tuned a lot for performance, and the performance gap is not as large as it was when our project started. Still we had to do a lot of optimization by hand to create acceptable speed, and the "interactive feeling", of *Cinderella*.

Sometimes you'll notice a short delay when you move a point. Do not blame your computer, Java or *Cinderella*. These delays are caused by extremely complex calculations which are necessary to get the correct result or the correct screen representation after a movement. The generation of correct loci is one reason for that; many intersections involving conicals are the other. We tried our best to speed up these calculations, but there is a (mathematical) limit where we do not want to sacrifice accuracy for speed.

Finally we want to mention the tools that helped us creating *Cinderella* and this documentation. First there is *XEmacs*, a powerful, extensible text editor, which is based on GNU Emacs, which in turn is a version of the original Emacs written by Richard Stallman in the seventies at MIT. It is definitely the best editor available, and we used it to write the whole program and all of the documentation.

The program itself was developed with the help of the *Java Development Kit* of Javasoft, a division of Sun Microsystems, in particular with the Linux port of it (see http://java.blackdown.org for more information on the Java-Linux porting project). *Linux* is a free, unix-like operating system originating from the work of Linus Torvalds, and is now continually improved by the effort of several hundred developers all around the world.

The parsing engine (used for loading saved constructions) was constructed with the help of *ANTLR 2.5.0*, a public-domain Java/C++ parser generator, written by Terence Parr of the MageLang institute.

Post-optimization and compression of the code was done with *Jax* from alphaworks, the research division of IBM. We want to thank the Jax team, in particular Frank Tip, for their help and IBM for the permission to use Jax commercially.

The *Concurrent Versions System (CVS)* by Cyclic Software did most of the version merging (and saved us from a lot of headaches). It is free software, too.

Thanks to the "browser war" between Microsoft and Netscape, the licensing terms for redistributing *Netscape Navigator* allow us to ship a Java-1.1 compatible browser with *Cinderella*.

The documentation of *Cinderella*, both the printed manual and the online version, were written with XEmacs in HTML. We used the same files for the printed version and the screen representation. The design of the web pages uses Cascading Style Sheets (CSS); the hardcopy was created using a whacked version of *html2ps* by Jan Karrman.

The icons and images used in *Cinderella* were designed by ourselves with *The GIMP* (GNU Image Manipulation Program), written by Peter Mattis and Spencer Kimball. In our view it is one of the most impressive freely distributed pieces of software. The additional figures in the documentation were created with *Cinderella*, of course, and *Povray*, a free 3D raytracing software, and some PostScript hacking.

Two people deserve special mention: *James Gosling*, the creator of the Java programming language, and *Jamie Zawinski*, responsible for the first Unix versions of Netscape Navigator. They are both connected to XEmacs in a special way: James Gosling did the first C-implementation of Emacs, known as GOSMACS, and Jamie Zawinski was the person responsible for XEmacs versions 19.0 to 19.10, which was at that time a collaborative work of Lucid (now out of business) and Sun Microsystems (sic!).

## 1.4 Installation and Updating

Cinderella is written in the Java programming language. On most computers the environment to run Java based software is pre-installed. This includes Mac OS X and most versions of Windows. On other operating systems, for example Linux, you might have to install a so-called Java Virtual Machine that is available for most platforms at http://www.java.com as a free download.

The Cinderella software itself is available as a free download from the Cinderella homepage at `http://cinderella.de`. Depending on your platform you either download an installer or the application itself.

### 1.4.1  Entering a License

Without further intervention, Cinderella will run in a mode that is prepared for the user that wants to have "just" a geometry program. All tools that are essential for doing geometry are unlocked. However, some of the advanced features like the script editor or the physics simulation and very few geometry modes are not enabled. By installing a valid license you can unlock the complete feature set (this means updating to the *Pro* version). A license can for instance be purchased at the Cinderella homepage. The license comes as a simple Cinderella file. Once the license file is opened, the personalized license is installed automatically. Licenses will be renewed automatically before they expire (no additional purchase necessary).

As the license is stored in the user preferences (the file cinderella2-user.properties that is located either in the installation folder or, on Mac OS X, in the Library/Preferences folder, you can safely install a newer version of Cinderella whenever it is available without having to enter the license again.

### 1.4.2  Network licenses

For network environments it is not convenient to have every user enter the license himself. Instead, it is possible to apply a license to the Cinderella installation as a whole, by copying the license key into a file called *cinderella2-global.properties*. This should be located in the same folder as the file *cinderella2.properties*, which contains the software's defaults. Usually, this is in the installation folder. On Mac OS X it is inside the application bundle: You can open the right location by ctrl-clicking on the application in Finder, selecting "Show Package Contents" and navigating to Contents/Resources.

### 1.4.3  Updating Cinderella

We are continuously updating Cinderella.2 with new features and, if necessary, bug fixes. Please make sure to always use the latest version, which is available from the home page at `http://cinderella.de/download`.

# Chapter 2
# Theoretical Background

## 2.1 Problems in Interactive Geometry

How should a system for doing interactive geometry behave when a user interacts with it? In a sense, the requirements are similar to the requirements for other programs:

- The program should be easy to use.
- The user should not be disturbed by unnatural behavior of the program.
- The user should not be tormented by being forced to make unnecessary input.
- The computed results should be correct.

Unfortunately, under these requirements interactive geometry turns out to be a difficult subject. There are two main reasons for this:

- There are problems that come from *special cases* that occur even in a static setup.
- There are problems that are of a *genuinely dynamic* nature.

### 2.1.1 Static Problems

Our usual "everyday geometry" is full of special cases. Two lines can intersect or they can be parallel. Two circles can intersect in one or two points or not at all. So, even for static constructions it is sometimes difficult to figure out what a correct and reasonable result for such a special case should be. For instance, what is the angular bisector of two parallel lines? Is it undefined? Can it be any line parallel to the two lines? Should it be a line that is equidistant to the two lines?

We could try to exclude all the special cases and not allow them at all. However, on the one hand, this would mean excluding nonesoteric cases such as parallel lines. On the other hand, when we allow points to be moved in a construction, it happens all the time that mutually dependent elements are forced into special cases. Observe that this is still a static problem!

Such static problems have been studied for a long time. The great geometers of the nineteenth century were aware of them, and it is due to their effort that most of them could be solved. The key to a solution is gradually to extend Euclidean geometry to a larger domain. First, the usual plane is extended by elements at infinity, leading to *projective geometry*. Then the underlying algebraic structure is extended to cover *complex numbers*. This essentially removes all special cases from geometry.

It was an exciting development in mathematics when finally, around 1870, these approaches led to a completely consistent system. This system explains the effects of Euclidean geometry as well as those of non-Euclidean geometries, such as hyperbolic geometry. Today it is called *Cayley-Klein geometry* in honor of two of its main contributors.

The mathematical background and the implementation of *Cinderella* rely on this general setup. In this way, *Cinderella* can deal with all special cases, and as an additional benefit it is able to do non-Euclidean as well as Euclidean geometry. It is an amazing fact that by using these general principles the program does not become more complicated. On the contrary, the exclusion of special cases allows a much simpler and straightforward program structure.

## 2.1.2 Dynamic Problems

For systems of interactive geometry there is a second class of problems, which are in a sense more subtle than the static problems. Unfortunately, they lead to even more drastic effects. Assume that you have created a construction that involves points, lines, and circles, and in particular the intersection of two circles (or of a circle and a line). When you move the mouse, the program has to decide, for every position, where the dependent elements are. However, there is a problem. Two circles do not have only one intersection. They have two, and we get both from our calculations. How should the system decide which of them is the one you "want"? When you construct the intersection, the answer to this question is easy: Take the point that is closest to the current mouse position. But when you start to move the construction, there is no obvious answer.

What would be most desirable is a *continuous* behavior of the program in the following sense: If you make a very small move with a free point, then the elements that depend on it should also move only a little bit.

At first sight it is not clear whether this requirement is satisfiable in general. Turn on your favorite system for doing interactive geometry or parametric CAD and make the following experiment: Draw a horizontal line and construct two circles of equal radius whose centers are constrained to slide along the line. Move the circles to a position in which they intersect, and construct the upper point of intersection of the two circles. Now move one circle so that its center passes through the center of the other circle. Most probably you will see that the point of intersection suddenly jumps from the upper intersection to the lower one. This is what has happened in all the systems we have tried so far. Such behavior runs counter to our requirement of continuity: You make a small move, and a dependent point suddenly jumps.

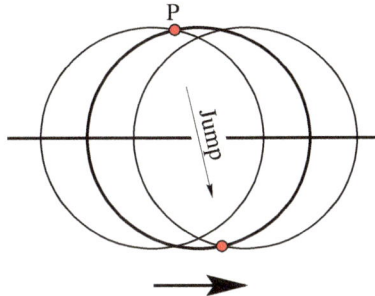

*This should not happen!*

At first, a single jumping point seems to be a mere curiosity that is tolerable. But what happens if large parts of a construction depend on this jumping point? Then these parts of the construction will jump too, without prior warning. Most systems for interactive geometry use heuristics based on orientation decisions that help to get rid of some of these jump situations. But still, in every system many cases remain unresolved. In fact, there is a proof that no heuristic based on orientations only will be able to resolve all of these dynamic problems [31, 20, 13]. In an article on dynamic geometry [24], Jean-Marie Laborde, the main designer of Cabri Géométrie, states this dilemma in the following way:

I think we need a real mathematical treatment of all consequences of stretching geometry in some way to a wider (dynamic) system. This system cannot be the projective one if we want to maximize the way the environment takes into account the special characteristics of non-static objects which are at the core of dynamic geometry.

*Cinderella* is the first program that is based on a theory that is capable of preventing dependent elements from jumping. This theory is also based on the use of complex numbers, which were used to solve the problems of static geometry.

The use of this theory has many benefits. For instance, it is the basis of the generation of correct loci. Consider the Three-Bar Linkage (p. 51) example of the second tutorial. The generation of the locus is based on the correct calculation of the points of intersection of two circles while a free point moves. In other systems for interactive geometry you will probably get only half of the figure-eight curve. The methods for automatic theorem proving, which are used internally throughout *Cinderella*, are based on this theory as well.

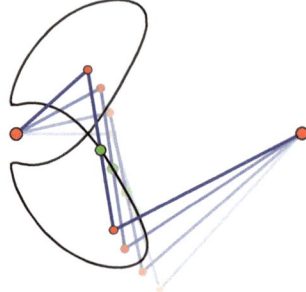

*The "three-bar locus" revisited*

The following paragraphs should give you an impression of the different mathematical methods and theories that form the basis of the implementation of *Cinderella*.

## 2.2 Projective Geometry

The first and perhaps most important step for a consistent geometric setup is to extend the usual Euclidean plane to contain elements at infinity. You have likely heard the phrase "parallel lines meet at infinity," and you might believe it when looking from a bridge along a very long and straight railroad track. This phrase is the key to projective geometry. The extension of geometry by infinite elements removes many of the special cases from usual Euclidean geometry.

Projective geometry has a very long tradition. Its historical origin can be traced back to the study of perspective undertaken by famous painters such as Albrecht Dürer and Leonardo da Vinci. Its mathematical origin is the work of *Gaspard Monge*, a French geometer, who developed, around 1795, a method called *descriptive geometry* for representing spatial configurations by planar perspective drawings. Monge observed that nontrivial facts about planar geometric configurations could be derived by considering these configurations as projections of configurations in space. The study of parallels in these projections was most elegantly accomplished by extending the plane by elements at infinity.

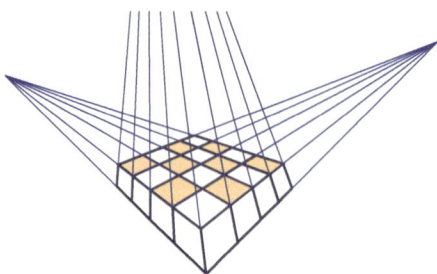

*In perspective drawings, parallels actually meet*

The *projective plane* consists of the points of the usual, Euclidean, plane together with one additional "infinite" point for every possible direction. The lines of the projective plane are the Euclidean lines together with one special "line at infinity." All infinite points lie on the line at infinity. The following nicely symmetric relations between points and lines hold:

- Every two distinct points have a unique connecting line (their *join*).

- Every two distinct lines have a unique point of intersection (their *meet*).

The first person who formalized these rules, around 1822, was *Victor Poncelet*, a student of Monge, who today may be considered the "father of projective geometry." In projective geometry there is no need to consider parallels as something special. They still have a point of intersection; it just happens to lie at infinity. For a readable introduction to projective geometry we refer to the books of H. S. M. Coxeter on that topic [4, 3].

## 2.3 Homogeneous Coordinates

On a computer we unfortunately do not have geometric objects as primitive data types. A point or a line has to be represented by numbers: the coordinates. Usually, a point in the plane is described by its $(x, y)$ coordinates. A line may be given by the three parameters $(a, b, c)$ of its defining equation $ax + by + c = 0$. However, when we want to do projective geometry, this turns out to be impractical. Each pair of coordinates $(x, y)$ represents a finite point, and there is no representation for the points at infinity. A solution to this problem gradually became clear in the first half of the nineteenth century. It started with Möbius's *barycentric coordinates*, via the refined setup of *homogeneous coordinates* given by Plücker, and finally led to Grassmann's setup of *multilinear algebra*.

The way out of the dilemma is as follows. For every point we use three instead of two coordinates, thereby introducing a third dimension. Consider the following scenario: The plane is embedded parallel to the $xy$-plane of three-space at a height of $z = 1$, so it does not pass through the origin $(0, 0, 0)$.

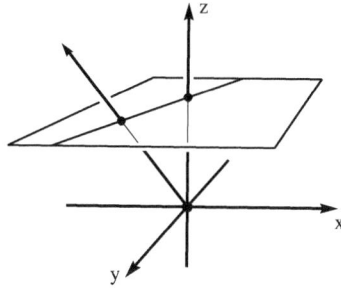

*Embedding the plane in space*

Every point $(x, y)$ is represented by its three-dimensional coordinates $(x, y, 1)$. These coordinates are the *homogeneous coordinates* of the point. What happens with the rest of the points in the three-dimensional space? Almost all of them will be interpreted as points in the original plane: we identify all three-dimensional points that differ by a nonzero multiple. For instance, (4, 6, 2) and (2, 3, 1) describe the same point. In general, a point with spatial coordinates $(x, y, z)$ is identified with the point $(x/z, y/z, 1)$ of the original plane. This process is called dehomogenization. In a way, every point of the original plane corresponds to the line spanned by this point and the origin in three-space.

However, there are points in three-space that do not correspond to points in the original plane. The points of the form $(x, y, 0)$ cannot be dehomogenized in the above way, since we would have to divide by zero. These points correspond precisely to the "points at infinity" of projective geometry. To see this, we study the behavior of a point that gradually moves to infinity in the original plane.

Assume that the moving point has coordinates $(r, r)$. As $r$ becomes larger and larger, this point gradually approaches a point at infinity in the 45° direction. Looking at its homogeneous coordinates, we see that they have the form $(r, r, 1) \sim (1, 1, 1/r)$. As $r$ increases, the contribution of the first two coordinates dominates the last coordinate more and more. In the limiting case of $r$ equals "infinity," the homogeneous coordinates are given by (1, 1, 0), an infinite point. You can also try to imagine the line through this point and the three-dimensional origin. As the point approaches infinity, this line becomes more and more horizontal, until, in the limiting case, it is contained entirely in the $xy$-plane.

A similar representation can be given for lines. For the line $ax + by + c = 0$ we take the parameters $(a, b, c)$ as the homogeneous coordinates of the line. As in the case of points, we identify nonzero multiples of such coordinates, since they do not alter the solution space of the corresponding equation. There is one set of parameters, (0, 0, 1), that does not correspond to a finite line. This is the line at infinity. The vector $(a, b, c)$ of a line is orthogonal to the plane spanned by the corresponding line and the origin of three-space. In particular, the vector (0, 0, 1) is orthogonal to the $xy$-plane, the "line at infinity."

In fact, the algebraic notion of homogeneous coordinates provides a complete symmetry between points and lines. Each point or line is represented by three homogeneous coordinates. A point $(x, y, z)$ lies on a line $(a, b, c)$ if and only if the scalar product $ax + by + cz$ is zero, which is simply rewriting the equation of the line. Geometrically, this means that the two corresponding vectors are orthogonal in three-space.

## 2.4 Complex Numbers

It is not only geometry that has been extended throughout the centuries. A similar process happened to numbers. Probably the first numerical concept considered by mankind was that of the positive integers: 1, 2, 3, .... From that beginning, it was reasonable to extend the number system gradually to encompass more powerful concepts. The *negative numbers*, the *rational numbers*, and the *real numbers* had to be invented to achieve a useful and self-contained system. The observation that there must be numbers that cannot be represented as quotients of two integers is of a geometric nature and dates back to approximately 600 B.C.E. It was observed by the Pythagoreans that there is no rational number measuring the length of a diagonal of a square with sides of length one. Through application of the Pythagorean theorem, this task is equivalent to finding a number $x$ such that $x^2 = 2$. This discovery led to a deep crisis in the foundations of ancient geometry.

However, the story of extending the number system does not stop at that point. One of the extensions, with perhaps the most drastic consequences, was the introduction of *complex numbers*. It was *Geronimo Cardano* in his 1545 work *Ars Magna* who was first to explicitly propose such an extension of the real numbers. He was led to his conclusions by the study of solutions of polynomials of degree three. Based on the work of other contemporary mathematicians, he discovered that a complete systematic representation of these solutions can be given only with the help of hitherto unknown values involving square roots of negative numbers.

A complex number is a number of the form $a + ib$, where the symbol $i$ satisfies the equation $i^2 = -1$, and $a$ and $b$ are real numbers. Clearly, the number $i$ cannot be a real number, since the square of a real number can never be negative. The system of complex numbers is, like the real numbers, closed under the four basic arithmetic operations of addition, subtraction, multiplication, and division (excluding, of course, division by zero). In other words, the sum and the product of two complex numbers, as well as their difference and quotient, can again be written in the form $a + ib$ for suitable parameters $a$ and $b$. However, unlike the real numbers, the system of complex numbers is also closed under the operation of finding solutions of polynomials. For instance, consider the polynomial

$$x^2 - 6x + 13 = 0$$

As you can easily check, it has no real solutions, but the complex numbers $3 + 2i$ and $3 - 2i$ do solve this equation. In fact, the following beautiful result is true: *Every polynomial equation with arbitrary real or complex coefficients has all its solutions again in the field of complex numbers.*

In a sense, the discovery of complex numbers is the starting point for most modern mathematics. Many mathematical theories find their broadest, most elegant, and most economical setting when they are formulated over the complex numbers. This happens also with geometry. Consider the situation of two circles. Depending on their position, they can have two, one, or no points of intersection.

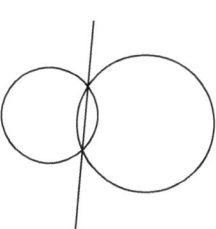

*Circles can intersect...*

*...or not.*

Finding the coordinates of the points of intersection amounts to solving a quadratic equation. Over the real numbers, this equation might have no solution. In this case, the circles do not intersect. Over the complex numbers, however, a solution always exists. So in the case of visually nonintersecting circles, we say that the intersections still exist, but that they have complex coordinates, and therefore we cannot see them in the real plane.

*Cinderella's* mathematical kernel is implemented entirely over the complex numbers. So when intersections visually vanish, *Cinderella* does not have to deal with special cases, and it can still continue calculating. The solutions just have complex coordinates.

What happens if two complex points are connected by a line? In general, this line will also have complex coordinates. However, if the points are so-called *complex conjugates*, which means that they differ only by the sign of their complex part, then their join is again real. Since the solutions of a quadratic equation with real coefficients are always complex conjugates, it follows that the intersection points of two circles are always complex conjugates. This is the reason why the line joining them is a real line, no matter where in the plane the circles are located, even if the intersection points are complex and therefore invisible. *Cinderella* will correctly calculate and draw this line, independent of the position of the circles. It may take a while to get used to the fact that intermediate results can disappear while some constructions depending on them remain visible. However, this is exactly what you should expect. Consider the case in which the circles have the same radius. The line is then the *perpendicular bisector* of the segment joining the two midpoints. If you include the complex situations, you are not compelled to consider so many special cases.

Another example of a theorem in which intermediate steps disappear is the following statement about three circles. Construct the line joining the two points of intersection of each pair of circles. The three chords that you obtain in this way meet in a point, no matter whether the circles intersect or not.

 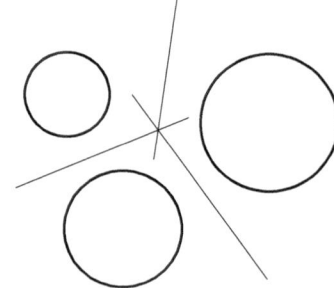

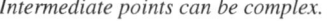

*Intermediate points can be complex...*                    *... but theorems remain true.*

So in *Cinderella* each point and each line is represented by *complex homogeneous coordinates*. This means that altogether, any point or line has a six-dimensional(!) internal representation in the mathematical kernel. This may sound crazy, but it is the most natural thing to do.

## 2.5 Measurements and Complex Numbers

If we were satisfied with projective incidence theory, then the system presented so far would be fairly complete. However, we want to be able to measure distances and angles, too. Measurements are in a sense the most fundamental geometrical operations. Unfortunately, at first glance, projective geometry does not appear to be capable of measuring, since under perspective transformations distances can change. Actually, for a long time mathematicians considered projective geometry a "nice toy" for doing incidence geometry, but not appropriate for the real stuff: measuring.

History proved them wrong. With the right setup, projective geometry is *the* universal system for doing measurements. This system unifies and explains different kinds of measurements. For example, it explains the relationship between Euclidean and hyperbolic geometry. However, it took a long time to finally find the algebraic setting in which Projective Geometry develops its full power. The key objects are called "Cayley-Klein geometries" in modern terms. It is an elegant and consistent mathematical approach to measurements that combines projective geometry and complex numbers.

### 2.5.1 Euclidean and Non-Euclidean Geometry

One part of this development started with the discovery of non-Euclidean geometries. Our everyday geometry is, with relatively great accuracy, described by Euclid's five postulates. He used these postulates to axiomatize geometry. This happened almost 2000 years ago. The last postulate, the so-called parallel axiom, plays a special role in the development of geometry. A way to formulate it is this: *"Whenever there is a line l in the plane and a point P not on l, then there is exactly one line through P that does not meet l."*

Euclid was very cautious about using the parallel postulate. Large parts of Euclid's elaborations, for instance the complete theory of congruence of triangles, were done without the explicit need for this axiom. Today we are relatively sure that Euclid himself believed that this axiom was a consequence of the other four axioms. But he could not prove this. After Euclid, many other mathematicians tried to do so, and some of them even presented proofs. But all these proofs were incorrect.

In the sixteenth to eighteenth centuries, mathematicians also found many equivalent formulations for the parallel postulate. One of the most prominent formulations is "The interior angles in a triangle sum up to 180°." If this statement could be derived from Euclid's first four axioms, then this would prove the dependence of the parallel postulate.

Proving that an axiom is dependent can be done by assuming its contrary and drawing conclusions until a contradiction is shown. Many people tried this, among them *C.F. Gauss*, *J. Bolyai*, and *N. Lobachevski*. They drew conclusion after conclusion, but to their surprise, instead of arriving at a contradiction, they found themselves developing a beautiful geometric system: *hyperbolic geometry*. There Euclid's parallel postulate is modified in the following way: *"Whenever there is a line l in the plane and a point P not on l, then there will be more than one line through P that does not meet l."* A consequence of this assumption is that the angle sum in a triangle is always less than 180°. Between 1815 and 1824, independently from each other, these three people, who are today considered the discoverers of hyperbolic geometry, came to the point at which they declared their system free from contradictions, simply because they could not find any. The system they developed was full of inner beauty, and it is a surprising fact that they could prove that under the assumption of the perturbed parallel postulate they end up with a unique theory (up to trivial isomorphisms).

It is worth mentioning that most probably Gauss was the first to arrive at these conclusions, around 1816. However, he did not dare to publish his results, since he was afraid of conflicts with the leading schools of Kantian philosophy at that time. They considered a straight line as the first thing whose nature is a priori clear.

If you are interested in the history of mathematics, we want to point you to the books of Bell [2, 1] and Struik [33]. As an introduction to hyperbolic geometry we recommend the book of Greenberg [8].

## 2.5.2 Cayley-Klein Geometries

For a long time it was not clear whether the system of hyperbolic geometry was indeed free of contradiction. What was missing was a model of this structure, a mathematical object that satisfied Euclid's first four axioms and the perturbed parallel postulate. With minor flaws, Beltrami was the first to construct such a model in 1868. However, the full beauty of a general theory was first seen when Felix Klein, a student of Plücker, presented his version of what we call "Cayley-Klein geometries"; see, for instance, [10]. What he gave was essentially a reduction of hyperbolic geometry to constructions of Euclidean geometry that implies the following: "If Euclidean geometry is free of contradictions, then so is hyperbolic geometry." This finally solved all problems surrounding Euclid's fifth postulate.

The idea behind Cayley-Klein geometries is to use the projective plane, and to distinguish a special conic, the "fundamental object." A special kind of global measurement is defined that depends only on the fundamental object. Depending on the type of the fundamental object you have chosen, you get different types of geometries: *Euclidean geometry*, *hyperbolic geometry*, *elliptic geometry*, *relativistic geometry*, and three other geometries of minor importance.

We will not go into the details of Cayley-Klein geometries, but we will present the major definitions and demonstrate some basic effects. We first need the concept of a *cross ratio*: For four points $A$, $B$, $C$, and $D$ on a line, the cross ratio is defined as the number

$$(A, B|C, D) = \frac{(A-C)(B-D)}{(A-D)(B-C)}$$

where $(A - C)$ denotes the usual "Euclidean distance" of the points $A$ and $C$. The cross ratio can also be defined without referring to the notion of Euclidean distance, which is important for a systematic treatment of geometry that is free from circular conclusions.

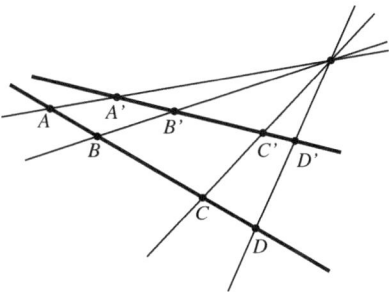

*The cross ratio*

The cross ratio is of remarkable value in projective geometry, since it is invariant under perspective transformations. So, if you have four points *A*, *B*, *C*, and *D* on a line and centrally project them to four points $A'$, $B'$, $C'$, and $D'$ on another line, then the cross ratios of the two quadruples of points are identical.

Similarly, the cross ratio of four lines through a point *P* can be defined to be the cross ratio of four points that are the intersections of each line with another distinct line not going through the point *P*.

Now the definition of a Cayley-Klein geometry is easy. Choose a quadratic form

$$ax^2 + by^2 + cz^2 + dxy + exy + fyz = 0$$

The zero set of this equation describes a (possibly complex) conic in the projective plane. This is the "fundamental object" of the geometry. Now measurements of angles and distances are defined as follows: For the distance between two points *A* and *B* take the line joining them. The intersection of this line with the fundamental object is two points *X* and *Y*. Calculate the cross ratio (*A*, *B* | *X*, *Y*). Take the logarithm of that number and call the result *distance*.

Angles are calculated in an analogous way. For the angle between two lines *L* and *M* first take their meet, that is, their point of intersection. The tangents through the meet that touch the fundamental object are two lines *P* and *Q*. Calculate the cross ratio (*L*, *M* | *P*, *Q*). Take the logarithm of that number and call the result *angle*. Usually, these two functions are multiplied by some cosmetic constants *r* and *s* in order to match the traditional definitions of measurements.

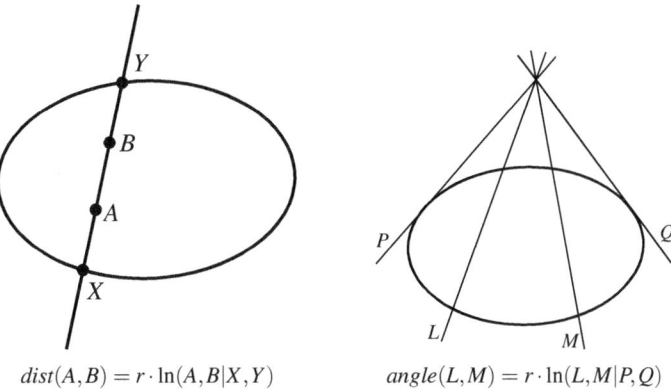

$$dist(A,B) = r \cdot \ln(A,B|X,Y) \qquad angle(L,M) = r \cdot \ln(L,M|P,Q)$$

It may sound like magic, but this is all you have to know. Depending on the type of fundamental object that you have chosen, you get different kinds of geometry. Up to isomorphism there are exactly seven different types of geometries you obtain that way. The three most important choices for the fundamental conic are the following:

- The circle given by $x^2 + y^2 - z^2 = 0$: The resulting measurement corresponds to hyperbolic geometry.

- The degenerate conic described by $x^2 + y^2 = 0$: The resulting measurement corresponds to usual Euclidean geometry.

- The equation $x^2 + y^2 + z^2 = 0$ with no real solutions: The resulting measurement corresponds to elliptic geometry.

Two things are worth mentioning:

- Distances in Euclidean geometry need a little twist. The formulas of Cayley-Klein geometry will always produce a zero distance. This is due to the fact that in Euclidean geometry there is no "absolute" notion of distance. Each length has to be compared to a unit length. The right formulas are obtained immediately in the limit situation.

- The intersections and tangents in the above constructions do not necessarily have real coordinates. For instance, in the case of elliptic geometry the fundamental object has no real points at all. In this case the intersections of the fundamental object with a line are always complex.

The metric part of *Cinderella* is based on Cayley-Klein geometries. All calculations of lengths, angles, orthogonality, circles, etc. refer to a fundamental object.

We finally want show at least one effect that is caused by this general theory. We do this in order to give you a feeling for what complex numbers, cross ratios, and projective geometry have to do with measuring.

We consider the case of Euclidean geometry. There the fundamental object has the equation $x^2 + y^2 = 0$. Using complex numbers, this quadratic form can be factored into two linear forms: $x^2 + y^2 = (1 \cdot x + i \cdot y + 0 \cdot z)$ and $x^2 + y^2 = (1 \cdot x + i \cdot y + 0 \cdot z)$. The points $I := (1, i, 0)$ and $J := (1, -i, 0)$ that occur in this formula play a special role in Euclidean geometry. They are not affected by any Euclidean transformation. In a very well defined way we can say that *"Euclidean geometry is projective geometry together with I and J."*

The points $I$ and $J$ are sometimes called the *imaginary circle points*, since they have a very special relation to circles: each Euclidean circle passes through $I$ and $J$. To see this consider a general circle equation in homogeneous coordinates:

$$ax^2 + by^2 + cz^2 + dxy + exy + fyz = 0$$

Now plug in the coordinates of $I$ and $J$. Using the rules for calculating with complex numbers we observe that the circle equation is obviously satisfied. Thus we can say that a circle is a special conic that passes through $I$ and $J$. With the notion of a circle it is easy to define what it means to have equal distances or equal angles. The remaining concepts of Euclidean geometry can be derived in a straightforward manner. A very good classical source on Cayley-Klein geometries can be found in [10]. A Modern treatment of these issues that also includes a broad introduction into projective Geometry may be found in [32] and [26].

## 2.6 The Principle of Continuity

It was mentioned in the preface of this manual that *Cinderella* uses some basic new methods to avoid inconsistent behavior. The geometric system we presented in the previous sections is a closed framework for doing geometry, including measurements. However, so far there is an element missing that is crucial for *Cinderella*: *dynamics*. Most other systems for interactive geometry, or parametric CAD, suffer from inconsistencies that come from an unsatisfactory treatment of the special effects of dynamics. For instance, consider the "theorem" that *the angular bisectors of the sides of a triangle meet in a point.* Every pair of lines has two angular bisectors that are perpendicular to each other. Depending on the choice of the angular bisectors, the above statement can be true or false. Now imagine you have constructed an instance of this "theorem" (i.e., you have chosen the correct bisectors). You start to drag the vertices, around and suddenly, without reason, one angular bisector flips to the other position and the "theorem" becomes false. Such a scenario can happen in any system that does not take extra measures to resolve the special problems from the dynamic aspects of geometry.

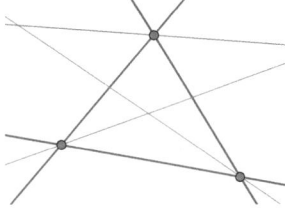

*The angular bisectors intersect. . .*                    *. . . or not.*

Let us consider another small construction. Take two circles and one of their intersections. While you drag elements around, *Cinderella* has to decide for every mouse move which intersection you "mean."

It is a good first attempt to "trace" this point of intersection using the following rule: *"Always take the intersection that is closest to the previous position,"* since this precisely reflects the definition of continuity. But how should we deal with vanishing intersections? Again it pays off having implemented everything in complex space. In *Cinderella* intersections never vanish; they can become complex, however. So we have to trace the intersections in complex space and use the above rule.

However, this is not enough. When you separate two circles that were previously intersecting, there is always a position in which the two points of intersection coincide. How can the points be distinguished in this situation? This time it is analytic function theory that rescues us. *If "detours" through complex space are allowed, then there is always a path that avoids all degenerate situations.* Again the whole approach is possible only because everything is embedded in a complex space.

Here is an approximate description of what happens while you drag the mouse from one position to another in the "move" mode of *Cinderella*. While you move a point from position $A$ to position $B$:

- *Cinderella* generates a path from $A$ to $B$ through complex space that avoids all degeneracies.

- The dependent elements are traced through complex space.

- The number of intermediate steps on the path is adjusted according to the required accuracy.

For the tracing, *Cinderella* uses an adaptive step-width algorithm. You can imagine that while you drag the construction elements the mouse pointer leaves the "real screen" and walks through complex space.

Why all this effort? With these methods we can be sure that elements do not "jump around" for no reason. So when you start with a correct drawing of the angular bisectors theorem you will never be able to move it to a false position. The theory also forms the basis for reliable randomized theorem checking and for the loci and animation functions of *Cinderella*.

# Chapter 3

# Quick-Start Tutorials

The following sections will guide you through large parts of *Cinderella's* functions. After working through the tutorials you should be able to use *Cinderella.2*. You can look up other functions in the reference part of this manual. Most of the functions described there follow patterns that are explained in the tutorials.

Here are a few general instructions on how to use these tutorials.

- Read the texts carefully. When you come to a new section of the tutorial, read it first completely. Then you should read it again, while you *perform the described actions.*

- Follow the instructions exactly. The tutorials ask you to create constructions in a certain order. If you do not follow the prescribed order, the labeling of your elements will be different from that in the tutorial.

- If you make a mistake in a geometric construction, you always have the opportunity to undo the last step by pressing the "Undo" button in the toolbar. You can undo as many steps as you want.

- The pictures in the tutorials are intentionally low-resolution screenshots. That should make it easy to match these drawings with the actual situation on your screen.

Each tutorial focuses on a particular topic. As the tutorials progress, the information provided by the text becomes less and less detailed. We assume that it is helpful in the very beginning to be told *precisely* what to do. The more advanced you become, the more freedom you will have to do the operations. There are three tutorials:

- *Pappus's Theorem* (p. 42): This tutorial introduces you to the basic construction principles of *Cinderella*. You will learn how to do a simple construction.

- *Three-Bar Linkage* (p. 51): Here you will learn how to use the animation features of *Cinderella*. You will also learn how to produce geometric loci.

- *Kepler Ellipses* (p. 56): Here you will learn how to use the physics simulation engine CindyLab in a very simple example.

- *Scripting the Seeds of a Sunflower* (p. 58): Here you will learn how to perform simple drawing operations with CindyScript. You will also learn how to combine CindyScript with the geometry part of Cinderella.

## 3.1 Pappus's Theorem

Pappus's theorem is one of the most fundamental theorems in projective geometry. In a certain sense it is the smallest example of an incidence theorem. In this step-by-step example we will construct an instance of the theorem that can be selected and dragged.

### 3.1.1 Drawing Your First Point

When you start *Cinderella.2*, the first window that appears is a "Euclidean view." This window has a large toolbar that is the key to most of *Cinderella's* functionality. Below this toolbar is the drawing surface on which you perform operations by constructing and dragging the elements you need.

You may notice that in the toolbar one button is slightly darker. This indicates the current mode of *Cinderella*.

All mouse actions in the drawing surface refer to the selected mode. For drawing a single point you have to switch to the add a point (p. 74) mode by clicking it. Now *move the mouse over the drawing surface and click the left button*. A new point is added and labeled with a capital letter.

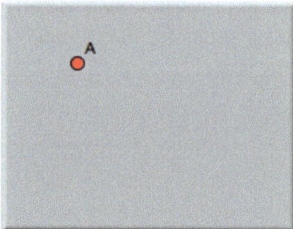

*Figure 1: Your first point*

Before you add a second point, continue reading. If you hold the left mouse button down while adding a point, you are able to move the point around. The definition of the point will be adapted to the geometric situation at the current location of the mouse pointer. (So far, there is not a lot of geometry in our drawing, but this will change soon.) *Move the mouse over the drawing surface, press the left button, and hold it. Now move the mouse.* You will notice that the new point sticks to the mouse pointer. Coordinates that indicate the current position are shown. When you approach an existing point (*try it*), the new point snaps to the old point. Only after you *release the button* is the new point added to the construction. If you release the mouse over an old point, no new point is added. *Play with these features and add a few new points.*

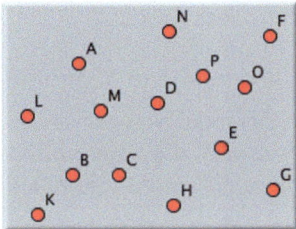

*Figure 2: Many points*

### 3.1.2 Undoing an Operation

Your screen may look a bit crowded now. There is an undo operation that inverts the actions you have performed. By pressing the undo button, ⬅ you will cause the last point you added to disappear. *Click the undo button until exactly two points remain on the drawing surface.* You can use the undo button whenever you make a mistake. You can undo as many consecutive operations as you want.

### 3.1.3 Moving a Point

We want to continue with our construction of Pappus's theorem. You should first move the two remaining points $A$ and $B$ to approximately the positions shown in Figure 3.

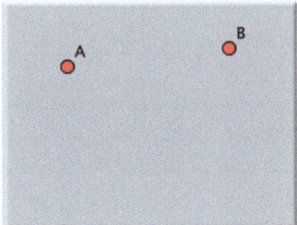

*Figure 3: Move to this situation.*

Most probably, your points are not already there. To move them, *select move mode* (p. 71) *by pressing the "move mode" button* 🖱 *in the toolbar.* Now mouse actions in the drawing surface no longer add points; instead, you can select points and move them around. *Move the mouse pointer over the point you want to move. Press the left mouse button. Hold it and drag the mouse.* The point follows the mouse pointer. You will also notice that in this mode a point does not snap to other points. When you *release the mouse button*, the point is placed. In general, move mode allows free elements of a construction to be moved. The rest of the construction will change accordingly.

### 3.1.4 Adding a Line

We are now going to add a line from *A* to *B*. *Switch to add a line* (p. 76) *mode by pressing the button* 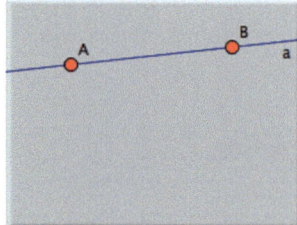. In this mode you can add a line between two points using a press–drag–release sequence with the mouse. *Move over point A. Press the left mouse button. Hold it and move the mouse over point B. Release the mouse.*

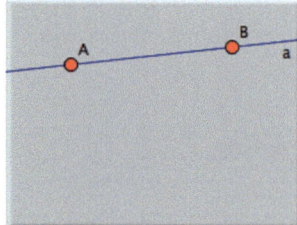

*Figure 4: Add a line*

This action should have created the desired line. You might have noticed a few things: When you pressed the mouse over point *A*, the point became highlighted. This means that you selected this point as the starting point of the line. While you drag the mouse, a second point is always present at the mouse position. When you release the mouse button, this second point will be added at the release point. When you approach another point, it will be highlighted, and the second point will snap to it. This is the action you used for attaching the line to point *B*.

If you did not notice these things or made a mistake, undo your operations and try again. Your final picture should look like Figure 4 before you proceed.

### 3.1.5 Adding More Lines

Now you are ready to add three more lines, in order to end up with Figure 5. You can do this with just three mouse operations. *First, move over point B and press–drag to the position of the not-yet-existing point C. Do another press–drag–release from C to D, and finally from D to E.*

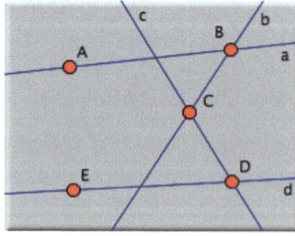

*Figure 5: More lines*

Observe that you have done all this without leaving "add a line" mode. You did some unnecessary work when you added the points *A* and *B* in add a point (p. 74) mode. You could have done this directly in add a line (p. 76) mode, since not only will the second point be created if necessary, but also the first one. The lines created so far were automatically labeled with the lowercase letters *a* to *d*.

### 3.1.6 Creating Points of Intersection

We will stay in add a line (p. 76) mode. Now we want to draw a line from point *E* to the intersection of the lines *a* and *c*. *Move the mouse over E. Press the button. Hold it and move to the intersection of a and c. Release the mouse.* .

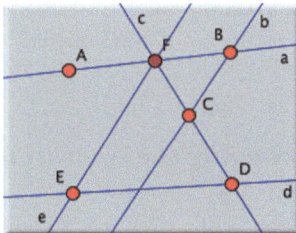

*Figure 6: Adding an intersection*

When you approached the intersection of the lines, they were highlighted and the endpoint of the new line snapped to the intersection. When you released the mouse, the new point was defined to be the intersection of the lines. If you move any one of the points *A* to *E*, the new line will follow the moves accordingly.

The new point *F* is drawn slightly darker. This means that it is not possible to move *F* freely in move mode. Thus, *F* is a "dependent" point. The other points in the construction are "free."

Two things are worth mentioning: With the same procedure you could also have added a "half-free" point that is bound to only one line. To do so, you need to release the mouse when you are over only one line. All these operations of creating intersections (dependent points), free, and half-free points work similarly in add a point (p. 74) mode. Many other modes, the interactive modes (p. 74), also come with this feature. Browse the reference guide of this manual for further information.

### 3.1.7 Finishing the Drawing

Now it should be easy for you to finish the drawing by adding four more lines to achieve the final configuration shown in Figure 7. *First add a line from A to the intersection of b and d. Then draw two lines from A to D and from B to E. Finally, draw a line from C to the intersection of e and f.*

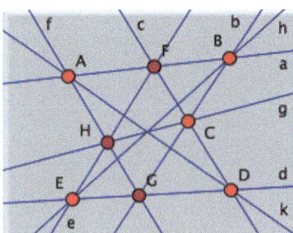

*Figure 7: Pappus's theorem*

You have created a configuration of eight points and nine lines. If you look at the configuration, you will notice that if everything was done correctly, the lines *g*, *h*, and *k* meet in a point. That this will always be the case in such a construction is the content of Pappus's theorem. *Switch to move mode (p. 71) by clicking* 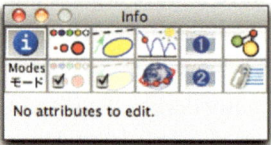 *in the toolbar, and drag around the free points of the construction to convince yourself of the truth of this theorem.* This theorem was known to the ancient Greek geometers and is named for Pappus of Alexandria, who flourished in the fourth century C.E. Later, the theorem turned out to be of fundamental importance for the theory of *projective geometry.*

### 3.1.8 Selecting and Changing the Appearance of a Drawing

You may not like the appearance of a drawing. The lines may be too thin, and perhaps you want to emphasize the final conclusion of the theorem. This can be changed by using the inspector (p. 153) of *Cinderella.2.* For this *choose the menu item "Edit/Information".* The following window will pop up:

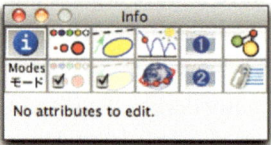

*Figure 8: The inspector*

There are several tabs in the inspector that refer to different aspects of the construction and the selected objects. To change the appearance of objects, go to the appearance tab by *clicking the button* 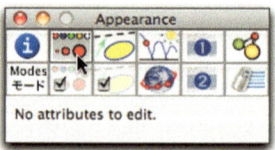. If no elements are selected, the inspector shows that there are no attributes to edit.

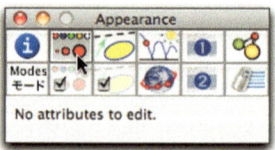

*Figure 9: Selecting the appearance tab*

As soon as an element is selected, the editable attributes of this elements are shown. *Now select all lines by clicking the button* ✕ *"Select all lines".* The inspector window then looks as follows:

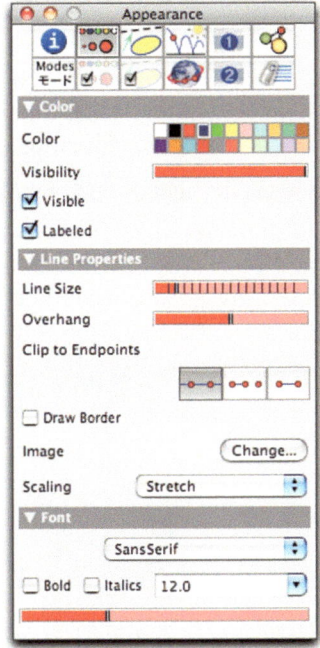

*Figure 10: The appearance tab*

Changes in the inspector are immediately applied to selected elements. If you now *move the slider "Line Size" in the inspector to its third tick*, all lines become thicker.

After this, *switch to move mode* . Now, when you click over one or more elements, those elements will be selected. If you hold the shift key while you click, the selection state of the element will be toggled.

*Click over line k, and then hold the shift key and click over lines h and g. The lines g, h, and k should be highlighted.* *Click the red color box in the inspector's color palette.* The color of the three lines changes from blue to red. You can also change the size of these lines by moving the "Line Size" slider to its fourth tick. Finally, *deselect everything* by clicking the button .

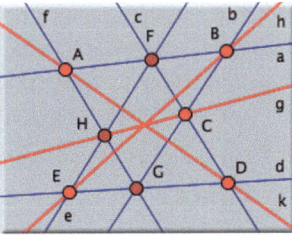

*Figure 11: Changed appearance*

### 3.1.9  Adding a Final Point and Proving a Theorem

Before continuing, open *Cinderella's* information window by choosing the menu item *Views/Information Window*. A console window pops up. *Cinderella.2* uses this console to report nontrivial facts about a configuration.

We will now create a final point at the intersection of the three red lines. This time we will add a point by *changing to intersection* (p. 92) *mode* 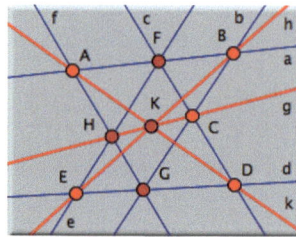. This mode expects you to mark two lines. A new point will be added at the intersection of these two lines. *Click two of the red lines, say g and h.* Their intersection point will be added.

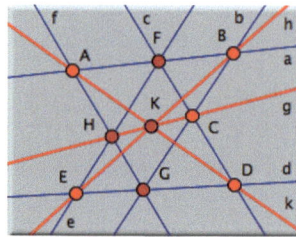

*Figure 12: The point of intersection*

Note that the newly added point *K* will always be incident to line *k* because of Pappus's theorem. The information console reports this remarkable fact automatically.

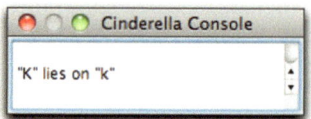

*Figure 13: A proven theorem*

You may wonder how this "theorem proving" works. *Cinderella* does not use symbolic methods to create a formal proof, but a technique called "randomized theorem checking." First the conjecture, "it seems that line *k* always passes through point *K*," is generated. Then the configuration is moved into many different positions, and for each of these it is checked whether the conjecture holds. It may sound ridiculous, but generating enough (!) random (!) examples in which the theorem holds is at least as convincing as a computer-generated symbolic proof. *Cinderella* uses this method over and over to keep its own data structures clean and consistent.

### 3.1.10  Moving Points to Infinity

Let us explore the symmetry properties of Pappus's theorem. For a nicer picture, we want to get rid of the line labels and make the points and lines a bit smaller. *First select all lines using* 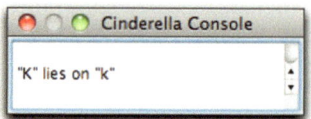. *Turn off the labeling by pressing the corresponding button in the inspector. Change the line size to "thin" (use the slider in the inspector). Then select all points using* . *Make them smaller using the "Point Size" slider in the inspector (observe that the inspector shows always information about the selected elements). Now choose the menu option "Views/Spherical View".* What you get will not look very instructive at first glance.

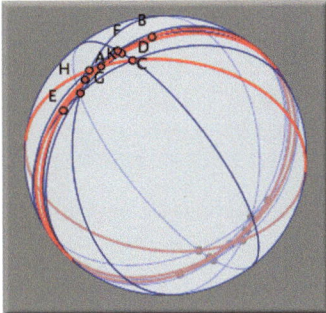

*Figure 14: A spherical view*

The spherical view shows a central projection of the plane to the surface of a ball. The projection point is the center of the sphere (see Figure 15). Every point is mapped to an antipodal pair of points, and each line is mapped to a great circle (a circle on the sphere dividing it into two equal hemispheres) of the ball.

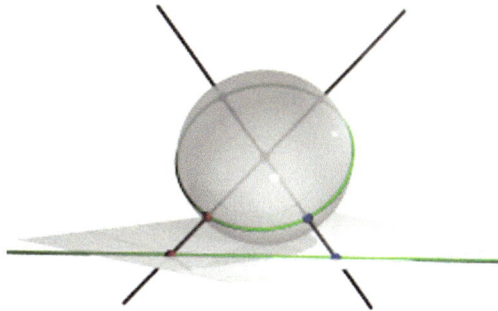

*Figure 15: Projection from the plane to the ball*

In spherical view there is a little red slider for controlling the distance of the ball to the original plane. Moving this slider corresponds to a kind of "zooming" operation on the sphere. Please *move the slider from its original position much more to the right.* Then the situation on the ball should become clearer. If you have adjusted the slider correctly, it should look somewhat like Figure 16. You should clearly recognize the original construction, now drawn on the surface of this ball.

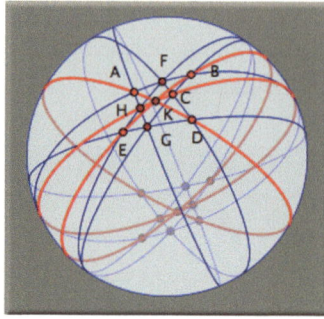

*Figure 16: Pappus's theorem on a ball*

The picture needs some explanation: For each point in the plane consider the line spanned by the point and the center of the ball. The intersection of this line with the surface of the ball gives the pair of antipodal points. For each line consider the plane spanned by the line and the center of the ball. The intersection of this plane and the surface of the ball is the great circle that represents the line in spherical view.

On the sphere there are points that have no correspondence in the Euclidean plane. If the ball touches the Euclidean plane at the "south pole," then the equator of the ball corresponds to the "points at infinity" in the usual Euclidean plane. In *Cinderella* it is possible to make manipulations in any currently open view. Therefore you can also move points in spherical view. The changes will be instantly reported to the Euclidean view. In particular, you can grab a point in spherical view and move it to what is really infinity in the Euclidean view.

We will do that now, in order to observe that our configuration has a nice threefold symmetry. *Select point A (in spherical view) and move it to the 11 o'clock position of the boundary.* This point is now really located "at infinity." Notice that in Euclidean view the lines passing through *A* became parallel, corresponding to the notion that "parallel lines meet at infinity." In a similar way, *move point E to the 7 o'clock position on the boundary and move point C to the 3 o'clock position.* Your spherical picture should then look like the ball in Figure 17. In Euclidean view you find three bundles of parallel lines, representing a kind of Euclidean specialization of Pappus's theorem. If you like, you can try to figure out the corresponding statement in terms of parallels and incidences.

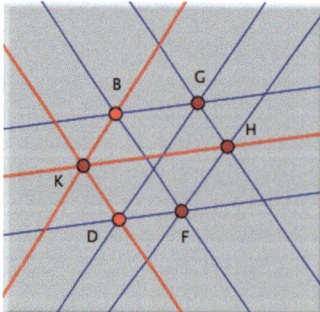
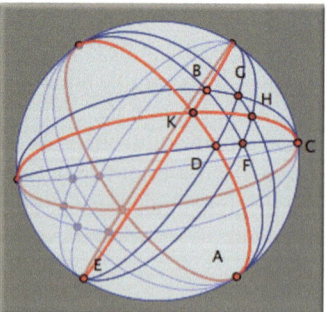

*Figure 17: A threefold symmetry*

It may happen that the drawing in Euclidean view is too big or too small. You can use the zooming tools to change this. There is a "Zoom-In" 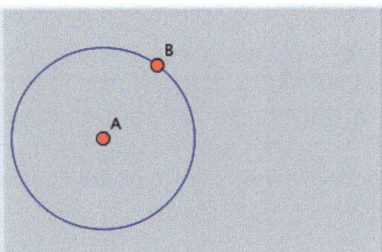 and a "Zoom-Out" tool below the Euclidean view. With a press–drag–release sequence you can mark the region to be zoomed in or out.

## 3.2  A Three-Bar Linkage

In the second tutorial we want to explore the dynamics of a little mechanical linkage. In particular, you will learn how to make bars of a fixed length. You will also learn how to produce a geometric locus of a point.

### 3.2.1  Making a Bar

After starting *Cinderella* or erasing a previous configuration, *switch to circle by radius (p. 82) mode by pressing the button*. This mode is used for working with circles whose radii remain constant when their centers move. *Move the mouse pointer over the view. Press the left mouse button. Hold the button while dragging the mouse. Release the mouse.* With these actions you have generated a circle. Switch to move mode (p. 71) to see how the circle behaves. When you select the center you can move the position of the circle, and the radius stays constant. You can also select the circle itself and move it. Then its center stays fixed and the radius of the circle changes.

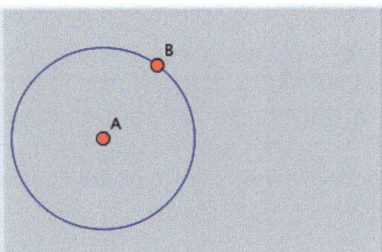

*Figure 1: A first circle*

We now want to add a point that is bound to the boundary of the circle. *Select add a point (p. 74) mode. Move the mouse over the view. With the button pressed, move the mouse pointer to the boundary of the circle. When the circle is highlighted, release the mouse.* You have now created a point that is on the boundary of the circle. You could alternatively have clicked the mouse over the circle boundary. However, that gives you a bit less control over whether you actually hit the circle. Your configuration should now look approximately like that in Figure 1.

*Switch to move mode (p. 71) to see how this new point behaves.* If you select it and drag the mouse, the point will never leave the circle; it attempts to remain as close to the mouse position as possible. When you move the center of the circle or change the circle's radius, the point also stays on the circle. The point furthermore keeps its relative angle to the center. *Choose the interactive add a segment (p. 84) mode by pressing the* 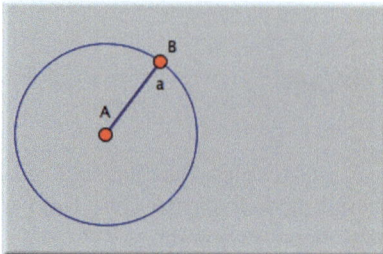 *and draw a segment from the center to the point on the perimeter* (see Figure 2).

*Figure 2: A bar*

This line now behaves like a bar of fixed length, given by the radius of the circle. The only way to change its length is to change the radius of the circle.

### 3.2.2 Adding Two More Bars

We now want to add two more bars to complete the three-bar linkage that was presented in the introduction (p. 5). The linkage is a chain of three consecutive bars that is pinned down at each of its ends. *Switch to circle by radius (p. 82) mode again and add a second circle to the right of the first one. Do not let the two circles intersect.*

The radius of the second circle represents the length of the third bar in the row (we have not created the second one yet). Before finishing the construction we should analyze the situation. If point $C$ in our construction is to be linked to point $B$ of the construction by two bars of given lengths, then there is not much freedom for the point that is common to those two bars. In fact, there are exactly two possible positions for this point. They are the two intersections of two circles: one circle around $C$, already drawn, and another, yet to be drawn, circle around $B$.

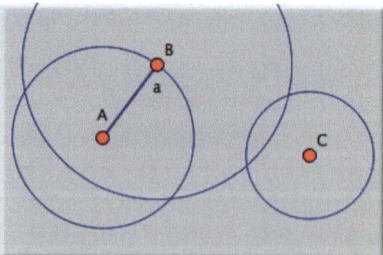

*Figure 3: Two more circles*

Therefore, as the next step, *draw a circle around point B in circle by radius (p. 82) mode. Make sure that its radius is large enough so the circle intersects the circle around C.* This stage of the construction is shown in Figure 3. Finally, draw a segment in segment (p. 84) mode ⧉ from *B* to one of the intersection points of the circle around *B* and the circle around *C* (this is an alternative to adding a line and clipping it to its endpoints).

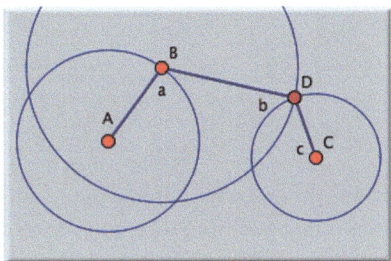

*Figure 4: The three bars*

This new point is automatically labeled *D*. Observe that this point is not a free point, since its position is determined, up to choosing the second intersection, by the positions of *A*, *B*, *C*, and the radii of the circles.

*Finish the construction by adding the third bar from C to D.* Most probably, the lines will be already clipped, since you made this choice in the appearance editor. If not, select them and use the appearance editor to clip them. Your drawing should now look like Figure 4.

### 3.2.3  Moving the Construction

The lengths of the bars are determined by the radii of the circles. *Select move mode (p. 71) again and play with the construction.* An interesting thing happens when you move point *B*, the point on the first circle.

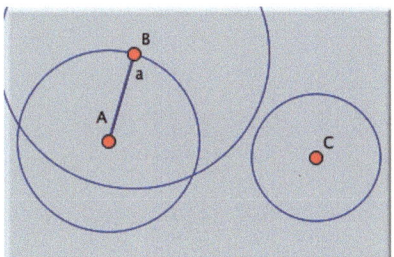

*Figure 5: The bars are too short*

First of all, you notice that the whole construction has exactly one degree of freedom when point *B* is moved. Moving it allows the whole construction to behave as if it were a mechanical linkage. We can consider the possibility of moving *B* as a kind of "driving input force"; that is what CAD people would call it. There are positions for point *B* for which the bars are too short (see Figure 5), which means that the two circles no longer intersect. When this happens, the lines and the intersection point disappear.

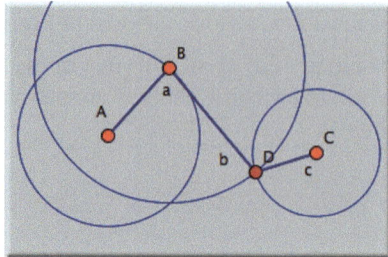

*Figure 6: After coming back*

One of the most interesting things, which you might initially overlook, happens if you return point *B* to the original position, where the circles once again intersect: the other point of intersection is chosen; the linkage takes the other possible position. Try this many times to get a feeling for it. *Move point B back and forth, so that the bars disappear and reappear.* This behavior seems a bit counterintuitive at first, but it is exactly the right thing to happen. Imagine that the bars were made from real matter, steel or wood, for instance. Then they would have a certain mass. What would happen if you pushed the whole construction and let it move freely, driven only from inertial forces? The two bars, those that disappear in the drawing, would come to a position in which they formed a straight line. Then point *D* would "sweep over", and the whole construction would move to the other possible position.

### 3.2.4 Starting an Animation

If you still do not believe that this is natural behavior, *choose animation* (p. 107) *mode by pressing* . The message line below the view asks you to choose a "moving element." In our case we want *B* to be the moving point: *Click B.* Since point *B* has only one degree of freedom, it is clear how to move it, and the animation is started immediately. Otherwise, you would have been asked to select a "road" on which *B* should move. In this construction the road is clear: it is the first circle.

An animation control panel

is added to the geometric view. It has buttons like a CD player to start, stop, and pause the animation, and it has a slider to control the animation speed. By pressing the play button you can start the animation.

For a moment enjoy watching the animation. Observe that point *B* moves only in a region in which the bars are long enough. It "knows" when to change the direction. *Cinderella* tries to model true physical behavior. It does not add masses to the moving elements, but it uses methods from complex analysis to determine the most reasonable thing to do. You can also move the free elements during the animation. Now exit the animation by pressing the "Stop" button in the animation controller.

### 3.2.5 Drawing a Locus

We now would like to know how the midpoint of the middle bar moves during the animation. We first add the midpoint. *Choose midpoint* (p. 83) *mode* ▢. By a press–drag–release sequence you can add the midpoint of two points: *Move the mouse over point D. Press it, and with the left button pressed move over point B. Now release the mouse button.* The midpoint is added.

Now *choose locus* (p. 103) *mode* ▢. This mode requires that you select a "mover," a "road," and a "tracer" in this order. The mover is the element that moves (the "driving force"). The road, which is the path along which the mover will move, must be incident to the mover. The tracer is the point whose trace will be calculated.

*Click point B (the "mover").* Cinderella recognizes that this point has a unique road and selects the circle as well. So finally, *point E has to be selected as tracer.* It takes a second, and the locus is generated automatically. You can switch to move mode and see how the locus changes when you move the free elements.

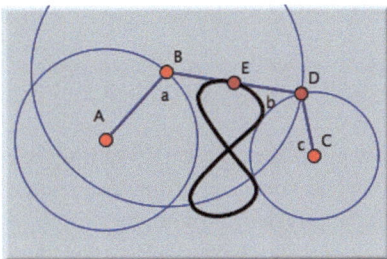

*Figure 7: Constructing a locus*

If you restart the animation by pressing the "play" button you will see how the locus was generated.

### 3.2.6 Experimenting

It is worth to experiment with the resulting construction. The constructed locus varies dynamically with movements of the free parameters of the construction (points A and C and the radii of the circles). The locus may take interesting and quite surprising shapes, as the following collection of pictures shows.

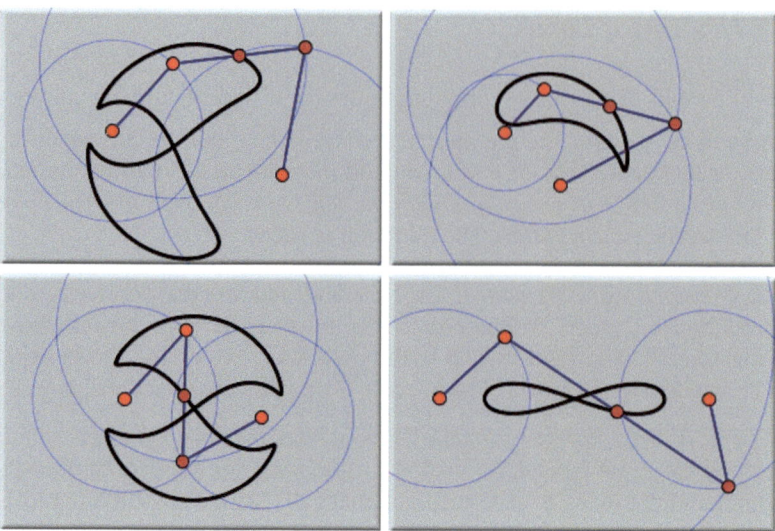

## 3.3 Kepler Ellipses

In this little tutorial we want to set up a very small physics simulation with Cindy-Lab. We want to model a planet orbiting a sun. You should learn in this tutorial how to set up and start running a physics simulation.

After starting Cinderella, *go to "add sun" mode by clicking the* ☀ *button in the toolbar*. Then add a sun at an arbitrary position. After the sun has been added, an animation control will appear. The animation control is always present when physics elements of CindyLab are present. The view will now look approximately like Figure 1.

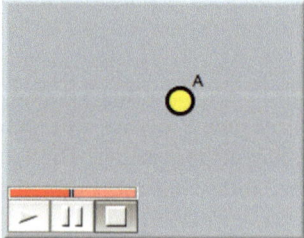

*Figure 1: A sun is born*

Now we add the planet that we want to orbit the sun. The planet needs an initial velocity perpendicular to the sun. Otherwise, it will be attracted directly toward the sun and end up in a singularity. For adding the point *go to "add velocity" mode by clicking* ➶ *, and perform a press–drag–release operation somewhere near the sun.* When the mouse is pressed the mass-point is created. Dragging creates the force, and releasing finalizes the situation. The picture should now look approximately like Figure 2. If not, you can change the position of the points and of the arrowhead in move mode.

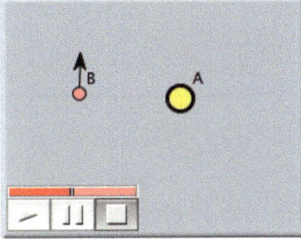

*Figure 2: A sun and a planet*

Now you can *start the simulation by pressing the play button*. Most probably, the animation will run too fast. You can change the animation speed with the slider in the animation control. The planet should describe an elliptical path (this is the essence of Keplers third law).

We now want to make the ellipse a little more visible. For this we want to create a trace of point *B*, the planet, while it is moving. Open the inspector and go to the "Special Appeareance" tab by clicking the button ▱. After point *B* has been selected, the inspector looks like Figure 3:

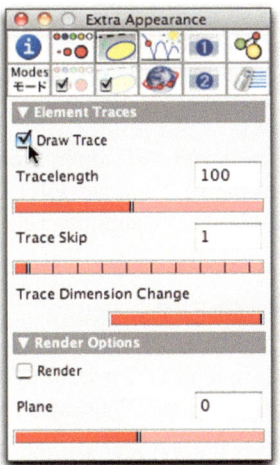

*Figure 3: Inspecting the planet*

*Check the "Draw Trace" checkbox*. This causes Point *B* to leave a trace as it moves. *Now restart the animation*. You will see the ellipse as a trace of the planet, as in Figure 4.

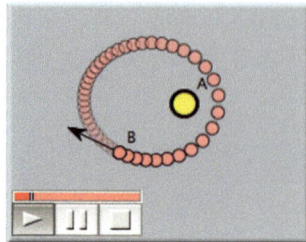

*Figure 4: The planet's orbit*

We recommend that you carry out several experiments with this configuration. Here are a few proposals:

- Vary the location of the planet.

- Vary the speed (the magnitude of the velocity).

- Vary the direction of the velocity.

- What happens when more than one sun is present?

## 3.4 Scripting the Seeds of a Sunflower

In this tutorial you will learn some elementary applications of CindyScript, the internal Programming language of Cinderella. You should be aware that this tutorial covers only a very small portion of the CindyScript functionality. Nevertheless, you will receive an impression of how CindyScript can be used.

The following example will be developed step by step. In each step we will add one conceptually new element. At the very end we will arrive at an interactive visualization of the packing pattern of seeds in a sunflower head.

### 3.4.1 A first line of CindyScript

We start with an extremely simple task. We want to draw one point by CindyScript that is bound to a given geometric point. We start by adding one point using the add a point mode (p. 74) . It is a good idea for the following example to also turn on the snap grid by pressing the button  in the button line below the drawing surface. After this step your drawing should look like Figure 1.

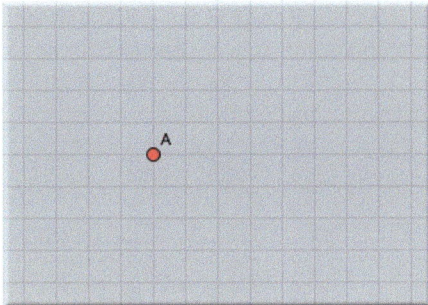

*Figure 1: A single point*

CindyScript is (among other things) capable of adding graphical elements to your construction. Now we want to draw a single point displaced by $(1,1)$ from point $A$. In order to achieve this we must open the CindyScript editor in the menu *Scripting –> Edit Scripts*. Then the window shown below will pop up. This is where CindyScript code is entered primarily.

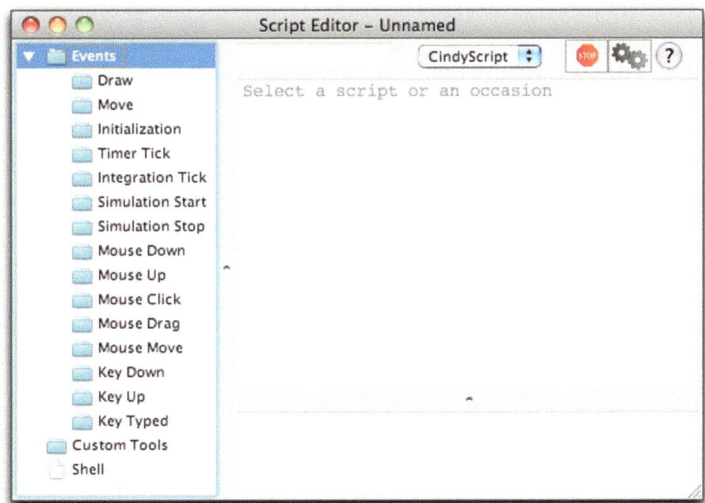

*Figure 2: The script editor*

You will notice that on the left side of this window there is a list of many occasions. Each piece of CindyScript is associated to one of these occasions and will be executed when a corresponding event occurs. We will add a piece of script in the *draw* event. This means that the script will be executed right before drawing. This is the right place to add code that produces additional graphical output. For adding the piece of code we click on the draw event (see Figure 3).

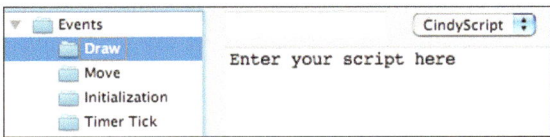

*Figure 3: Entering code in the draw event*

After this we can add a piece of code in the main part of the editor. To draw the desired point the code snippet shown in Figure 4 is already sufficient.

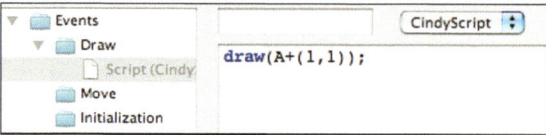

*Figure 4: A first line of CindyScript*

We will comment on this line of code briefly. The `draw(...)` statement accepts a two-dimensional vector. It will draw a point at the corresponding position on the drawing surface. By default this point will be green and small. (It is also possible to use the draw statement with two vectors as arguments. Then a segment between the two corresponding points will be drawn.) In our code the argument is `A+(1,1)`. This summation statement will automatically use the *xy*-coordinates of point *A*. To these coordinates we add the vector '(1,1)'. In arithmetic expressions geometric elements are automatically cast into their coordinates.

After entering the line of code you must press the script start button (the one with the two gears) in the script editor. After this your drawing should approximately as shown in Figure 5.

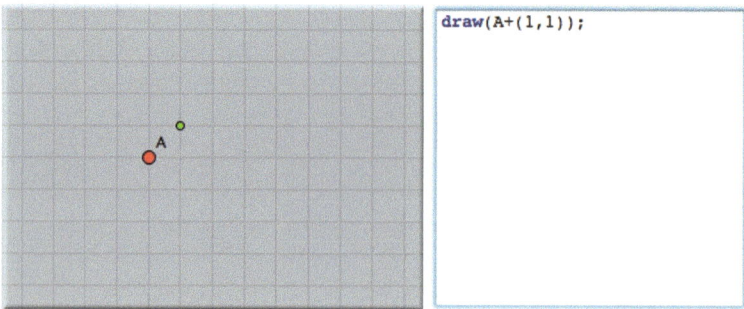

*Figure 5 : Drawing a point*

### 3.4.2 Loops

Now we will add a simple control structure to our script. Instead of one single point we want to draw a whole sequence of points. For this we will use the `repeat(...)` statement to generate a simple loop. `repeat(...)` can be used with three arguments. The first one is the number of loops. The second is the name of the running-variable. The last argument is the piece of code that will be executed during the loop. The arguments are separated by commas. In our case we will use the piece of code shown in Figure 6. (Remember to press the start button again to activate the changes to your script.)

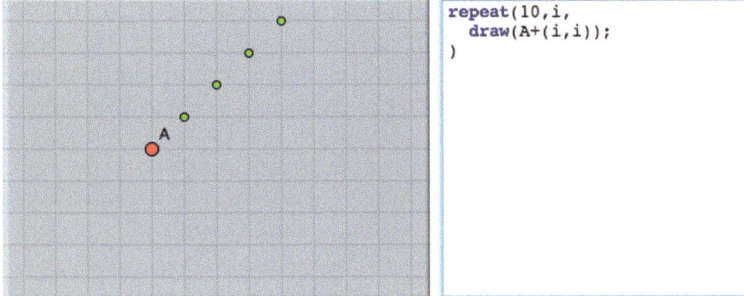

*Figure 6 : A loop that draws points*

This code draws all together 10 points. The variable `i` is used as running-variable. During the loop it assumes values `1,2,3,4,5,6,7,8,9,10`. In contrast to the last step we now calculate the vector `A+(i,i)`. All together this creates a diagonal chain of points.

In the next step we want to place the point cyclically around point *A*. The distance from *A* should be 1 and the points should have an angular displacement of $10°$. For this we define a new variable `w` that plays the role of the angle. We add the line `w=i*10°` to our loop. This automatically creates the variable `w` (there is no need of an explicit typing) and associates the angle to it. Then we change our code by adding the vector `(sin(w),cos(w))`. The result is shown in Figure 7 below.

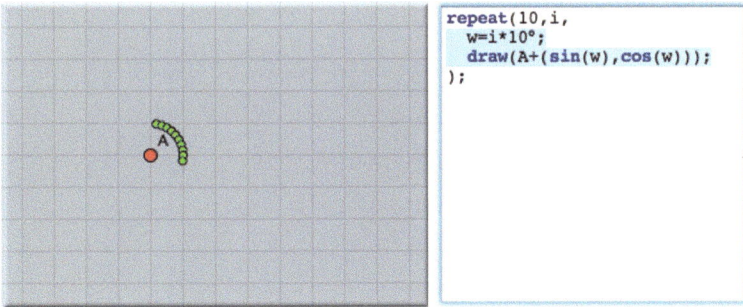

*Figure 7 : Drawing points in a circle*

### 3.4.3 Adding a slider

We now want to interactively control the number of points added by a slider in the drawing surface. Sliders are constructed explicitly as geometric objects. For this one usually adds a line segment and places a point on it. However also other types of sliders (for instance circular sliders) are possible variants. We add a simple segment by using the add a segment (p. 84) ![icon] mode. The endpoints of this segment will be *B* and *C*. Then we add a point on this segment by using the add a point (p. 74) ![icon] mode. This point will be point *D*. We will now use the distance from *C* to *D* as a controlling entity for the number of points. The distance can be measured by `|C,D|`. We multiply this value by 10 and round it to get the value `n` of number of points to be drawn. This can be done using the simple expression `n=round(|C,D|*10)`. Again observe that no explicit declaration of variables or typing is necessary. Variables like n are simply containers for values, no matter what these values may be. We further-more have to change the number of loops from the value 10 to n. The corresponding piece of code together with a picture of the current state of our drawing is shown in Figure 8. In this drawing the number of points can be controlled interactively by moving point *B*.

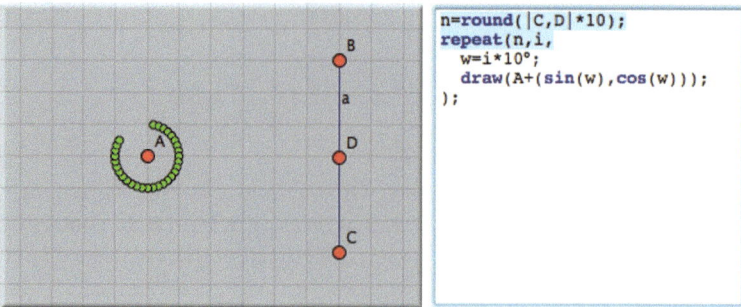

*Figure 8 : Adding a slider*

### 3.4.4 A strange angle

We will now do something the importance of which may not be clear immediately. We change the angular displacement to a value of 137,508°. The result of this action is shown in Figure 9 below.

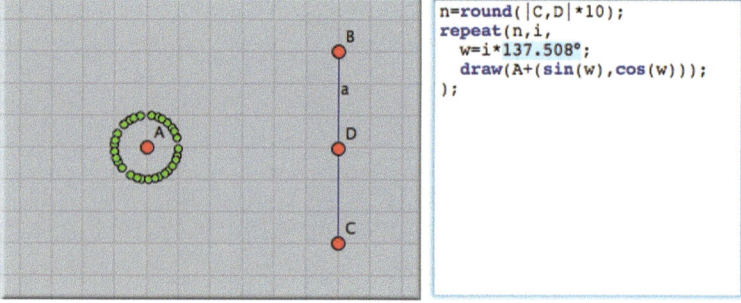

*Figure 9 : A "strange" angle*

The points seem to be distributed almost randomly. In fact the value of 137,508° has a certain mathematical meaning. It is 360° times the square of the golden ratio. It has the strange property that it scatters the points in an kind of optimal way. Each new point is placed in a position that is as far away from the previous points as possible. We can see this by changing the distance of each point to the center *A*. A good choice of this radial displacement is *sqrt(i)\*0.2*. We do so by multiplying the vector we add by a factor r that we set to this value. The result is shown in Figure 10. Observe that the points in the drawing are placed in a way that each of the points has approximately the same amount of free space around itself.

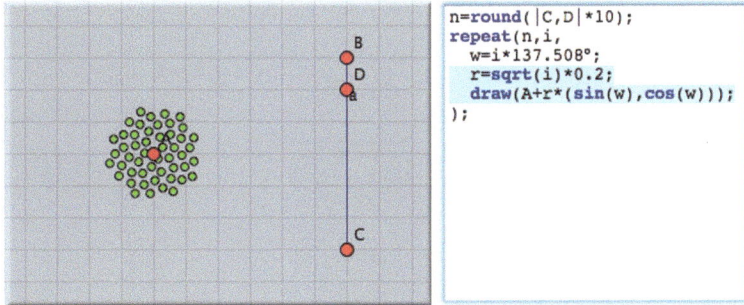

*Figure 10 : Adjusting the radii*

Let us be ambitious and increase the number of points significantly. We change our factor 10 in the first line to a value of 80. We recommend to move the points *A* and *D* to see the speed of the visualization. A Situation with roughly 400 points is shown in Figure 11.

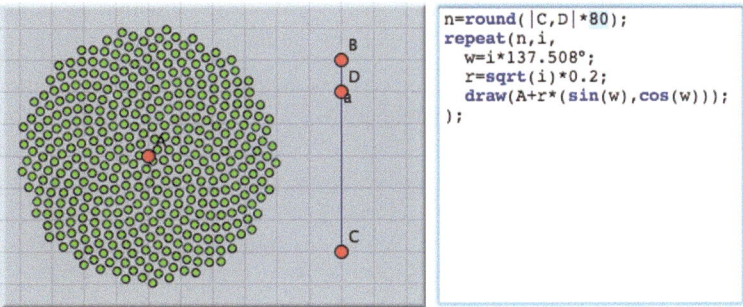

*Figure 11 : Much more points*

### 3.4.5  Spirals in a sunflower

The points we drew are roughly positioned like the seeds in the head of a sunflower. You will also observe the typical spiral structures that one sees in a sunflower. The angular displacement factor of 137,508° is crucial for this pattern. We will exemplify this by slightly disturbing this angle and seeing what happens. For this we add another slider with which we can control the amount of disturbance. We again draw a segment with a point on it and add define a (small!) disturbance value by `dd=|E,G|*0.1°`. We add this value to the angular displacement, by changing the angle to `w=i*(137.508+dd)`. The effect of such a small disturbance is shown in Figure 12. We observe the ever so slightly changing the angular displacement results in the appearance of obvious spiral structures in the placement of the seeds.

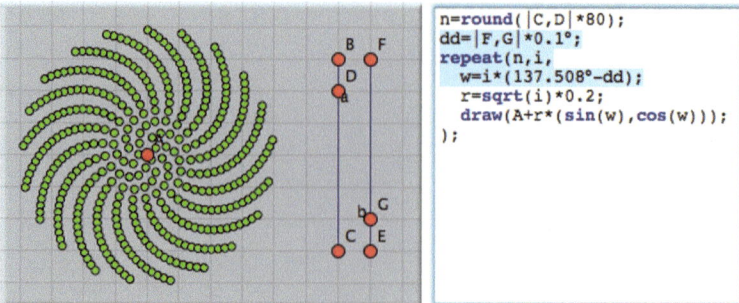

```
n=round(|C,D|*80);
dd=|F,G|*0.1°;
repeat(n,i,
    w=i*(137.508°-dd);
    r=sqrt(i)*0.2;
    draw(A+r*(sin(w),cos(w)));
);
```

*Figure 12 : Disturbing the angle*

The number of spirals is intimately related to the golden ratio and its development by continued fractions. We will not elaborate on this (quite deep) mathematical structure. We just mention that the so-called Fibonacci numbers *1,1,2,3,5,8,13,21,34,...* come into play. Here each number is the sum of its two predecessors. Dividing two consecutive Fibonacci numbers by each other gives a better approximation of the golden ratio the larger these two numbers are. Amazingly the number of prominent spirals will be a Fibonacci number (pick a real sunflower and check it!).

We will exemplify this by changing the color of the points in our visualization. Many CindyScript functions (and in particular `draw(...)`) allow for the use of so called *modifiers*. By using them one can for instance change the color, size or opacity of the element that is drawn. In our particular example we will add the following piece of code to the draw statement: `color->hue(i/21)`. The function `hue(...)` creates a rainbow color with full saturation. While the argument runs from 0.0 to 1.0 the color will proceed a full cycle in the color circle. By setting the color to `hue(i/21)` we paint the seeds in colors that are cyclically with period 21 (a Fibonacci number). We visually observe that the disturbed pattern has 21 prominent spirals.

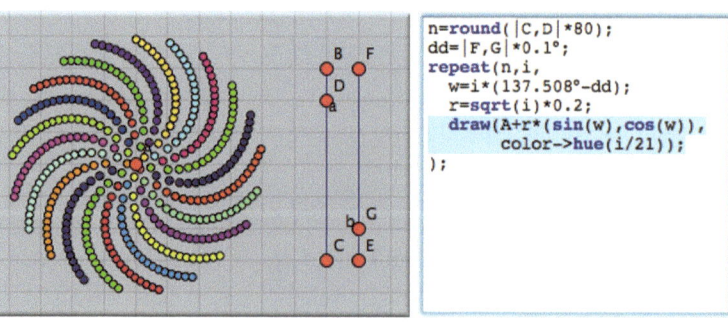

```
n=round(|C,D|*80);
dd=|F,G|*0.1°;
repeat(n,i,
    w=i*(137.508°-dd);
    r=sqrt(i)*0.2;
    draw(A+r*(sin(w),cos(w)),
        color->hue(i/21));
);
```

*Figure 13 : Coloring the spirals*

# Part II
# Cinderella Reference

# Chapter 4

# Interactive Geometry with Cinderella

The primary and most fundamental usage of Cinderella still is interactive geometry, often also called *dynamic geometry*. With a few mouse clicks you can generate constructions from elementary geometry, projective geometry or even hyperbolic geometry and fractals. The main feature of Cinderella is the so called move mode, sometimes also called drag mode. This mode gives you much more than an ordinary geometric drawing tool. You can drag the base elements of your construction and while you do this all the dependent elements follow accordingly. Using move mode you can explore many geometric constructions and observe effects and theorems on an experimental level.

The following explanations give you a detailed guide through all the different construction tools and possibilities that are available for doing interactive geometry. After a description of the elementary construction modes an explanation of various views and types of geometry is given.

As a novelty, *Cinderella.2* lets you access various geometric transformations as geometric objects. Using these transformations it is also possible to generate transformation groups and fractals easily.

## 4.1 The Main Menu

All the functionality of Cinderella is accessible through the main menu. You will get to know the menu items and how to use them while reading this manual. This little section provides only a rough overview of the different parts of the program that are accessible through the main menu.

*The main menu*

### 4.1.1 File

In the file menu you can find the usual operations for loading and saving constructions. Unlike the old version of Cinderella, *Cinderella.2* is able to handle several constructions in the same application.

In the file menu you also have access to various ways of exporting a drawing. You can generate different kinds of image formats, send your drawing to a printer, or create an interactive html page from it.

### 4.1.2 Edit

This menu contains tools for copying and pasting parts of a construction. Furthermore, several selection devices are offered. Here you will also find access to the inspector (p. 153).

### 4.1.3 Modes

The modes menu contains a collection of all items that can be constructed. Here you will find construction tools for all geometric operations, as well as construction tools for the CindyLab (p. 167) objects. The most important ones of these modes are also accessible via toolbars (p. 69).

### 4.1.4 Views

Here you can open different views on your construction. For details consult the section Views (p. 116). Here you can also open the Information Window.

### 4.1.5 Scripting

This menu item gives you access to the various ways to attach CindyScript (p. 219) code to your construction.

### 4.1.6 Configuration

Here you can choose the type of geometry you are working in (the default is *Euclidean geometry*, but you can also work in *hyperbolic geometry* or *elliptic geometry*). For details on geometries, please consult the section Geometries (p. 114). Also you can choose the default formats in which units and equations should be displayed.

### 4.1.7 Help

References to online help, updates, and licensing issues.

## 4.2 The Toolbars

The toolbar offers you a fast way to access commonly-used modes and actions. The main toolbar is located above the main construction area. Actually, Cinderella.2 has two toolbars – there is a secondary one below the construction area.

### 4.2.1 Selecting toolbars

You can change between different versions of the toolbars. To do so, either choose "Customize Toolbars" from the "Configuration" menu, or right-click in an empty area in the toolbar to access a context menu that allows you to invoke the "Choose toolbars"-dialog. The bottom toolbar can be changed similarly.

In this dialog you can select one of the currently available toolbars. When you open Cinderella for the first time, the *default toolbar* will be selected. It looks approximately as shown in the following picture.

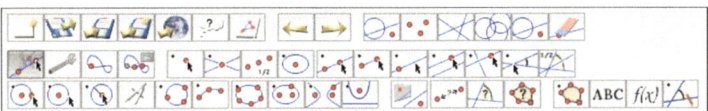

*The default toolbar*

It contains a collection of a lot of useful tools. You should keep in mind that there are even more construction tools available via the menu. Cinderella provides the following standard toolbars:

- *empty toolbar:* Contains no tools at all
- *default toolbar:* A collection of useful geometry tools
- *school toolbar:* A collection of tools for ruler and compass constructions
- *full toolbar:* Contains all available buttons for professional usage
- *physics toolbar:* Mainly focuses on CindyLab (p. 167) construction tools

### 4.2.2 Saving toolbars with a construction

Cinderella.2 can save the currently active toolbar along with the construction. To do so, you enable the option "Save toolbars into file" in the "file" tab of the Inspector. When that option is not active, Cinderella stores only the construction. This feature is useful to create files with a non-confusing tool selection for beginning learners.

### 4.2.3 Customizing Toolbars

Currently it is not possible to customize toolbars beyond the possibilities described above. A feature that will provide this possibility will be added in future releases. Still it may be desirable to create interactive worksheets that focus only on a specific collection of tools. It is possible to include such customized toolbars in an exported HTML page by using a suitable CindyScript command. This can be done via the `createtool(...)` (p. 392) operator. This method is also a useful method for creating Interactive Exercises (p. 433) as described at the end of this manual.

## 4.3 General Tools

### 4.3.1 File Operations

The file operations are rather standard and similar to corresponding operations in other programs. Constructions are stored in a special file format. The construction files have the extension `.cdy`.

- *New*: Opens a new and empty construction window.

- *Load*: Loads a construction.

- *Save*: Prompts for a filename if needed and saves the construction.

- *Save As*: Prompts for a new filename and saves the construction.

**Export Tools**

- *Export to HTML*: This operation automatically generates an interactive HTML page.

- *Print All Views*: Prints all currently open views. This operation makes use of the Java printing routines. The output is inferior to the output produced using the "Print PDF" command of a view, due to the limits of the Java printing interface. If you have the facilities to print and preview PDF files, it is much better to use this print format.

**Undo/Redo/Delete**

- *Undo*: This operation undoes the last performed action. You can undo as many consecutive operations as you want. The following actions are undoable:

  - Construction steps

  - Movements

  - Appearance changes

   – Zooming, translating, and rotating views

   – Deletion of elements

- ⟶ *Redo*: Redoes the last undo operation. You can redo as many consecutive undo operations as you want.

- *Delete*: Deletes all selected elements and all elements that depend on them. If you delete elements by accident, just use "Undo" to restore them.

### 4.3.2  Selection Tools

- *Select All*: Selects all geometric elements.

- *Select Points*: Selects all points.

- *Select Lines*: Selects all lines.

- *Select Conics*: Selects all conics and circles.

- *Deselect*: Clears the current selection.

## 4.4  Move Mode

This mode is probably the most important feature of *Cinderella*. It allows objects to be selected and to be moved around.

### 4.4.1  Moving Elements

Usually, move mode is used to move the free elements of a construction. To do this, place the mouse over a movable element, select the element by pressing the left mouse button, and drag it (with the button still pressed) to the desired position. After the button is released, another element can be moved around.

*Cinderella* distinguishes two types of elements: *movable elements* and *fixed elements*. As the name indicates, movable elements are those that you can move around in this mode. The fixed elements are those whose position is already completely determined by the rest of the construction. If, for instance, a point is defined to be the intersection of two lines that have already been defined, then the point is no longer movable and becomes a fixed element. You can visually distinguish movable points from fixed points by their appearance: movable points appear brighter.

There are seven types of movable elements:

- *Free Points:* These are points that are not dependent on other elements of the construction. These points can be moved about freely.

- *Points on a line:* Points may be defined always to be incident to a certain line. During a move these points slide along the line.

- *Points on a circle:* Points may be defined to be always incident to a certain circle. During a move these points slide along the circle.

- *Points on a conic:* Similar to points on a circle, since circles are special conics.

- *Line with a slope:* A line may be defined to be always incident to an already constructed point, with no further restrictions. Such a line can be selected directly for a move. During the move the line rotates about the point.

- *Circles by center and radius:* A circle that is defined by its center and its radius may be scaled in move mode. Select the boundary of the circle with the mouse and drag the mouse. The radius changes according to the current mouse position. You can move the whole circle by dragging its center point.

- *Texts and Measurements:* Texts and measurements may be also dragged in move mode. You select the text and move it to the desired position. Texts and measurements may be placed everywhere in the view. The boundary of the view has snap points that support the creation of nice layouts. Texts and measurements may also be "docked" to points of the construction.

There is one more reason to use the move mode: you can use it to *change the position of labels*. To do this, press the *shift key* on the keyboard, select the label with the mouse, and move it to the new position. However, labels cannot be moved to an arbitrary position. They must be close to their corresponding element. Movements that drag the label too far away are automatically blocked.

### 4.4.2 Selecting Objects

Move mode is also used to select one or more geometric elements with the mouse. You can easily recognize selected elements, since they are highlighted in all views. There are four reasons why you might want to select elements:

- The most common use of select mode is to individualize the appearance of geometric elements. All changes that you make in the inspector (p. 153) are immediately applied to all selected elements. If, for instance, several lines are selected and you move the size slider in the appearance editor, then the thickness of the selected lines changes.

- Selected elements may be deleted by using the Delete (p. 71) action.

- Selected elements may be moved simultaneously.

- Elements may be selected to define "Input," "Output," and "Hints" for an exercise. For details on this, consult the section about creating interactive web pages (p. 419) and exercises.

You can select elements in the view by clicking on them. Depending on what you do, several different behaviors are possible:

- When you click with the mouse somewhere on a view, then precisely those elements hit by the mouse pointer will become selected. All other elements will be deselected.

- When you hold the shift key down while clicking the mouse over an element, the selection state of the element will toggle. The selection state of elements not hit by the mouse will remain the same. You can use "shift-clicking" to select more than one object.

We strongly recommend playing with the different possibilities of select mode.

### 4.4.3 Moving Several Elements

It is also possible to move several free points at once. To do this, one first has to select the points that are to be moved. This can be done by pressing the shift key and clicking all points to be selected. If after this selection one of the selected points is clicked on and dragged, all other selected points will move as well. The shape and orientation of the selected points will remain constant during the move.

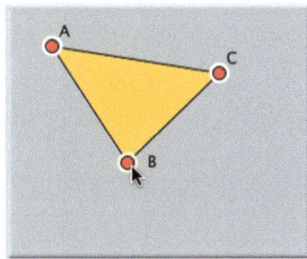
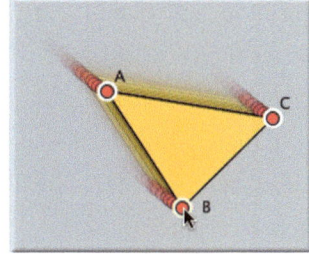

*Select several points...*          *...and move them.*

The possibility of moving many selected points is particularly useful after a "copy and paste" action (or the use of a user-defined tool). The pasted objects will be selected after the pasting operation. With move mode they can be moved all at once to a different place by selecting one of the selected points.

**Synopsis**

Move elements by dragging. Select elements by clicking.

**See also**

- Add a Point (p. 74)
- Add a Line (p. 76)
- Line through a Point (p. 77)
- Circle by Radius (p. 82)
- Text (p. 97)
- Multitouch Support (p. 438)

## 4.5 Interactive Modes

The interactive modes are very powerful construction tools. They take full advantage of the mouse. A single click–drag–release sequence constructs several elements. Moreover, the definitions of the new elements are adapted according to the mouse position. For instance, in "add a line" mode, the line together with its endpoints is generated. The definition of the start point and endpoint depends on the mouse position when the mouse button is pressed and when it is released. As another example, when the mouse pointer is clicked on the intersection of two lines, the new point will be defined as the point of intersection of those lines.

If elements are already present in the construction, they will not be added a second time. This refers not only to the definition of elements, but it is guaranteed by *Cinderella's* automatic theorem checking facilities.

Interactive modes can easily be recognized in the toolbar since their icon contains a little mouse pointer symbol.

Add a Point (p. 74)

Add a Line (p. 76)

Line through a Point (p. 77)

Add a Parallel (p. 78)

Add a Perpendicular (p. 80)

Add a Line with a Fixed Angle (p. 80)

Add a Circle (p. 81)

Circle by Radius (p. 82)

Circle by Fixed Radius (p. 83)

Midpoint (p. 83)

Segment (p. 84)

In addition to the standard modes, that are provided with a button, there are also many new modes in *Cinderella.2* that are accessible only via the menu. For these, please consult the sections Polygons (p. 112), Conic Operations (p. 110), and Transformation Modes (p. 122).

### 4.5.1 Add a Point

Usually, a single new point will be added when this mode is used. The mode is designed to be multipurpose and easy to use. Pressing the left mouse button generates a new point. When you drag the mouse with the left button pressed, you can change the position and definition of the newly added point. When you finally release the mouse button, the point is fixed. Its definition depends on the position at which the mouse was released. Sometimes, it can be more convenient to use the more powerful interactive modes such as add a line (p. 76), add a parallel (p. 78), and add a circle (p. 81), which generate points together with other geometric elements.

As already mentioned, constructing a point can be described as a three-step procedure:

- *Pressing the left mouse button* generates the point.

- *Dragging the mouse* moves the new point close to the mouse position. You will see that the point snaps to existing elements. Whenever this happens, the definition of the point is adapted to the current situation and the defining elements are highlighted. In particular, the point snaps to intersections of existing elements, and it snaps to existing points.

- *Releasing the mouse button* fixes the definition of the point. If at the moment of release the mouse is . . .

    - over no element at all, then a free point is added with no additional constraints. In "move mode" this point can be freely moved.

    - over an already existing line, then the new point will always be incident to the line. In "move mode" this point can only slide along the line.

    - over an already existing circle, then a point that is always incident to this circle is added. In "move mode" this point can only slide along the circle.

    - over the intersection of two elements (line, circle, or conic), then that intersection of these elements is added. Also, such a point will appear slightly darker than the free points. It is no longer freely movable in "move mode."

    - over an existing point, then no element will be added.

The elements that lead to the definition will always be highlighted.

The figures below show the three main situations: a "free point," a "point on a line," and an "intersection point." Notice that while the mouse is dragged, the coordinates that correspond to the point's position are displayed.

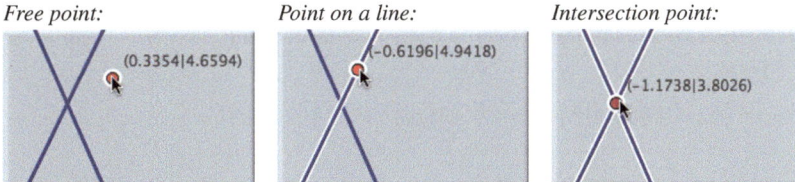

*Free point:*          *Point on a line:*          *Intersection point:*

A point can also be added with a simple mouse click at the position at which the point should be placed. The above-mentioned three-step procedure collapses then to a single mouse click. However, it is often more convenient to take full advantage of the three steps mentioned above:

*Press the mouse button down somewhere, drag the point to the desired position, and release the mouse.*

**Synopsis**

Add a point mode creates a new point with a press–drag–release sequence. The definition is automatically adapted to the situation.

**Caution**

There are two situations in which it is not appropriate to add a point with this mode:

- If you want to add an intersection of two lines that cross outside the visible area of the view, then you should use intersection mode (p. 92).

- If you want to add a point at a position at which more than two elements cross, then you should do one of the following:

  - Perturb the situation slightly using move mode (p. 71) if possible.

  - Zoom in (p. 117) in to get a higher resolution.

  - Use intersection (p. 92) mode to define exactly what intersection is meant.

**See also**

- Move Mode (p. 71)

- Add a Line (p. 76)

- Add a Circle (p. 81)

- Intersection (p. 92)

### 4.5.2  Add a Line

This mode allows a line through two points to be added. The mode is powerful enough to generate the line together with the two points with just one mouse action. When the mouse is pressed, the first point is added. Dragging the mouse generates the second point and the line. When the mouse is released, the position of the second point is frozen, and the construction is finished. The logic behind this mode is very similar to the logic behind other interactive modes, such as add a parallel (p. 78), add a perpendicular (p. 80), and add a circle (p. 81).

Constructing a line in this mode is a three-step procedure:

- *Pressing the left mouse button* generates the first point. The definition of this point depends on the position of the mouse at the moment the button is pressed:

  - If the mouse pointer is over an existing point, then this point is taken as the first point.

  - If the mouse pointer is over the intersection of two elements (line, circle, or conic), then this intersection is automatically constructed and taken as the first point.

  - If the mouse pointer is over just one element (line or circle), then a point is constructed that is constrained to this element. This point is taken as the first point.

  - Otherwise, a free point is added.

- *Dragging the mouse* generates the line and the second point. The definition of the second point depends on the mouse position. As in add a point (p. 74) mode, the second point snaps to existing elements. The choice of definition is completely analogous to the choice for the first point. The elements that define the second point are always highlighted.

- *Releasing the mouse button* freezes the definition of the second point. The construction is then finished.

The figures below show the three stages during the construction of a line. Here the first point is a free point and the second point is the intersection of the two existing lines.

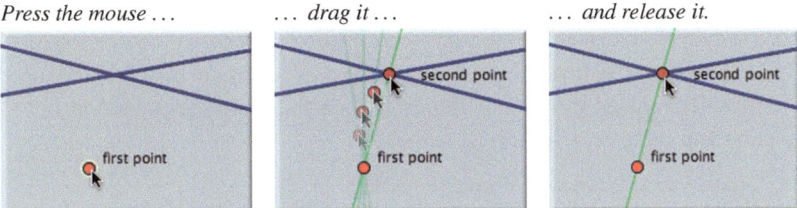

*Press the mouse ...*          *... drag it ...*          *... and release it.*

**Synopsis**

Add a line mode creates a line connecting two points by a press–drag–release sequence.

**See also**

- Add a Point (p. 74)

- Line through a Point (p. 77)

- Add a Parallel (p. 78)

- Add a Perpendicular (p. 80)

- Add a Circle (p. 81)

### 4.5.3  Line through a Point

This mode creates a line through a point with a certain slope. When the point is moved, the slope of the line stays constant. However, in move mode it is also possible to select the line and change its slope. So this mode could also be called "line by slope." The mode generates the line together with the point through which it passes by a single press–drag–release sequence. When the mouse is pressed, the point is added. Dragging the mouse generates the line. It is always attached to the mouse pointer. When the mouse is released, the position of the line is frozen, and the construction is finished. More precisely:

- *Pressing the left mouse button* generates the point. The definition of this point depends on the position of the mouse at the moment the button is pressed:

- If the mouse pointer is over an existing point, then this point is taken.

- If the mouse pointer is over the intersection of two elements (line, circle, or conic), then this intersection is automatically constructed and taken as the point.

- If the mouse pointer is over just one element (line or circle), then a point is constructed that is constrained to this element. This point is taken as the point.

- Otherwise, a free point is added.

- *Dragging the mouse* generates the line. The slope can be adjusted freely. The line also snaps to existing points if they are selected by the current mouse position.

- *Releasing the mouse button* freezes the definition of the line. The construction is then finished. Depending on the final position of the mouse pointer, two things can happen:

  - If the mouse pointer is over an existing point, then this point is used as a second point on the line. The line is then the "join" (connecting line) of the first and the second points.

  - Otherwise, a "line with slope" is added.

**Synopsis**

Line through point mode creates a line through a point by a press–drag–release sequence.

**See also**

- Add a Line (p. 76)
- Circle by Radius (p. 82)

 ### *4.5.4 Add a Parallel*

With this mode you can construct a line through a point parallel to another line with a press–drag–release sequence. The point through which the parallel should pass can also be generated in this mode. Constructing the parallel is a three-step procedure.

- *Move the mouse over the line* for which you want a parallel. Press the left mouse button. This creates the parallel and the point through which it should pass.

- *Hold the left button and drag the mouse.* This moves the parallel and the new point to the desired position.

- *Release the mouse.* Now the construction is frozen. Depending on the position at which you release the mouse, the definition of the new point is adapted:

  - If the mouse pointer is over an existing point, then this point is taken.

  - If the mouse pointer is over the intersection of two elements (line, circle, or conic), then this intersection is automatically constructed and taken as the new point.

- If the mouse pointer is over just one element (line or circle), then a point is constructed that is constrained to this element. This point is taken as the new point.

- Otherwise, a free point is added.

The figures below show the three stages during the construction of a parallel. Here the new point will be bound to the existing point *P*.

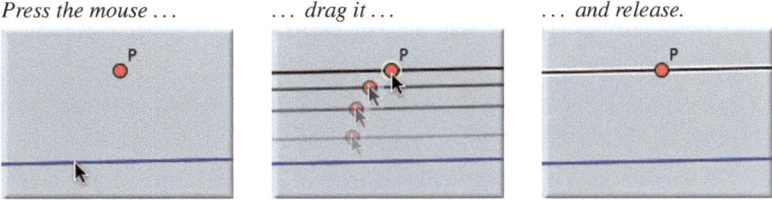

*Press the mouse ...*       *... drag it ...*       *... and release.*

**Synopsis**

Add a parallel mode creates a parallel with a press–drag–release sequence.

**Caution**

The behavior of this mode is dependent on the geometry that is chosen. While in Euclidean geometry there is always exactly one parallel, in non-Euclidean geometries the number of parallels is subject to the definition of parallelity. Depending on the underlying "philosophy," in hyperbolic geometry there can be infinitely many parallels, from an incidence-geometric viewpoint, or there can be exactly two parallels, from an algebraic or measurement-based point of view. *Cinderella* takes the algebraic point of view: *A parallel to a line L is a line that creates a zero angle with L.* So in hyperbolic geometry the mode produces exactly two parallels.

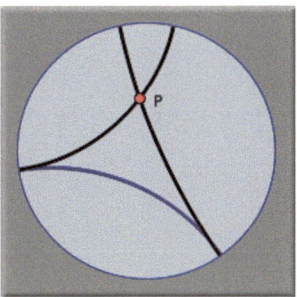

In elliptic geometry the usual viewpoint is that there are no parallels at all. However, from an algebraic standpoint there exist such parallels. It is just that they have complex coordinates. In other words, they will never be visible. *Cinderella* constructs these parallels, and they are invisible but are nevertheless present in the Construction Text (p. 121) view. From there they can be accessed for further constructions.

**See also**

- Add a Line (p. 76)

- Add a Perpendicular (p. 80)

### 4.5.5  Add a Perpendicular

This mode allows a perpendicular to a line and through a point to be constructed with a press–drag–release sequence. The use of the mode is exactly analogous to add a parallel (p. 78) mode. The point through which the perpendicular should pass can also be generated in this mode. Constructing the perpendicular is a three-step procedure.

- *Move the mouse over the line* for which you want a perpendicular. Press the left mouse button. The perpendicular and the point through which it should pass are created.

- *Hold the left button and drag the mouse.* You can move the perpendicular and the new point to the desired position.

- *Release the mouse.* The construction is frozen. The definition of the new point is adapted according to the position at which you release the mouse:

    - If the mouse pointer is over an existing point, then this point is taken.

    - If the mouse pointer is over the intersection of two elements (line, circle, or conic), then this intersection is automatically constructed and taken as the new point.

    - If the mouse pointer is over just one element (line or circle), then a point is constructed that is constrained to this element. This point is taken as the new point.

    - Otherwise, a free point is added.

**Synopsis**

Add a perpendicular mode creates a perpendicular with a press–drag–release sequence.

**Caution**

The definition of perpendicularity depends on the chosen geometry.

**See also**

- Add a Line (p. 76)
- Add a Parallel (p. 78)

### 4.5.6  Add a Line with a Fixed Angle

This mode allows a line to be constructed with a fixed numerically given angle to another line. The new line and a point through which it should pass can be added with a press–drag–release sequence. The point through which the new line should pass can also be generated with this mode.

When you select this mode, the property inspector pops up, asking you to specify the desired angle.

Otherwise, the use of this mode is analogous to add a parallel (p. 78) mode.

- *Move the mouse over the line* to which you want the new line to make an angle. Press the left mouse button. The new line and the point through which it should pass are created.

- *Hold the left button and drag the mouse.* You can move the new line and the new point to the desired position.

- *Release the mouse button.* Now the construction is frozen. The definition of the new point is adapted according to the position at which you release the mouse button.

**Synopsis**

This mode creates a line with fixed angle with respect to another line with a press–drag–release sequence.

**Caution**

The definition of angle depends on the chosen geometry.

**See also**

- Add a Line (p. 76)

- Add a Parallel (p. 78)

### 4.5.7 Add a Circle

This mode allows the creation of a circle given by its center and a point on its circumference. The mode generates the circle together with the two points with just one mouse click. When the mouse is pressed, the center is added. Dragging the mouse generates the circumference point and the circle. When the mouse button is released, the position of the circumference point is frozen, and the construction is finished. The logic behind this mode is analogous to that in add a line (p. 76) mode.

Constructing a circle in this mode can be described as a three-step procedure:

- *Pressing the left mouse button* generates the center point. The definition of this point depends on the position of the mouse at the moment the button is pressed, as in add a line (p. 76) mode.

- *Dragging the mouse* generates the circle and the circumference point. The definition of the circumference point depends on the mouse position. As in add a point (p. 74) mode, the second point snaps to existing elements. The definition of the second point is analogous to that of the first point. The elements that define the circumference point are highlighted.

- *Releasing the mouse button* freezes the definition of the perimeter point. The construction is then finished.

**Synopsis**

Add a circle mode creates a circle with a press–drag–release sequence.

**Caution**

The "shape" of a circle depends on the choice of the geometry.

**See also**

- Add a Line (p. 76)

- Circle by Radius (p. 82)

- Circle by Fixed Radius (p. 83)

- Circle by Three Points (p. 89)

- Compass (p. 87)

 ### 4.5.8  Circle by Radius

This mode allows a circle defined by its center and its radius to be added. When the center of such a circle is moved, the radius remains constant. However, in move mode it is also possible to select the boundary of the circle itself and change the radius.

This mode generates the circle together with its center with just one press–drag–release sequence. When the mouse is pressed, the center is added. Dragging the mouse generates the circle. Its boundary remains attached to the mouse pointer. When the mouse is released, the position of the circle is frozen, and the construction is finished. The logic behind this mode is analogous to that in Line through a Point (p. 77) mode.

Constructing a circle in this mode can be described as a three-step procedure:

- *Pressing the left mouse button* generates the center. The definition of this point depends on the position of the mouse at the moment the button is pressed, as in add a line (p. 76) mode.

- *Dragging the mouse* generates the circle. The radius can be adjusted freely. The circle also snaps to existing points if they are selected at the current mouse position.

- *Releasing the mouse button* freezes the definition of the circle. The construction is then finished. Depending on the final position of the mouse pointer two things can happen:

  - Either a circle by radius (p. 82) is added,

  - or, if the mouse pointer is over an existing point, then this point is used as a circumference point for the circle. The circle is bound to that point.

**Synopsis**

Circle by radius mode creates a circle with freely chosen radius with a press–drag–release sequence.

**Caution**

The actual "shape" of a circle depends on the choice of the geometry.

**See also**

- Add a Circle (p. 81)
- Circle by Fixed Radius (p. 83)
- Line through a Point (p. 77)

### 4.5.9 Circle by Fixed Radius

This mode allows a circle that has a fixed, numerically given, radius to be created. The new circle and its center are added with a single press–drag–release sequence.

When you select this mode, the property inspector pops up, asking you to specify the desired radius, as in add a line with a fixed angle (p. 80) mode. After that the use of this mode is analogous to that of add a point (p. 74) mode.

**Synopsis**

Circle by fixed radius mode creates a circle with numerically given fixed radius with a press–drag–release sequence.

**Caution**

The "shape" of the circle depends on the choice of the geometry.

**See also**

- Circle by Radius (p. 82)
- Add a Line with a Fixed Angle (p. 80)

### 4.5.10 Midpoint

In this mode, the midpoint of two points can be constructed with a press–drag–release sequence. As in add a line (p. 76) mode, the two points can also be added.

- *Pressing the left mouse button* generates the first point. As in the other interactive modes, the definition of this point depends on the position of the mouse at the moment the button is pressed.

- *Dragging the mouse* generates the second point together with the midpoint. The second point also snaps to existing points or intersections if they are selected at the current mouse position.

- *Releasing the mouse button* freezes the definition. The construction is then finished.

**Synopsis**

Midpoint mode creates two points and their midpoint using a press–drag–release sequence.

**Caution**

This innocent-looking mode becomes interesting in non-Euclidean geometries. While in Euclidean geometry there is always a unique finite midpoint, in hyperbolic or elliptic geometry there are always two. Both points will be generated in this mode. In the Poincaré disk model of hyperbolic geometry, the second midpoint of two points in the disk always lies outside the disk. The presence of this point does not much matter, but there are circumstances in which one should be aware of this fact.

**See also**

- Add a Line (p. 76)

### *4.5.11  Segment*

This mode is completely analogous to add a line (p. 76) mode. By a press–drag–release sequence you add two points and a segment joining them. In addition, segments can also be supplied with arrows at one or at both ends. The arrow type can be chosen in the inspector tab for special appearance.

In contrast to the old release of Cinderella, segments are now fully active geometric elements. They can be used like any other geometric object. Intersections with segments disappear when the segment is not hit by the intersecting object, as do all elements depending on these intersections. This creates the possibility of allowing geometric objects to appear and disappear depending on certain geometric relations. However, it is much more convenient to create these kinds of optical effects using CindyScript (p. 219).

**See also**

- Add a Line (p. 76)
- CindyScript (p. 219)
- Traces and Arrows (p. 160)

## 4.6 Definition Modes

The *definition modes* use a simple "define by selection" mechanism to construct new elements. You choose a certain definition mode by clicking on its icon in the toolbar. Then you are asked to select a certain number of elements in the view. After a sufficient number of elements are selected, the newly defined element is added to the construction. Although these modes are slightly inconvenient compared to the interactive modes (p. 74), using these modes is sometimes unavoidable. There are four major circumstances in which the definition modes should be used:

- A certain geometric operation is provided only as a definition mode. (This is for instance the case for the following modes: angle bisector (p. 86), compass (p. 87), circle by three points (p. 89), center (p. 86), conic by five points (p. 89), polar of a point (p. 91), polar of a line (p. 91), and polygon (p. 91)).

- A point of intersection cannot not be reached in the visible area of a view. For instance, two lines can be almost parallel, so that their point of intersection lis far outside the viewing region in the window. Then the usual "add a point" or "add a line" modes are not applicable. But you can still select the lines and apply the "intersection" mode.

- An element could be invisible or complex. It can still be selected (e.g., in a "construction text" view), and a definition mode can be used.

- A situation could be ambiguous, such as three lines passing through a point. Then the interactive "add point" mode cannot be used to add a point of intersection (since all three lines would be close to the mouse pointer position). In this case, use "intersection" mode and select the lines whose intersection is desired. In the case of a *geometric theorem* in which three lines always meet, such as the altitudes of a triangle, *Cinderella's* theorem checking mechanism ensures that the added point is considered incident to all three lines.

A few things are common to all definition modes and should be known in advance:

- Elements are selected by clicking them with the mouse pointer. The selection can be made in any of the geometric views, in particular in the construction text (p. 121), where all elements are always visible and selectable).

- Selected elements are highlighted in the view.

- Preselected elements are often considered when a definition mode is chosen. So if two lines are selected and you press the button that chooses "meet" mode, then the point of intersection is instantly added.

- If you made a mistake in selecting an element you can deselect it with a second click.

- You can also select the elements by moving the mouse while holding the mouse button down. All elements that are touched by the mouse pointer will be selected (or deselected).

- Selected elements that do not contribute to the required selection of the mode are ignored. For instance, if you choose circle by three points (p. 89) mode, all selected lines, circles, and conics are ignored.

- Selection modes communicate with the user using the message line. There you can find messages that tell you about the next expected input.

In addition to the standard modes that are provided with a button, there are also many new modes in *Cinderella.2* that are accessible only via the menu. For these please consult the sections Polygons (p. 112), Conic Operations (p. 110), and Transformation Modes (p. 122).

### 4.6.1  Center

This mode constructs the center of a conic. In particular, you can use this mode to construct the center of a circle. For general conics the center is the intersection of its axes of symmetry. The center of a conic is sometimes useful for constructing a point on a movable conic.

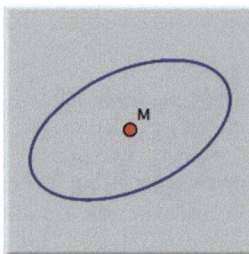
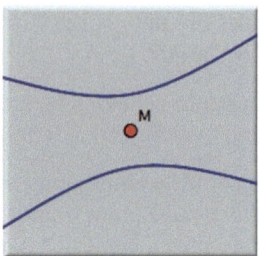

Center of an ellipse       Center of a hyperbola

**Synopsis**

Select a conic and construct its center.

**Caution**

This mode is not properly supported in non-Euclidean geometries. It always constructs the Euclidean center.

### 4.6.2  Angle Bisector

This mode is used to construct the angular bisector of two lines. The application of this mode requires a little care, since two lines do not have only one angular bisector; *they have two!* To take this fact into account, this mode is provided with a position-sensitive selection mechanism.

To define the angular bisector, two lines have to be selected. In order to indicate which angular bisector should be chosen, three points are relevant: the click point of the first selection, the click point of the second selection, and the intersection of the two selected lines. Imagine a triangle formed by these three points. The inner angle at the intersection point of the lines will be bisected. This is what you would intuitively expect.

To make the selection process a bit simpler, *Cinderella* gives graphical hints as to which angle will be bisected.

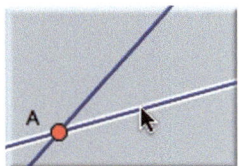 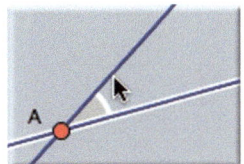 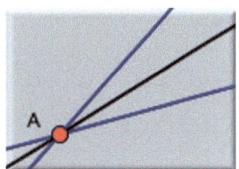

**First selection:**
*A line is highlighted.*
*The selection point*
*is memorized.*

**Moving the mouse:**
*An indication of*
*the chosen angle*
*is given.*

**Second Selection:**
*The angular bisector*
*is added.*

### Synopsis

Angle bisector mode selects two lines and constructs their angular bisector.

### Caution

The definition of angular bisector depends on the type of geometry (Euclidean, hyperbolic, or elliptic). In hyperbolic geometry, angular bisectors can have complex coordinates.

## 4.6.3 Compass

The compass is a very useful tool for transferring the distance between two points to some other location. *Cinderella's* compass works exactly like a real compass. You select a first point by clicking (i.e., you poke the needle into the first point). You select the second point by clicking (i.e., you adjust the compass to the distance between the first and the second points). Then you are ready to transfer the distance to another location. When you click on a third point, a circle with the specified distance around this point is added to the construction. Each single click may as well create a new point with the same synopsis as in the add a point (p. 74) mode.

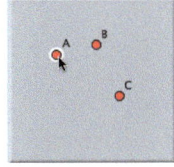 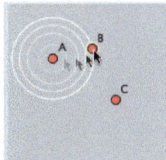 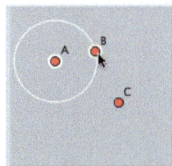 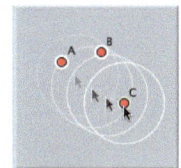 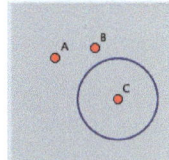

**First selection:**
*The first point*
*is highlighted.*

**Moving:**
*Hints are shown*
*for the distance.*

**Second Selection:**
*The distance*
*is fixed.*

**Moving:**
*Hints are shown*
*for the position.*

**Third Selection:**
*The construction*
*is finished.*

### Synopsis

In Compass mode you select two points whose distance is the radius of a new circle centered at a third point.

### Caution

The definition of a circle changes with the type of geometry (Euclidean, hyperbolic, or elliptic).

**See also**

- Add a Circle (p. 81)

- Circle by Radius (p. 82)

### 4.6.4 Mirror

The mirror is a multipurpose tool for creating reflections at points, lines, or circles. The first mouse click selects the "mirror." The following clicks select the elements that should be reflected. Reflected elements are either points, lines, or conics. You can deselect the mirror by clicking on it a second time. Depending on the choice of the mirror, different actions are performed:

- *If the mirror is a line* then the usual reflection is taken. The mirror image of a point with respect to a line is a point that has the same distance to the line as the original point and lies on the perpendicular of the line that goes through the original point.

- *If the mirror is a point* then the reflection "at this point" is taken. The mirror image of a point with respect to a point is a point that has the same distance to the mirror-point as the original point and lies on the join of the mirror-point and the original point.

- *If the mirror is a Euclidean circle*, then the inversion at that circle is taken. The inverse of a point with respect to a circle is a point that lies on the join of the original point and the center of the circle. The distance of the inverse point to the center of the circle is such that the product of this distance and the distance from the center to the original point is the circle's radius squared.

Reflections of lines or conics are considered as pointwise reflections.

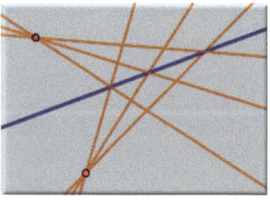
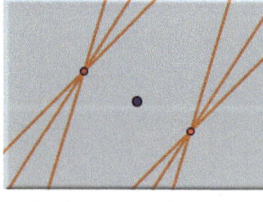

*Reflection at a line*          *Reflection at a point*          *Inversion at a circle*

**Synopsis**

First select a mirror. Then select elements that should be reflected.

**Caution**

The definition a mirror image heavily depends the choice of the underlying geometry.

**See also**

● Transformation Modes (p. 122)

### 4.6.5 Circle by Three Points

This mode is for constructing a circle that passes through three points. In Euclidean geometry such a circle is always uniquely defined to be the circumcircle of the triangle defined by the three points. You choose the points one after another. The definition phase does not supply graphical hints.

**Synopsis**

Select three points. Then their circumcircle is constructed.

**Caution**

This mode is available only in Euclidean geometry. In other geometries (hyperbolic, elliptic) there is no unique circle with this property.

**See also**

● Add a Circle (p. 81)

● Conic by Five Points (p. 89)

### 4.6.6 Conic by Five Points

This mode constructs the unique conic that passes through five points. It is one of the basic modes for constructing a general conic.

**Synopsis**

Select five points to create a conic.

**Caution**

If four of the five points are collinear, the conic is no longer unique and a null element is computed.

**See also**

● Circle by Three Points (p. 89)

### 4.6.7 Ellipse by Foci

This mode constructs an ellipse defined by two foci and one perimeter point. After selecting the mode one has to click focus1, focus2, perimeter point in this order.

**Synopsis**

Ellipse by foci mode constructs an ellipse from two foci and a perimeter point.

**Caution**

Due to the continuity behavior of *Cinderella* it may happen that while free elements are being moved, an ellipse transforms continuously into a hyperbola. This may happen if one of the foci passes through infinity.

**See also**

- Hyperbola by Foci (p. 90)

- Parabola by Focus and Directrix (p. 90)

 ### 4.6.8  Parabola by Focus and Directrix

This mode constructs a parabola defined by its focus and directrix. To define the parabola simply a point and a line must be selected.

**Synopsis**

A parabola is constructed from a focus and directrix.

**See also**

- Ellipse by Foci (p. 89)

- Hyperbola by Foci (p. 90)

 ### 4.6.9  Hyperbola by Foci

This mode constructs a hyperbola defined by two foci and one perimeter point. After selecting the mode one has to click the first focus point, the second focus point, and a perimeter point in this order.

**Synopsis**

A hyperbola is constructed from two foci and a perimeter point.

**Caution**

Due to the continuity behavior of *Cinderella* it may happen that while free elements are being moved, a hyperbola transforms continuously into a ellipse. This may happen if one of the foci passes through infinity.

**See also**

- Ellipse by Foci (p. 89)

- Parabola by Focus and Directrix (p. 90)

### 4.6.10 Polar of a Point

This mode constructs the polar of a point with respect to a conic. The point and the conic can be selected in any order. The polar, which is a line, is then constructed.

This mode has a few very interesting special cases that deserve some extra attention. If the point is on the conic itself, then the unique *tangent* to the conic that passes through that point is constructed. Even more special, if the conic is a circle and the point lies on the circle, then the *tangent* to the circle that passes through the point is constructed. Although this mode gives access to general polars, this special case will most probably be its main application.

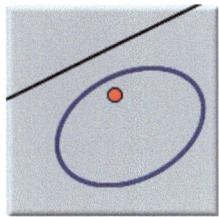
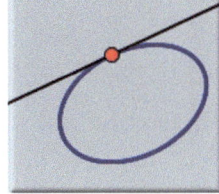
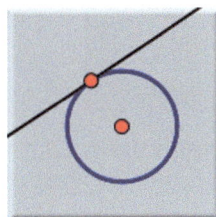

| *General Polar* | *Tangent to conic* | *Tangent to circle* |

**Synopsis**

The polar line of a point with respect to a conic is constructed by selecting a point and a conic.

**See also**

- Polar of a Line (p. 91)

### 4.6.11 Polar of a Line

This mode constructs the polar of a line with respect to a conic. The line and the conic can be selected in any order. This mode can be used to construct the point where a tangent line touches the conic.

**Synopsis**

The polar point of a line with respect to a conic is constructed by selecting a line and a conic.

**See also**

- Polar of a Point (p. 91)

### 4.6.12 Polygon

In this mode a polygon can be constructed from a sequence of vertices. Select the vertices in the order in which they should appear on the boundary of the polygon. Unlike the other definition modes, this mode does not expect a fixed number of input selections. You finish the creation of a polygon by selecting the first point a second time to close the polygon. For instance, if you want to construct a polygon through vertices A, B, C, and D, you have to select A, B, C, D, and A in this order.

If you made a mistake in selecting a point, you just click the point again to delete it from the boundary of the polygon. However, only the last point in the sequence can be deselected in this way.

Graphical hints guide you through the definition phase. They always resemble the part of the polygon that has been constructed so far.

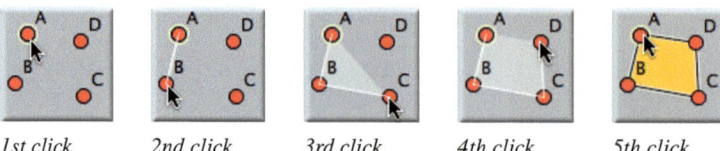

|   1st click   |   2nd click   |   3rd click   |   4th click   |   5th click   |

For construction regular polygons there are also a variety of other modes described in the section Polygons (p. 112).

**Synopsis**

Select a sequence of points to define a polygon. To finish the definition select the first point again.

**Caution**

This mode is supported only in the Euclidean, spherical, and textual views. The hyperbolic view ignores polygonal objects. The orientation of the polygon is important when one measures its area.

**See also**

- Area (p. 96)

- Polygons (p. 112)

### 4.6.13  Intersection

In this mode the point of intersection of two geometric objects (line, circle, or conic) is constructed by selecting them. The mode generates all intersections of the objects. Two lines have a unique intersection. A circle and a line have two intersection points. Two conics may have four intersections.

This mode may at first sight seem to be unnecessary. Usually, points are added using add a point (p. 74) mode or as a byproduct of some other interactive mode. However, under certain circumstances it can be unavoidable to use intersection (p. 92) mode. For instance, when the intersection of two lines is not reachable in a usual view, the lines are still listed in the Construction Text (p. 121), where they can be selected and used for "intersection" mode. Another case in which meet mode is a good choice is that of ambiguities, for example when three lines pass through one point. Then meet mode is a secure way of adding a particular point of intersection.

**Synopsis**

Select two objects and create their intersection.

**See also**

- Add a Point (p. 74)

### 4.6.14 Arc

This mode is very similar to circle by three points (p. 89) mode, except that here, the three points are not used to define a circle. They are used to define an arc.

Three clicks are necessary to define the arc. The first and the last click define the start and end of the arc. The second click defienes a point that should lie between the other two on the arc.

**Synopsis**

An arc by three clicks that select three points.

**See also**

- Circle by Three Points (p. 89)

### 4.6.15 Angle Mark

This mode creates a graphical mark for an angle. The mark itself is not geometrically acrive and cannot be used as input for other constructions.

Three clicks at three points are necessary to define the angle mark. The middle point marks the one at which the angle mark is drawn. The other two points define the start and the end of the angle mark.

**Synopsis**

An angle mark by three clicks that select three points.

**See also**

- Angle (p. 95)
- Angle Bisector (p. 86)

## 4.7 Measurements

Measuring is an important part, perhaps even the origin, of geometry. *Cinderella* has modes for measuring distances, angles, and area. Their behavior, at least in Euclidean geometry, is relatively straightforward:

- Select two points and obtain the distance between them.

- Select two lines and obtain the angle between them.

- Select a polygon, circle, or conic and obtain its area.

For most purposes that is all you have to know. However, as usual, there are many fine details that sometime make things more difficult. If you deal with non-Euclidean geometries, their crucial defining property is a strange way to measure things. What goes on exactly is described in more detail in the section Theoretical Background (p. 27). For now, it is sufficient to know that in non-Euclidean geometries measurements differ from the usual Euclidean measurements. Values of distances or angles may even become complex numbers. These values can be calculated with the theory of *Cayley-Klein geometries*.

The treatment of measurements in such a general way is one of the core features of *Cinderella*. You should not worry about the strange behavior of measurement in non-Euclidean geometries. Perhaps the best way to understand it is to play with different constructions to get a feeling for its unique aspects. Allowing people to get an intuitive feel for non-Euclidean geometries is one of the main goals of *Cinderella*.

Another fine point comes from the measurements of areas. *Cinderella* is able to measure the area of polygons and conics. In both cases there are a few things that deserve some extra attention. It is easy to define the area for a polygon that does not intersect itself. But what happens if it does? The approach chosen in *Cinderella* is to use a general and consistent formula for area. Areas are counted with respect to an orientation. How much a point inside the polygon contributes to the area depends on its winding number with respect to the boundary.

The area of conics is also a delicate topic. It is easy to define the area of an ellipse. But what is the area of a hyperbola? Is it infinite? Is it undefined? Is it something completely different? *Cinderella* chooses an algebraic approach that tries to use only one formula for all the different cases. It turns out that the area of a hyperbola is most reasonably described by a complex number. So, if you make measurements of areas of conics do not be surprised if complex numbers sometimes show up.

In particular, the measurement modes are

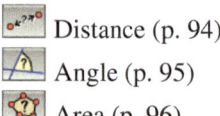 Distance (p. 94)

Angle (p. 95)

Area (p. 96)

### 4.7.0.1 Remark

The measurements are shown as texts. They can be used as "text objects," so they can be dragged around and be repositioned. Consult the description of text (p. 97) and function (p. 99) for these features.

## *4.7.1 Distance*

Measuring distance in *Cinderella* is very much like drawing a line in add a line (p. 76) mode. You have to select the two points whose distance from each other you want to measure. This is done with a single mouse press–drag–release sequence.

- Move the mouse pointer over the first point and press it.

- Drag the mouse, with the left button pressed, to the second point. While you drag, a ruler with the actual distance from the first point to the mouse pointer is shown.

- The ruler snaps to selected points. When the second point is selected, you can release the mouse button.

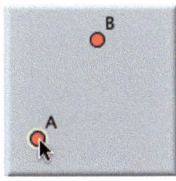 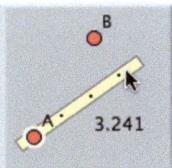 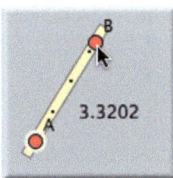 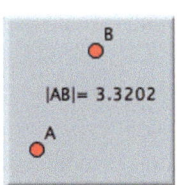

**Mouse press:**          **Mouse drag:**          **Mouse drag:**          **Mouse released:**
*A point is selected.*    *A ruler is displayed.*  *Second point selected.* *Measurement added.*

### Synopsis

Measure the distance between two points with a press–drag–release sequence.

### Caution

The definition of distance changes with the type of geometry (Euclidean, hyperbolic, or elliptic). In hyperbolic geometry, distances may even be complex numbers.

## 4.7.2 Angle

This mode is used to measure the angle between two lines. The application of this mode requires a little care. Between two lines there is not only one angle: *there are two*, an angle and its supplementary angle. To take this fact into account, "angle" mode is provided with a position-sensitive selection mechanism. To measure the angle, two lines have to be selected. In order to determine which angle will be chosen, three points are relevant: the click point of the first selection, the click point of the second selection, and the intersection of the two selected lines. Imagine a triangle formed by these three points. The inner angle at the intersection point of the lines will be measured. This is what you would intuitively expect.

*Cinderella* simplifies the selection process with graphical hints that show the angle that will be measured.

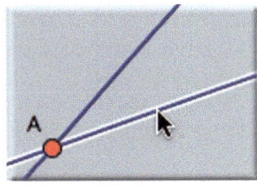 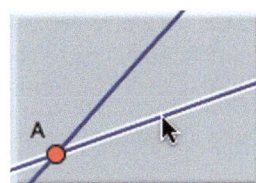 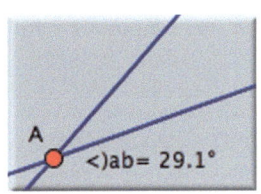

**First selection:**       **Moving the mouse:**      **Second selection:**
*A line is highlighted.*   *Hints are shown that*     *The angle is*
*The selection point*      *indicate the angle*       *measured.*
*is memorized.*            *that will be measured.*

**Synopsis**

Angle mode measures the angle between two lines.

**Caution**

The definition of angle depends on the type of geometry (Euclidean, hyperbolic, or elliptic). In hyperbolic geometry angles may even be complex numbers.

### 4.7.3 Area

In this mode you can measure areas of polygons and of conics. To measure the area of a polygon you simply click inside the polygon. To measure the area of a conic, you simply select it. In particular, you can use this mode to measure the area of a circle.

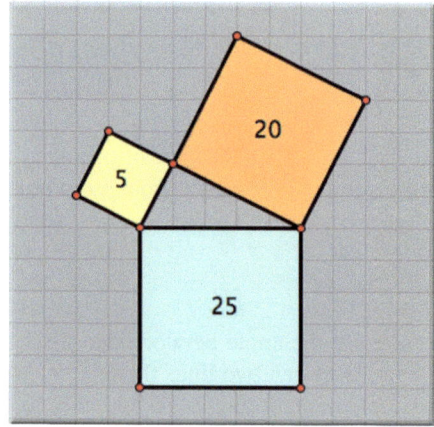

*The Pythagorean theorem with area measurements*

**Synopsis**

Select a polygon or a conic and obtain its area.

**Caution**

The area of a hyperbola is a complex number.

The contribution of a point to the area of a polygon is calculated with respect to its "winding" number. In particular, areas may become negative if the polygon is oriented clockwise. If the polygon is self-intersecting, areas may be zero.

You can measure the area of triangles in CindyScript (p. 219) using the area (p. 351) function.

## 4.8 Special Modes

In this section we summarize modes that are extremely powerful but do not fit into any of the categories presented elsewhere. Each of these modes encapsulates powerful featuers, and the wise use of these modes can enhance the use of *Cinderella* considerably.

### 4.8.1 Text

ABC

"Add text" mode is a multipurpose mode for almost everything related to displaying text in a view. Here are the main applications of this mode:

- *Add plain text:* Click at some position in a view. Then an input dialog pops up asking you to input the desired text. When you exit the window, the text is displayed at the position where you clicked.

- *Edit text:* When you click on an existing text you will get an editing window that shows the old text. This text can now be changed. After the input window is exited, the old text is replaced by the new text.

- *Changing a label:* When you click on a geometric element you will also get an editing window, now showing the label of that element. You can now input a new label for the element. However, there is a restriction: The label of an element must be unique in the construction. If you try to input an existing label, a warning message is shown and the label remains unchanged.

#### 4.8.1.1 Docking and Referencing

Texts are more flexible objects than you might at first think. This comes from two important features: *docking* and *referencing*.

- *Docking:* Usually, a text is located at a certain position relative to the coordinate system of the drawing. When you zoom or translate the view, the text will follow the zooming. This is sometimes desired, but sometimes you might want a different effect. Imagine that you have a descriptive text that should always be shown in the upper left corner of the view. Then you should use "docking." In move mode you select a text and move it around by dragging the mouse. If you come to a position close to the boundaries of the window, you recognize that the text snaps to predefined docking positions. When the text is docked, it stays in its position relative to the window.

It is also possible to dock a text to a point. Drag the text close to the point until the point is highlighted. The text will then consistently stick to the point.

- *Referencing:* Often it is necessary to have variable parameters inside a text. Imagine a text that reads, "The distance between A and B is 25 cm." In such a text you usually want to refer to the actual distance between "A" and "B" and not to a fixed string "25". You can do this by referencing the value of a geometric element. If "A" is the geometric element, then @#A references its value. Moreover, you usually want to mark the fact that the "A" and the "B" in your text are labels. This forces further changes of element names to be reported to the text object and that the displayed text be changed accordingly. You can use @$A to refer to a label. If `dist` is the label of the distance object, then the above effect would be generated by the following input string:

```
The distance between @$A and @$B is @#dist cm.
```

In the case of points, lines, and conics, the "@#" operator refers to the coordinates. So, you can write

```
The coordinates of @$A are @#A.
```

Altogether you have three possibilities for referring to the data of a geometric element. If `A` is the label of the element, then you access

- the *label* of the element by @$A,

- the *defining algorithm* of the element by @@A, and

- the *value or position* of the element by @#A.

The texts that are then generated are precisely the ones you can see as well in construction text (p. 121) view. The exact representation of the element's value or position can be influenced by the relevant settings in the "Format" menu.

If you want to include Greek characters, you can do so by using @ and the name of the character, for example, @alpha or @Omega.

Furthermore, you have the ability to evaluate CindyScript (p. 219) code within a text and display the result. The syntax for this is @{...}, where ... refers to an arbitrary piece of CindyScript code. For instance, @{sin(A.x)} produces the sine of the 'x'-coordinate of A.

Furthermore it is possible to produce formula type setting by inserting TeX code in a text object. For instance it is possible to use Text like `The point $P_1$ has coordinates $({x \over y},x2)$` to create the text

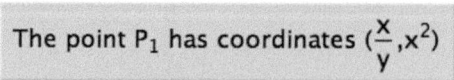

*Formula typesetting in texts*

For details on formula typesetting please refer to TeX Rendering (p. 313).

**Click Referencing**

As in function mode (p. 99) it is possible to obtain the reference of a geometric element by simply clicking on it in an arbitrary view. This simplifies the process of entering a text line in the dialog box. Clicking on another text produces a copy of this text.

**Attaching CindyScript**

You can transform a text into a clickable button using the info block in the Inspector (p. 153). A button can be a push button or a toggle button. A toggle button has a state that can be read using a property (p. 234). You can attach CindyScript code (p. 219) to both types of buttons, as described in the corresponding section of this manual (p. 395).

**Synopsis**

Texts can be added and edited.

**See also**

- Measurements (p. 93)

- Function (p. 99)

- CindyScript (p. 219)

## 4.8.2 Function

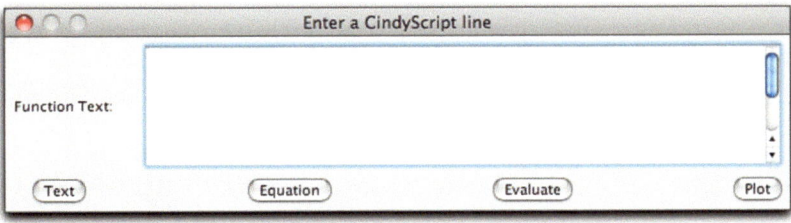

Function mode provides an extremely powerful tool that can be used in many circumstances. Briefly, function mode provides a one-line interface to the CindyScript language. This one line can be either simply evaluated (so that the result is displayed) or displayed as an equation or evaluated with all side effects. Since CindyScript is a powerful language and even one line of CindyScript may encode very high level interactions, we will here demonstrate a few uses of function mode by means of a couple of examples.

When you are in function mode and click somewhere in the window, the following window will pop up:

*The "enter a function" dialog*

There you can enter one line of CindyScript code. The buttons allow you to determine how this line should be processed.

- *Text:* calculates the result of the code and displays it.

- *Equation:* displays the line entered literally, followed by an equation sign and the result of the evaluation. So a line "4+7" becomes the displayed equation "4+7 = 11".

- *Evaluate:* evaluates the line and performs all side effects of the line (such as setting variables, drawing, or plotting).

We will present a few examples to illustrate the use of functions.

**Calculations in Constructions**

Consider the following drawing of a rectangle:

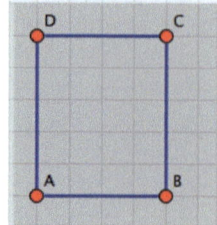

*A rectangle*

We want to calculate and display the area of this rectangle. A suitable CindyScript expression for calculating the area of this rectangle is |A, B| * |A, C|. We enter this function into the function dialog.

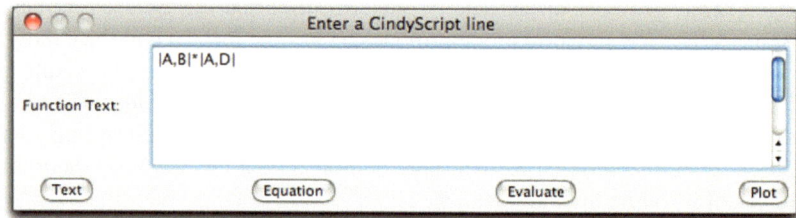

*Entering the text*

If we now press the "Text" button, we get the following picture on the left; if we press the "Equation" button, we get the picture on the right.

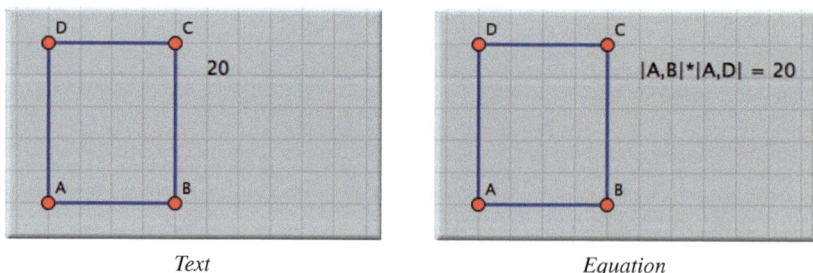

*Text*                                        *Equation*

If we want to display the result in a custom text, we may achieve this by entering the following code and hitting the "Text" button:

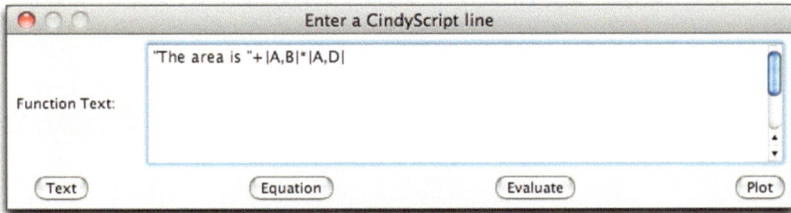

*Creating a custom text*

In this case, the line consists of a string to which the result of the calculation is appended. The result is again a string, which contains the desired text. The result in the window looks as follows:

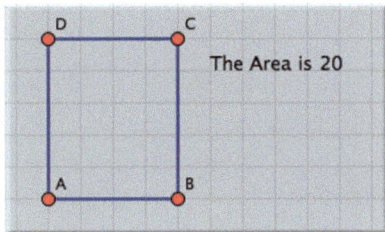

*Result of creating a custom text*

### Evaluation with Side Effects

Pressing the evaluate button allows for the evaluation of side effects. We will illustrate this feature by a slightly more sophisticated example. In the picture below the rectangle has been constructed in a way such that its perimeter remains constant when point C is moved. We will use this construction to analyze for which position of point C we obtain the largest area of the rectangle. Entering the line

```
F.xy=(C.x,|A,C|*|A,E|)
```

into the function dialog will move the point F to a position whose x-coordinate is the same as that of point C. The y-coordinate of F will be the area. Thus while we move C we can watch point F and determine for which position of C it assumes its maximal y-value. In the figure on left a locus is constructed (mover = C, tracer = F) that shows the areas for all possible positions of C.

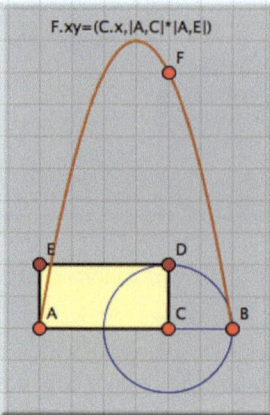

*Combining geometry and functions*

## Plotting Functions

As a final application we shall demonstrate how to use function mode for generating a plot of a function. For this we simply use the plot function of CindyScript and evaluate it via the function mode dialog. If the plot contains parameters that depend on the data of the elements in the drawing, then it is automatically updated as the elements are moved. In the example below, two sine waves are superimposed whose frequencies depend on the position of A. The code line is simply evaluated in the function mode dialog.

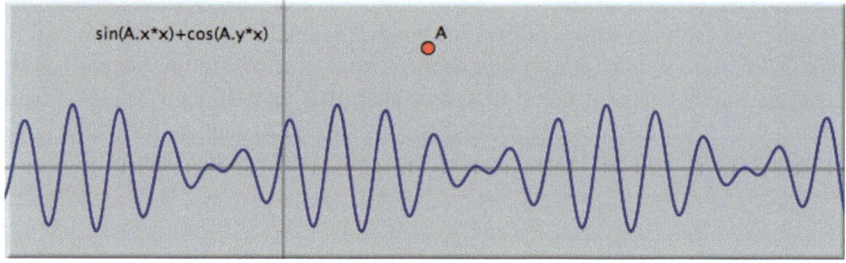

*Plotting a function*

## Click Referencing

As in text mode (p. 97) it is possible to obtain the reference of a geometric element by simply clicking on it in an arbitrary view. This simplifies the process of entering a formula in the dialog box. Clicking on a text that contains a measured number (for instance a distance, an angle, or an area) produces a reference to this number. Clicking on the text of another function reproduces the defining text of this function.

Furthermore, it is possible by a press–drag–release operation with the mouse to measure the distance between two points directly. If, for instance, A and B are two points in a geometric view, then pressing the mouse over A, dragging to B, and releasing it will produce the text |A,B| in the function dialog.

**Synopsis**

Function mode allows for the calculating, evaluating, and plotting of functions via a one-line interface to CindyScript.

**See also**

- CindyScript (p. 219)

- Text (p. 97)

- Measurements (p. 93)

## *4.8.3 Locus*

### 4.8.3.1 Constructing Loci

A locus is the trace of a point under the movement of another point. A locus is defined by three objects:

- The *mover*, a free element whose movement drives the generation of the locus.

- The *road*, an element incident to the mover. The mover will be moved along the road.

- The *tracer*, the element whose trace is calculated and presented as a geometric locus.

The mover, road, and tracer must be selected in this order. However, if the mover is either a "point on a line," a "point on a circle," or a "line through a point," then *Cinderella* automatically recognizes that there is a unique road and selects it for you automatically. You should watch the message line to check what input is required at each stage of a construction. Currently, the following combinations of mover and road are supported:

- *mover = point, road = line:* The point moves along the line.

- *mover = point, road = circle:* The point moves along the circle.

- *mover = point, road = segment:* The point moves along the segment.

- *mover = point, road = arc:* The point moves along the arc.

- *mover = line, road = point:* The line rotates around the point.

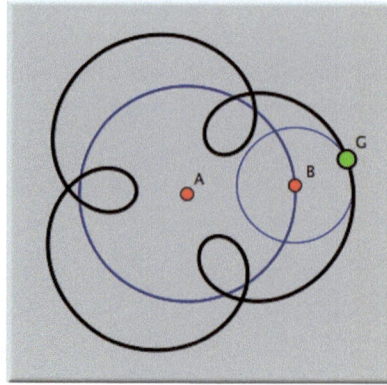

*A cycloid*

*Cinderella* also supports the selection of a "line" as the tracer. In this case the envelope of the moving line is calculated.

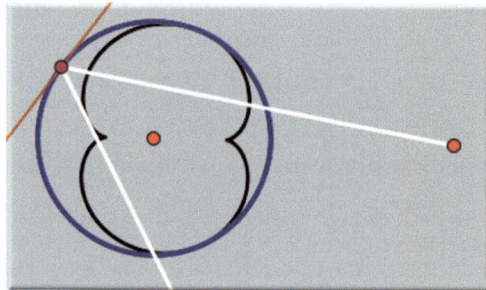

*The envelope of light rays in a circular mirror*

Loci of elements are real branches of algebraic curves. *Cinderella* will always try to generate an entire branch.

### 4.8.3.2 Inspecting Loci

Compared to *Cinderella 1.4*, loci have become much more flexible and powerful. Most of the new functionality is accessible through the inspector (p. 153). Here we give only a rough overview of the possibilities.

In the appearance tab of the inspector a selected locus offers the following control access:

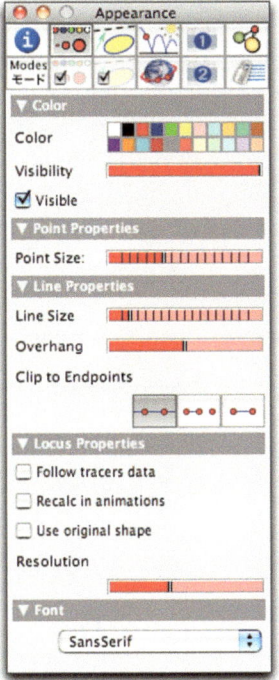

*Appearance of a locus*

Usually, loci are drawn as lines, so only the color, visibility, and line size controls will be relevant. However, in *Cinderella.2* it is also possible to present a locus as sample snapshots of the trace element. In this way, the locus of a moving point is drawn as a sequence of points. Hence in this case, the appearance controls for points will also be relevant. The behavior of the locus can be influenced by the "Locus Properties" section of at the bottom of the inspector. The precise meanings of the three checkboxes are as follows:

- *Use original shape:* If this box is checked, then instead of a line, a trace of images of the original object is drawn.

- *Follow tracers data:* This box ensures that the original data of the tracer are taken; in particular, appearance information is also represented by the trace.

- *Recalc in animations:* This selection forces that in animations the locus is always updated with the actual data and position of the mover.

The resolution slider allows for control of the number of sample points in a trace.

To get a slight impression of the possibilities of the new locus features, consider the example given by the following construction:

| | Who? | What? | Where? |
|---|---|---|---|
| ● | A | Point(0.08\|3.28) | (0.08\|3.28) |
| ○ | C0 | Circle(A;4.0613) | $(x - 0.08)^2 + (y - 3.28)^2 = 4.0613^2$ |
| / | a | LineThrough(A;−0°) | $y = 0x + 3.28$ |
| ● | B | PointOn(C0;34.2°) | (3.4407\|5.5604) |
| / | b | Perpendicular(a;B) | vertikal |
| / | c | Join(A;B) | $y = 0.6786x + 3.2257$ |
| ● | C | Meet(a;b) | (3.4407\|3.28) |
| / | d | Perpendicular(c;C) | $y = -1.4737x + 8.3504$ |
| ● | D | Meet(c;d) | (2.3811\|4.8415) |
| | E0 | Locus(B;C0;D) | – |

*Construction text*

The two pictures below show two different ways of presenting the locus.

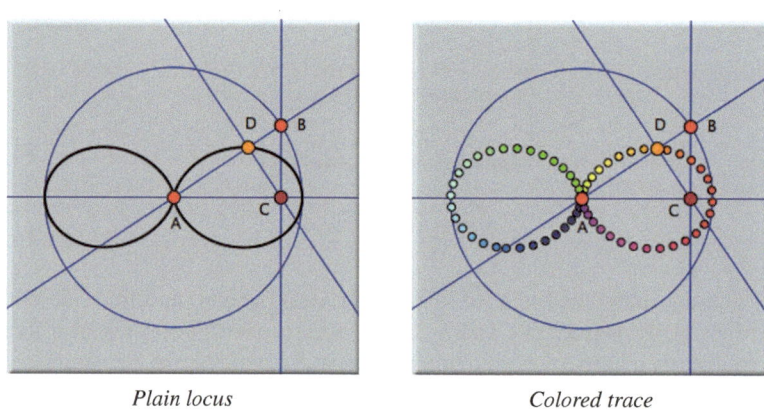

*Plain locus*                                        *Colored trace*

The first picture shows the plain locus. After adding the line

```
D.color=hue(B.angle/pi/2)
```

in the CindyScript panel, the tracer changes its color in accordance with the angle of the mover. If one then checks the "Use original shape" box and the "Follow tracers data" box, the resulting image is the one on the right. Observe that in this way, one recognizes, the nontrivial way in which the tracer follows the figure eight in this example.

**Synopsis**

Generate a geometric locus by selecting a mover, a road, and a tracer.

**Caution**

Sometimes you will notice a short delay when a locus is being calculated. Do not blame your computer or *Cinderella* (or, even worse, the authors). These delays can be caused by extremely difficult calculations that are necessary to get the correct result (a complete branch of the curve) or the correct screen representation after a movement. We have tried our best to speed up these calculations, but there is a (mathematical) limit: We do not want to sacrifice accuracy for speed.

**See also**

- Animation (p. 107)

- Traces (p. 160)

### *4.8.4  Animation*

In an animation you do not move the points yourself; *Cinderella* does it for you. An animation is defined by a "mover" and a "road," as in the definition of a locus.

- The *mover* is a free element whose movement drives the animation.

- The *road* is an element incident to the mover. During the animation, the mover will be moved along the road.

Either you select the mover and the road in this order, or you click on a locus, which will select the mover and road of the locus. If the selected mover is either a "point on a line," a "point on a circle," or a "line through a point," then *Cinderella* automatically recognizes that there is a unique road and selects it for you. Currently, the following combinations of mover and road are supported:

- *mover = point, road = line:* The point moves along the line.

- *mover = point, road = circle:* The point moves along the circle.

- *mover = line, road = point:* The line rotates about the point.

After you have defined an animation, control devices are added to the main windows. The following picture shows Cinderella after an animation of a point on a circle road was added and started:

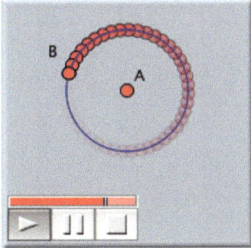

*Animation of a rotating point*

#### 4.8.4.1  Global Control of Animations

The main control device consists of three buttons and a speed slider in the lower left corner of the window. These controls refer to all animations. The precise meaning of the buttons is as follows:

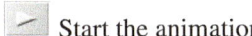 Start the animation

Pause the animation

Stop the animation

If you stop a running a animation, the picture will return to its position before the animation started. If you stop a paused animation, you are able to continue from the position of the paused animation. The slider can be used to adjust the speed of an animation.

There are a few significant differences regarding animations compared to the older version *Cinderella 1.4*.

- After defining an animation, you have to explicitly start it by pressing the start button.

- While an animation is running you can still move the free elements of a construction.

- You can have more than one animated element. Each of the animations can be started or stopped individually by pressing the port buttons in the upper right of the picture.

### 4.8.4.2 Individual Control of Animations

The port button associated with an animation is also a handle for selecting the animation. For selecting an animation you must press and hold the shift key and click the associated port button. Selected animations are indicated by a highlighted border of the port button.

The properties of a selected animation can be changed in the inspector (p. 153). In particular, the "Info" panel of the inspector can contain slots where you can adjust the speed of an animation. You can thereby adjust the relative speed of different animations. Animation speeds can even be set to negative values. In this way you can change the direction of an animation.

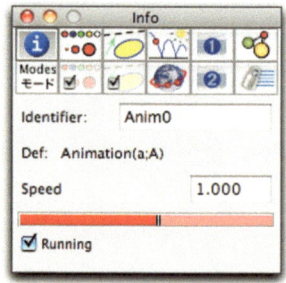

*Inspecting animations*

In the inspector you can also start and stop an individual animation by checking the "Running" box.

### 4.8.4.3 Tracing Elements

It will often be desirable to emphasize the trace of moving elements visually while an animation is running. To do this, consider the section on tracing (p. 160) elements, which shows you how to create stunning visual effects. Tracing is often a very interesting alternative to creating an explicit locus (p. 103).

### 4.8.4.4 Animations and CindyScript.

You can also control the speed of an animation via the CindyScript (p. 219) programming language. By default, the labels of the are "Anim0," "Anim1," .... If `Anim0` is the name of an animation, you can access the speed and running flags via the fields `Anim0.speed` and `Anim0.run`. Thus a CindyScript code fragment like

```
Anim0.run=(A.x>0);
```

forces "Anim0" to run only if the x-coordinate of A is positive. Similarly, the fragment

```
Anim0.speed=B.x;
```

allows you to control the speed of "Anim0" via the x-coordinate of B.

### 4.8.4.5 Animations and CindyLab.

The main control panel (play/pause/stop) of animations is identical to the control panel of simulations in CindyLab (p. 167). Thus all physics simulations are linked and synchronized with animated elements. Using this feature one can easily use animations to add motor-like devices to a physics simulation. The picture below, for instance, shows a simulated rubber band that is periodically driven by a point that moves up and down. The movement of the point is derived from an animated point that moves along a circle.

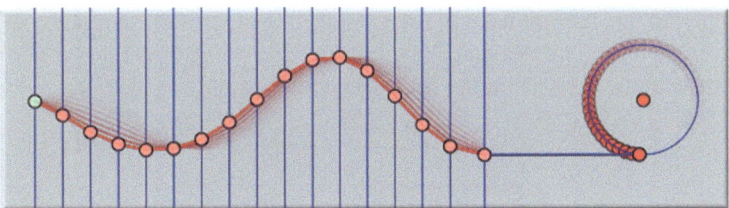

*Animation of an oscillating wave*

### 4.8.4.6 HTML Export

In contrast to the old version of Cinderella, exported animations generate full access to the control panel. Thus the user can start, stop, and pause animations freely on an HTML page. One is also allowed to move free elements while an animation is running. For a detailed treatment of this issue we refer to the section HTML Export (p. 419).

**Synopsis**

Start an animation by selecting a mover and a road.

**Caution**

The are many changes compared to *Cinderella 1.4*.

**See also**

- Locus (p. 103)

- CindyLab (p. 167)

- CindyScript (p. 219)

- Traces (p. 160)

- Inspector (p. 153)

- HTML Export (p. 419)

## 4.9 Conic Operations

There are four conic operations that are accessible by buttons:

Conic by Five Points (p. 89)

Ellipse by Foci (p. 89)

Parabola by Focus and Directrix (p. 90)

Hyperbola by Foci (p. 90)

Besides these four operations, there are also several others that are accessible by the menu only. In this section we will briefly sketch the use of these modes. All of these modes are definition modes that ask for a selection of input elements. The complete conic mode menu consists of the following entries:

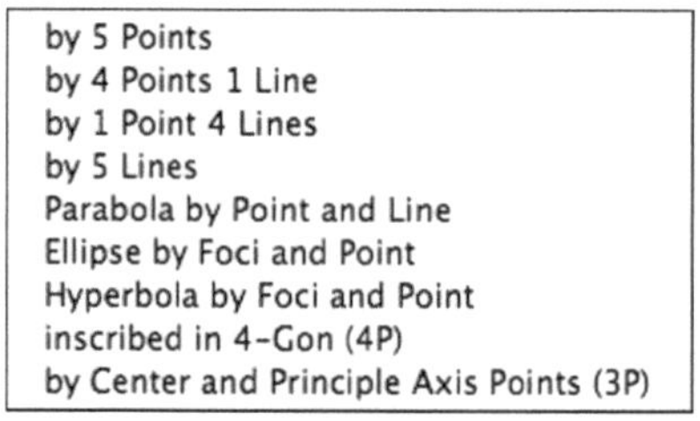

```
by 5 Points
by 4 Points 1 Line
by 1 Point 4 Lines
by 5 Lines
Parabola by Point and Line
Ellipse by Foci and Point
Hyperbola by Foci and Point
inscribed in 4-Gon (4P)
by Center and Principle Axis Points (3P)
```

*Conic mode menu*

### 4.9.1 Conic by Five Lines

Just as one can show that a conic is uniquely defined by five of its points, one can also show that a conic is uniquely defined by five tangent lines. This mode makes this functionality accessible. To use it, simply select the five lines before or after invoking the mode.

### 4.9.2  Conic by Four Points and a Line

There are two conics that simultaneously pass through four points and are tangent to a line. This mode allows one to construct these conics by selecting the defining elements.

### 4.9.3  Conic by Four Lines and a Point

There are also two conics that simultaneously are tangent to four lines and pass through one point. This mode allows one to construct these conics by selecting the defining elements.

The pictures below give examples for the three above-mentioned modes:

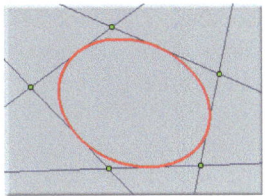  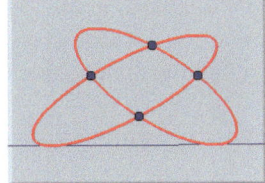  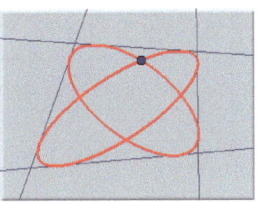

*Five lines*                 *Four points / one line*                 *One point / four lines*

### 4.9.4  Conic Inscribed in a Four-gon

This mode is very convenient for constructing perspective drawings. It generates a conic that is inscribed in a four-gon. The conic is constructed in such a way that the final picture corresponds to the perspective drawing of a circle inscribed in a square. To define it one has to click the four points in cyclic order.

### 4.9.5  Conic by Center and principal Axes Points

A conic is very often characterized by its principal axes and a point on each of them. This mode allows you to construct an ellipse by the center and two points on the principal axes through which it should pass.

The pictures below give examples of the two above-mentioned modes.

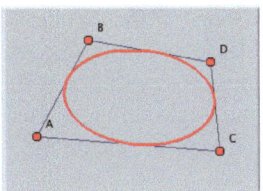       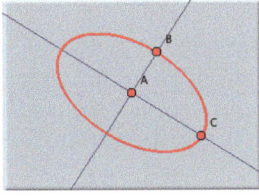

*Inscribed conic*                       *Conic by principal axes*

**See also**

- Conic by Five Points (p. 89)

- Ellipse by Foci (p. 89)

- Hyperbola by Foci (p. 90)

- Parabola by Focus and Directrix (p. 90)

- Definition Modes (p. 84)


## 4.10 Polygons

In *Cinderella.2* various methods are provided to generate regular polygons. They are accessible via the "Polygon" submenu in the "Modes" menu. These modes can be subdivided into two different classes. One class consists of "regular polygon" mode only; the other consists of all the remaining modes in this menu.

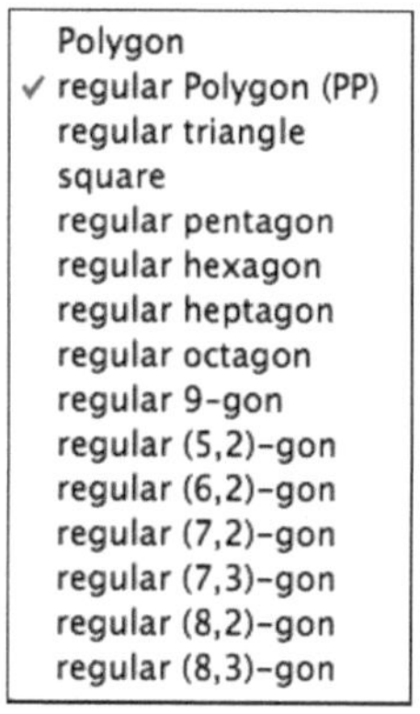

```
Polygon
✓ regular Polygon (PP)
  regular triangle
  square
  regular pentagon
  regular hexagon
  regular heptagon
  regular octagon
  regular 9-gon
  regular (5,2)-gon
  regular (6,2)-gon
  regular (7,2)-gon
  regular (7,3)-gon
  regular (8,2)-gon
  regular (8,3)-gon
```

*Polygon mode menu*


### 4.10.1 Regular Polygon

This mode can be used if two points of one edge of a regular polygon are already present. First one clicks on these two points. Then, while the mouse is being dragged, a hint is displayed that indicates the shape of the regular n-gon that will be generated. By adjusting the mouse one can choose whether one wants a triangle, a square, a pentagon, and so forth. Clicking a third time will freeze the position of the polygon, and the vertices and segments for the edges are generated. They are then available in the construction as fully featured geometric elements.

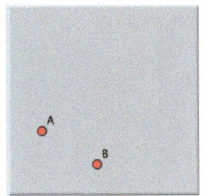 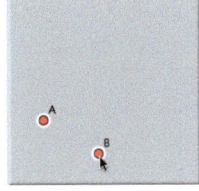 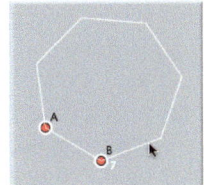 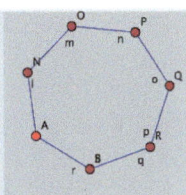

*The two points*     *Selecting them*     *Hints while dragging*     *Situation after the third click*

This mode is also available in hyperbolic geometry, in which case a regular *hyperbolic* polygon is created. This is a polygon all of whose sides have equal lengths and all of whose angles are equal.

**Caution**

This mode is currently not supported in spherical geometry. In hyperbolic geometry, in contrast to Euclidean geometry, the angle of an n-gon is not determined by the number n.

## 4.10.2 Regular Polygons by Center and Vertex

The remaining modes for regular polygons are very similar to "add a circle" mode. They are defined by a press–drag–release sequence. Pressing the mouse creates the center of the polygon, and dragging creates the polygon. Releasing freezes the construction.

For each regular $n$-gon from $n = 3$ to $n = 9$ a specific mode is provided. Furthermore, there exist several modes for regular star polygons. The following picture shows all regular polygons constructible by these kinds of modes.

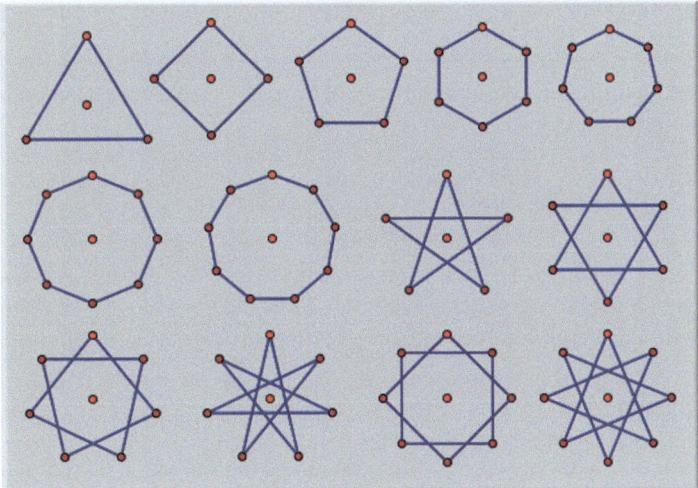

*A sampler of regular polygons*

This mode is also available for hyperbolic geometry. There another important feature comes into play. A regular n-gon in hyperbolic geometry is a polygon with identical angles at each vertex and with all edges of equal length. In contrast to Euclidean geometry, the number n does not determine the angle. Depending on the size of the polygon, the angle can take any value between $0°$ and $\frac{n-2}{n} \cdot 180°$. Thus there exist, for instance, regular pentagons for which all angles are exactly $90°$. Four such pentagons fit around a vertex, like squares in Euclidean geometry. Usually it is difficult to construct a polygon of the right size for studying this effect. The regular polygon modes of *Cinderella.2* provide a method for generating these polygons in a controlled way. While the mouse is being dragged, snap points are indicated, and a small number is displayed that indicates how many similar polygons fit around a corner.

**See also**

- Polygon (p. 91)

- Add a Circle (p. 81)

- Transformation Groups (p. 145)

- Hyperbolic Geometry (p. 114)

# 4.11 Geometries

It is one of *Cinderella's* main features that it supports different kinds of *geometries*. If you are not accustomed to the idea that there are "different kinds of geometry," this may sound confusing. You should read the section Theoretical Background (p. 27) to get a feeling for the underlying ideas and how they are implemented in *Cinderella*. Users who already have some basic knowledge about Euclidean and non-Euclidean geometries might be satisfied reading this reference part only.

Another warning message: Do not confuse *geometries* with views (p. 116); they both come in similar flavors, but the first defines the behavior of elements, while the second describes the presentation of elements.

## 4.11.1 Types of Geometry

In each main window of *Cinderella* you will find three buttons for choosing the type of geometry. In the present version, Cinderella provides three different kinds of geometry: *Euclidean geometry*, *hyperbolic geometry,* and *elliptic geometry.* You can switch among these three geometries by pressing the button $\boxed{\text{Euc}}$ for Euclidean geometry, $\boxed{\text{Hyp}}$ for hyperbolic geometry, or $\boxed{\text{Ell}}$ for elliptic geometry.

The choice of a new geometry does not affect the behavior of the elements you have already constructed. However, every newly added element is interpreted with respect to the new geometry. You can think of each element as having an entry that tells it to which geometry it belongs. The basic notions affected by the choice of the geometry are the measurements of *distances* and *angles*.

However, also other constructions are influenced by this choice. For instance, the *angle bisector* is defined to be a line whose angles to two other lines are equal. If the measurement of angles has changed, then the definition of angle bisector has to change, too. Similar things happen to "parallels" and "perpendicular lines." The definition of a circle is also influenced by the geometry. A circle is the set of all points that have the same given distance to the center. If the notion of "distance" is changed, the concept of "circle" changes as well.

Other operations are not affected by the choice of the geometry at all: the line joining two points will always be the same no matter in which of the above geometries you are.

The following list collects all constructions that are influenced by the choice of the geometry. Observe that the position, as well as the number, of elements that are constructed can change.

- *Distance:* The notion of distance depends on the geometry. It can even happen that in hyperbolic geometry distances of real points become complex numbers, for instance when the line joining both points lies completely outside the horizon.

- *Angle:* The notion of angle depends on the geometry. As in the case of distances, angles can also become complex numbers.

- *Circle:* The exact notion of circularity depends on the definition of "distance," which changes in each geometry. This influences all construction modes for circles. In the Euclidean view, hyperbolic or elliptic circles can look like arbitrary conics. The picture is clarified in the other views. In the hyperbolic view (Poincaré disk), hyperbolic circles really look like circles. In the spherical view, elliptic circles look like circles on the surface of a ball.

- *Mirror:* The notion of reflection depends on distances and angles and thus it depends on the geometry. All kinds of mirrors are influenced this way.

- *Angle bisector:* In all three geometries, you have two angle bisectors for a line. However, the exact position depends on the choice of the geometry. In hyperbolic geometry, angle bisectors of real lines can become complex.

- *Midpoint:* The midpoint of two points depends on the definition of "distance." In Euclidean geometry, there is exactly one such midpoint. In hyperbolic and elliptic geometry there are two such midpoints (points of equal distance to the defining points). *Caution:* If you are in the "hyperbolic view," only one of these points will be visible, since the other one lies outside the horizon.

- *Line with fixed angle:* This construction is influenced by the choice of the geometry because angles are involved.

- *Perpendicular:* The notion of a perpendicular depends on the notion of "angle" and is influenced by the choice of the geometry.

- *Parallel:* In *Cinderella* parallels of a line $L$ are defined as lines that have an angle of zero to $L$. In Euclidean geometry there is a unique parallel to $L$ through a point. However, in hyperbolic and elliptic geometry there are two such parallels in general. The parallels in elliptic geometry will usually have complex coordinates, so that you will see them only in the "construction text" view.

In the present version of *Cinderella* some operations are not supported in all geometries. These operations are Circle by Three Points (p. 89), Area (p. 96), and Center (p. 86), where the Euclidean result is always calculated.

### 4.11.2 Views and Geometries

Although every geometry is usable together with every view, a few combinations are a bit more common than others. Here is a short list of what these common choices represent.

*Euclidean view* in *Euclidean geometry*: This may be the most common choice. The geometric elements behave like "usual elements" in a "usual plane."

*Spherical view* in *Euclidean geometry*: This choice gives you control over the behavior "at infinity" of the Euclidean plane. The spherical view represents a double cover of the Euclidean plane. Each line is mapped to a great circle and each point is mapped to an antipodal pair of points. The boundary of the nonrotated view corresponds to the "line at infinity" of the Euclidean plane.

*Euclidean view* in *hyperbolic geometry*: What you see here is the "Beltrami-Klein" model of hyperbolic geometry. In this model hyperbolic lines are really straight. Measurement is done according to the definitions of the Cayley-Klein geometry. In the Euclidean view the horizon of hyperbolic geometry is shown as a thin circle.

*Hyperbolic view* in *hyperbolic geometry*: This is what is known as the Poincaré disk. Hyperbolic lines are represented by circular arcs that cross the boundary of the disk at right angles. The Poincaré disk distorts the usual plane in a way that hyperbolic angles between lines correspond to "Euclidean" angles between the corresponding circular arcs. In mathematical terms, the Poincaré disk is a conformal representation of the hyperbolic plane. In this picture hyperbolic circles really look circular.

The whole disk represents only a part of the full plane of the corresponding Cayley-Klein geometry. The part that is shown corresponds to the region inside the circle shown in the Euclidean view. The measurement of distances is such that the distance from any interior point to any point on the boundary is equal to infinity.

*Spherical view* in *elliptic geometry*: The spherical view is the natural view for elliptic geometry. The angle between two lines corresponds to the spherical angles of the corresponding great circles. Measurement of distances corresponds to geodesic measurement of distances on the surface of a ball. Elliptic circles correspond to circles on the surface of the ball.

However, one has to be a bit careful. Elliptic geometry is not equivalent to spherical geometry (geometry on the ball). This comes from the fact that in elliptic geometry antipodal points of the ball are identified with each other.

## 4.12 Views

The entry "Views" in the menu bar offers items for opening windows containing different views of a geometric construction. Each view is a kind of "projection" of the abstract configuration to some visible part of the computer screen. Usually you can create constructions and manipulations in any of the views. The changes are reported to the other views immediately. In particular, you can have several views of the same type (for instance two Euclidean views with different scales).

The views are also related to the different kinds of geometries (p. 114). Which view is appropriate for which geometry is discussed in the section Views and Geometries (p. 116).

### 4.12.1 Euclidean View

The Euclidean view is the usual drawing surface. When *Cinderella* is started you will get a window containing a Euclidean view. It is the natural window for doing Euclidean geometry.

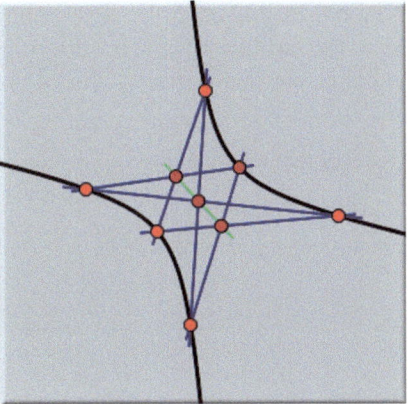

*Pascal's theorem in a Euclidean view*

The Euclidean view has view-specific control buttons. They can be used for zooming and translation. Furthermore, they are used to control grids and snap points.

*Translate mode:* This mode allows the entire coordinate system to be translated. After you have selected this mode, you can move the view around while pressing the left mouse button.

*Zoom in:* With this mode you can zoom into the drawing. The operation is inverse to the zoom out operation. There are two ways of doing this:

- With a press–drag–release sequence you mark a region. The view will be zoomed to show exactly this region.

- You can click the left mouse button over any position in the view. The view will be zoomed around the click point. The factor of zooming is 1.4. You can also click the right button (or shift-click the left button) to get the inverse of this zooming operation.

*Zoom out:* With this mode you can zoom out the drawing. The operation is inverse to the zoom in operation. There are two ways of using this mode:

- With a press–drag–release sequence you can mark a region in the view. The view will be appropriately zoomed so that the currently visible part zooms to that region.

- You can click the left mouse button at any place in the view. The view will be zoomed around the click point. The factor of zooming is 0.7. You can also click the right button (or shift-click the left button) to get the inverse of this zooming operation.

*View all points:* Pressing this button adjusts the current zoom settings in order to show all points of the construction.

*Toggle square grid:* Shows/hides a quadrangular grid on the view.

*Toggle triangle grid:* Shows/hides a triangular grid on the view.

*Toggle axes:* Shows/hides a coordinate system within the view.

*Snap:* Toggles "snap" mode. In this mode the grid points are magnetic and will attract the mouse. This is the ideal tool for exact drawings. When you first select this mode, the grid and axes are automatically shown. You can hide them again individually.

*Fine grid:* Gives the grid a higher density.

*Coarse grid:* Gives the grid a lower density.

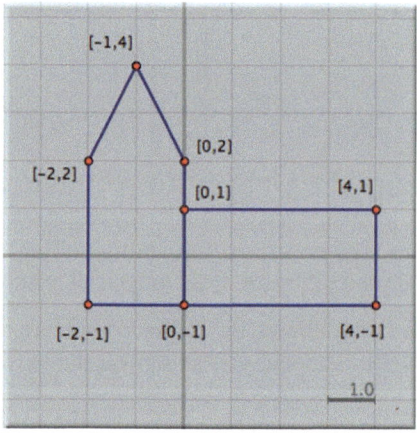

*Using "snap" mode for exact drawings*

### 4.12.2 Spherical View

The spherical view arises from a projection of the Euclidean plane onto the surface of a ball. The center of the projection is the center of the sphere. The plane does not pass through this center.

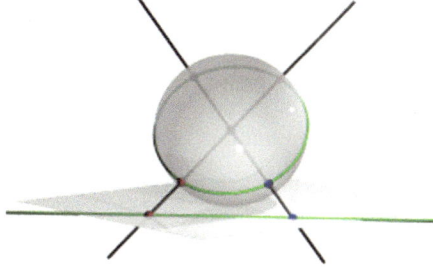

*Projection from the plane onto the ball*

This projection maps each point to an antipodal pair of points on the sphere. Each line is mapped to a great circle (an equator) on the sphere. The incidence structure is preserved. Working with the spherical view allows elements at infinity to be manipulated. They lie on the boundary of the image of the unrotated sphere.

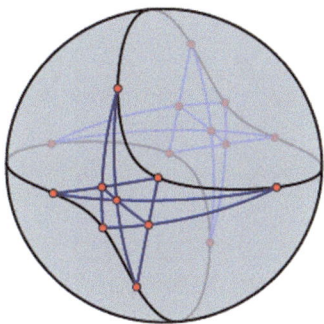

*Pascal's theorem in a spherical view*

The spherical view is a natural way to represent elliptic geometry. Measurement of angles between lines corresponds to measuring angles on the sphere. Measuring distances corresponds to the usual geodesic measurement of distances on the sphere, keeping in mind that antipodal points are identified with each other.

*Rotate:* This mode allows the spherical ball to be rotated (after the projection). In this mode you can rotate the view by moving the mouse while keeping the left button pressed.

*Home:* This action resets the rotation of the spherical view to its original position. After a reset, the visible boundary of the ball corresponds to the line at infinity again.

*The scale slider:* This slider lets you control the distance from the sphere to the Euclidean plane. You can use this slider to find the proper magnification of the drawing.

### 4.12.3 Hyperbolic View

The hyperbolic view is the natural view for hyperbolic geometry. In fact, doing hyperbolic geometry is the main reason for opening a hyperbolic view. The hyperbolic view represents an implementation of the Poincaré disk model of hyperbolic geometry. In this model the (finite part of the) hyperbolic plane is represented by a disk. Each line is represented by a circular arc that is orthogonal to the boundary of this disk. The measurement of angles between lines is conformal. This means that you can read off angles by measuring Euclidean angles between the circular arcs. The measurement of distances is such that the elements on the boundary are "infinitely far away" from any other point on the disk. If you "walk" in hyperbolic unit steps in one direction, you will never reach the boundary. In the disk, the steps seem to become smaller and smaller (in Euclidean measurement).

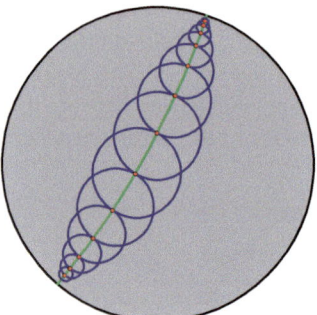
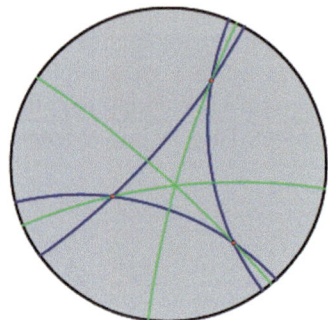

*Hyperbolic circles of equal size*          *Hyperbolic altitudes meet in a point.*

### 4.12.4 Polar Euclidean and Spherical Views

Polarity is an important concept of projective geometry. Because of the complete symmetry between lines and points, it is possible to turn every statement about incidences of points and lines into a corresponding "polar statement" in which the roles of points and lines are interchanged.

*Cinderella* offers two polar views for visualizing polarity. The polarity implemented in *Cinderella* is the polarity with respect to the identity matrix. In algebraic terms, we use the homogeneous coordinates of a point and interpret them as a line, and vice versa. Geometrically, this finds its easiest interpretation in the spherical view. Whenever you have a point, consider it as a "north pole"; the corresponding "equator" is its polar line. Whenever you have a line, consider it as an "equator"; the corresponding "north pole" is its polar point. The figure below shows a configuration in the spherical view and in the polar spherical view.

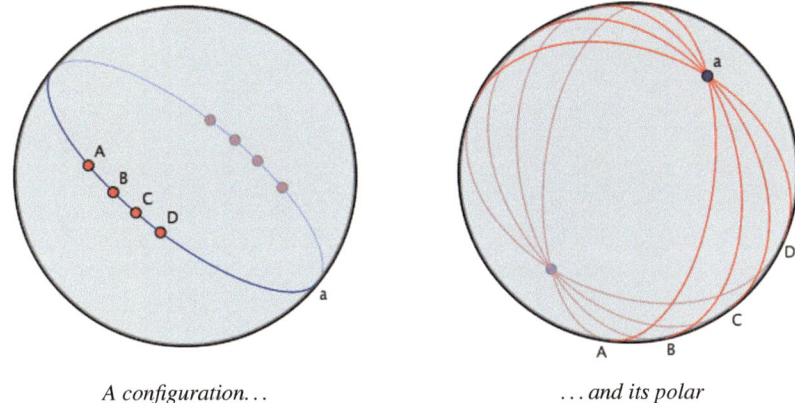

*A configuration...*                    *...and its polar*

Polar view elements are selectable, but moving them is disabled. If you want to move elements, you have to control them in a primal view.

### 4.12.5 Construction Text

The construction text is a textual description of the construction steps. Each element of the geometric construction is represented by a row in the construction text window.

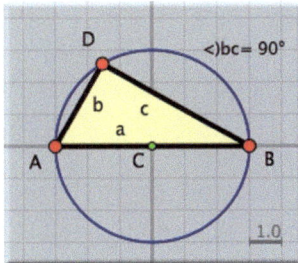

*A picture of Thales' theorem*

Each row shows a little icon that resembles the element it refers to. The icon is shown in the size, color, and shape of the element. This makes it easy to identify elements quickly.

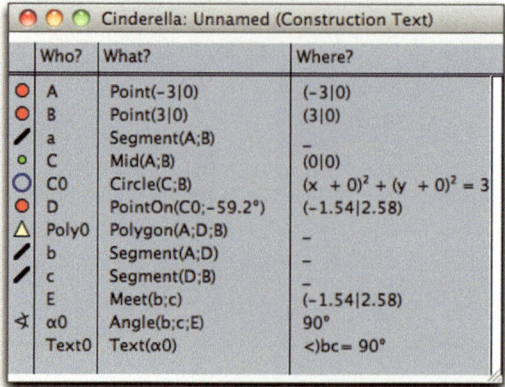

| Who? | What? | Where? |
|------|-------|--------|
| A | Point(−3\|0) | (−3\|0) |
| B | Point(3\|0) | (3\|0) |
| a | Segment(A;B) | − |
| C | Mid(A;B) | (0\|0) |
| C0 | Circle(C;B) | $(x + 0)^2 + (y + 0)^2 = 3$ |
| D | PointOn(C0;−59.2°) | (−1.54\|2.58) |
| Poly0 | Polygon(A;D;B) | − |
| b | Segment(A;D) | − |
| c | Segment(D;B) | − |
| E | Meet(b;c) | (−1.54\|2.58) |
| α0 | Angle(b;c;E) | 90° |
| Text0 | Text(α0) | <)bc = 90° |

*The construction steps for Thales' theorem*

The construction text consists of four columns. The first column shows the icons. The second column shows the labels of the geometric elements. These labels are unique identifiers within the construction. The third row explains how the element is defined. The fourth row represents the current value of the element. In most cases this is the present location with respect to the coordinate system. With the menu "Format" you can change how the locations or values of elements are shown.

The texts that appear in the "construction text" are exactly the three entries that can be referenced in text (p. 97) mode.

The four columns are separated by vertical lines. You can select these lines with the mouse to control the width of each column. If the text has too many rows for the window, a scrollbar at the right side of the window is available for scrolling.

### 4.12.6 General Functions

In each view, a toolbar at the bottom displays view-specific operations. Those that are common to all views are described below.

*Generate PDF:* If you press this button, the contents of the view will be exported to a PDF file.

*Choose the geometry:* With these buttons you change the current geometry. All geometric constructions refer to this geometry. For a detailed discussion of geometries consult the chapters Geometries (p. 114) and Theoretical Background (p. 27).

## 4.13 Transformation Modes

Transformations are one of the most powerful new geometric enhancements in *Cinderella.2*. With transformations it is easy to transfer parts of a construction to another place, reflect or rotate them, or even apply a perspective transformation. The use of transformations often allows significant simplifications of constructions even in unexpected circumstances.

Transformations are conceptually slightly different from usual geometric construction modes in *Cinderella*. While a usual geometric construction generates a visible geometric object (such as a parallel or an intersection point), a transformation creates an action button that permits the application of the transformation to other objects.

The transformation mode menu collects all possible types of transformation operations:

```
Reflection (old)
Rotation P|Ang
Rotation LL
Reflection
Translation
Similarity
Affine Transformation
Projective Transformation
Moebius Transformation
Add Function
Inverse
Composition
Map Element
```

*Transformation mode menu*

### 4.13.1  General Use of Transformations

We will demonstrate the use of transformations by a simple translation. The generation of a transformation consists of two steps:

- First, a transformation has to be defined by a suitable transformation mode.

- Second, once a transformation has been defined, it can be applied to arbitrary geometric objects.

During a move in move mode, the parameters of a transformation become updated automatically such that the images of the transformed objects move accordingly.

#### 4.13.1.1  Defining a Transformation

As with the usual geometry modes, how to define a transformation depends on the exact type of transformation you want to define. A *translation* is defined by clicking two points *A* and *B*. This translation corresponds to the geometric shift that moves *A* to the position of *B*.

Consider the following example: Assume that points *A* and *B* are already present in the drawing. To define a translation you have to choose the mode *Translation* from the *Mode -> Transformation* menu. After point *A* is clicked, an arrow appears as an optical hint for the construction of the translation. Then you click point *B*. After this, a little rectangle, a so *View Button*, appears that serves as a link to the translation further on.

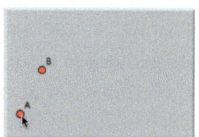 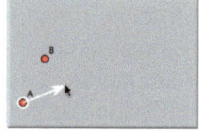  

| **First selection:** | **Moving the mouse:** | **Move to second point:** | **Second selection:** |
|---|---|---|---|
| *The first point of the translation is selected.* | *An arrow hint is shown.* | *The arrow helps you.* | *The translation is created.* |

The view button serves a kind of representative for the transformation and acts almost like an ordinary geometric mode. By pressing it you apply the transformation to the currently selected elements.

### 4.13.1.2 Applying a Transformation

After you have defined the transformation, you can use it. For this you have to go to Move Mode (p. 71), which at the same time serves as the selection mode. Now you select one or several elements by clicking them. If you want to have a multiple selection, you must hold down the shift key. After you have selected the elements you want to map via your translation, you simply press the view button of your transformation. In this way the mapped elements are created. They will automatically have the same appearance settings as their preimages. Furthermore, they will automatically become the active selection, so that you can easily apply a transformation to them again.

 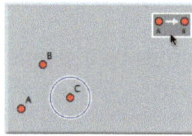 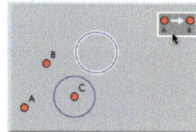 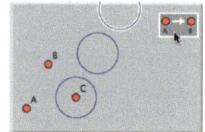

| **Select the preimage:** | **Move the mouse:** | **Click the trafo:** | **Another click:** |
|---|---|---|---|
| *You must do this in move mode.* | *If you move over a transformation, it is highlighted.* | *This maps the element.* | *Maps the mapped elements again.* |

### 4.13.1.3 Changing the Parameters of a Transformation

A transformation is a dynamic object. If the position of the preimages changes, then the mapped elements become updated accordingly. If in our example *A* or *B* is moved, then the mapped circles will move as well.

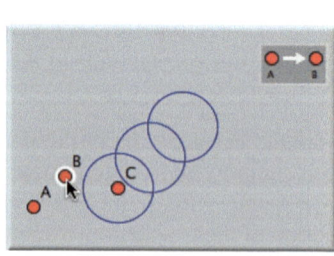

*Changing the translation*

#### 4.13.1.4  Selecting a Transformation

Sometimes, it may be necessary to *select* a transformation. This can also be done in Move Mode (p. 71). However, for selecting the transformation you have to press the shift key while you click it. Selected transformations may be used as input parameters for transformation groups (p. 145) or iterated function systems (p. 141).

**See also**

- Move Mode (p. 71)

- Iterated Function Systems (p. 141)

- Transformation Groups (p. 145)

### *4.13.2  Translation*

A translation is a shift in a certain direction by a particular length. A translation is defined by selecting two points that serve as a sample of a preimage and an image. Defining the translation requires two mouse clicks.

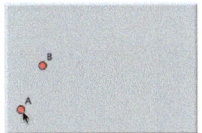 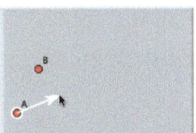 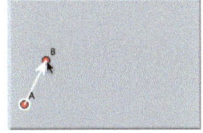 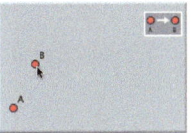

**First selection:**          **Moving the mouse:**      **Move to second point:**   **Second selection:**
*The first point of the*      *An arrow hint*            *The arrow helps you.*       *The translation*
*translation is selected.*    *is shown.*                                             *is created.*

A detailed description of defining a translation can be found in the section General Use of Transformations (p. 123).

Applying a translation to a selection of many objects will map them all simultaneously. In this process all appearance information is transferred from the preimages to the images. The following pictures demonstrate how appearance information is transferred.

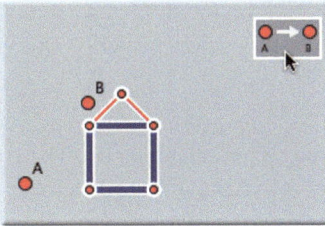                  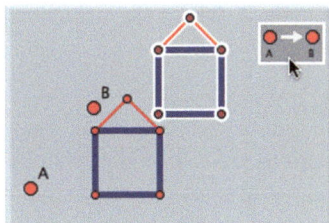

    *The preimage*                      *Situation after mapping*

One use of translations is to create regular ornamental patterns. The following picture shows a pattern that is generated by the repeated application of two translations in two different directions.

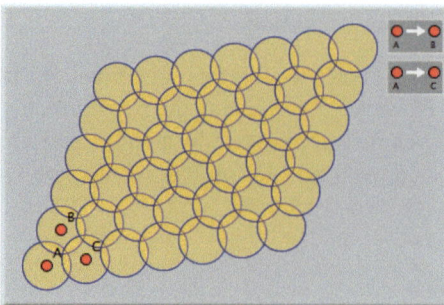

*A regular pattern of circles*

Transformations may also be used in order to study a certain situation that occurs locally around a moving point. In such a situation one may introduce another point as a kind of fixed base point, define a translation that maps the point of interest to this point and finally maps the point whose relative movement is to be studied with this translation. Since this application of transformations may appear slightly advanced, we give a little example of this technique. Consider the image below. There a rotating red point $D$ and a rotating green point $C$ are shown. The green point rotates three times as fast as the red one. What is the trace of the green point viewed from the rotating red point. For this, a transformation is defined that maps the red point $D$ to some fixed point $E$. Then the map is applied to the green point. The trace of the mapped point is shown as a dark-green trace. This is how $D$ sees $C$.

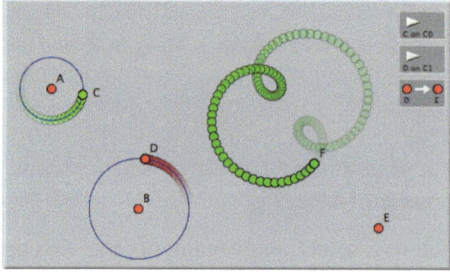

*Analyzing from different perspectives*

The same technique has been used in the section Geometry and CindyLab (p. 208), in Example 2, where the behavior of the velocity vector in planetary motion was analyzed.

**Caution**

Translations are also available in hyperbolic and spherical geometry. There they are defined as the transformations that map a point $A$ to a point $B$ while leaving the line joining them invariant.

**See also**

- Similarity (p. 128)
- Rotation (p. 127)

- Reflection (p. 128)
- General Use of Transformations (p. 123)

### *4.13.3 Rotation*

A rotation can be defined in two different ways: One way is by choosing the mode *Rotation P|Ang,* which defines a rotation by a rotation center and an angle (which is given by a number). To apply this mode one has to click the center of a rotation and a text in the drawing that represents the requested angle (for instance a measured angle). Alternatively, one can use *Rotation LL* mode. For this one has to click two different lines that serve as a sample for an image and a preimage.

The mode *Rotation P|Ang* can be particularly useful if the rotation angle is an already measured angle or if the rotation angle must satisfy a certain predefined value. The mode *Rotation LL* is useful whenever a line and its image under the rotation are already available.

The picture below shows an iterated application of a rotation to a circle.

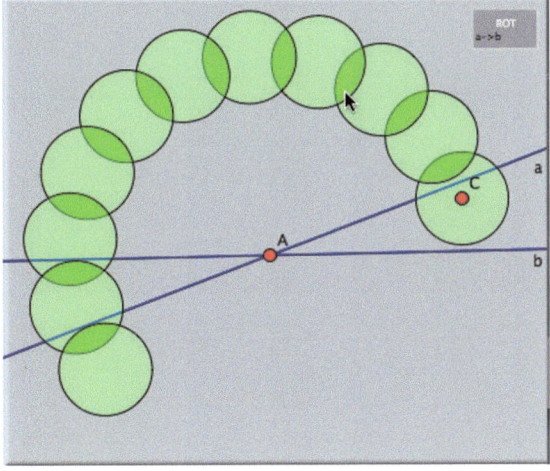

*Rotating a circle*

**Caution**

Rotations are also defined in hyperbolic and elliptic geometry.

**See also**

- Translation (p. 125)
- Similarity (p. 128)
- Reflection (p. 128)

### 4.13.4 Reflection

Defining a reflection can be dome with just one mouse click. One simply clicks the element that should serve as a mirror. As in mirror (p. 88) mode, reflections can be created with respect to various objects. Reflection in a line gives the usual mirror symmetry. Reflection in a point gives point symmetry, and reflection in a circle generates an inversion in the circle. For a detailed geometric description of these reflections please refer to mirror (p. 88) mode.

The picture below shows different kinds of mirrors applied to a drawing of a house.

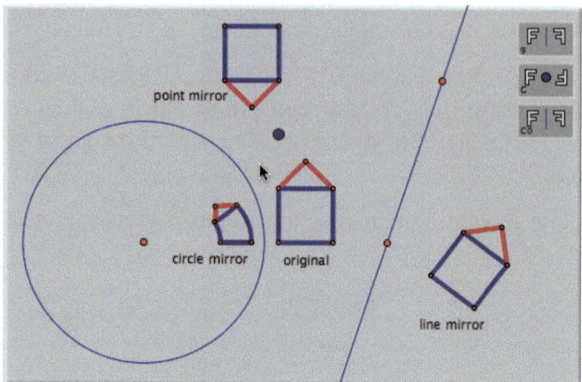

*Different kinds of mirrors*

**Caution**

Translations are also available in hyperbolic or elliptic geometry.

**See also**

- Similarity (p. 128)
- Rotation (p. 127)
- Translation (p. 125)
- Mirror (p. 88)

### 4.13.5 Similarity

Similarities are among the most useful transformations available in Cinderella. They can be considered as a composition of a rotation and a scaling. With a suitably chosen similarity one can map an arbitrary (ordered) pair of points to another one.

A similarity can be defined by specifying the preimage and image for two pairs of sample points. Thus a similarity is defined by choosing an (ordered) collection of four points.

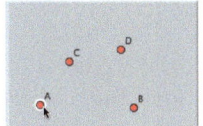 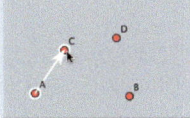 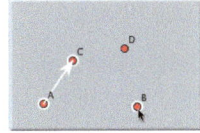 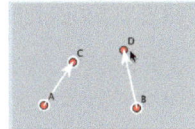

**First click:**          **Second click:**        **Third click:**          **Fourth click:**
*The first preimage*      *The first image*         *The second preimage*     *The second image*
*is defined.*            *is defined.*             *is defined*              *is defined.*

After this sequence of four clicks, the similarity has been defined completely. In this example the similarity will be uniquely defined by the facts that *A* is mapped to *C* and that *B* is mapped to *D*.

Applying a translation to a selection of many objects will map them all simultaneously. The following pictures demonstrate the effect of a similarity transformation.

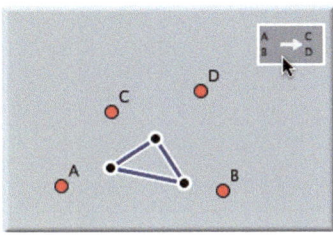 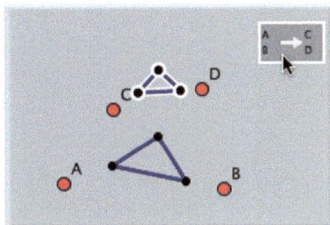

*The preimage*                              *The situation after mapping*

If one iterates a similarity, the sequence of mapped pictures will automatically form some kind of logarithmic spiral. The picture below demonstrates this effect.

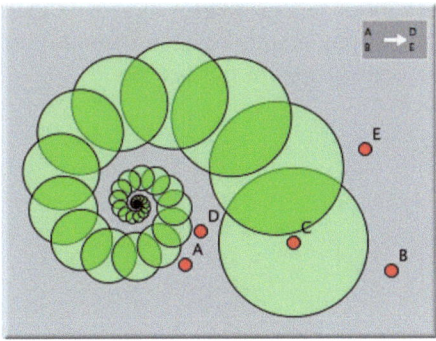

*A logarithmic spiral*

In the definition of a similarity it is also possible to use points more than once. Thus it is possible to map a point to itself by selecting the same point as image and preimage. We get several interesting special cases of similarity:

- *A → A and B → B:* The *identity transformation*, which leaves all points invariant

- *A → A and B → C:* A *similarity*, with fixed point *A*

- *A → B and B → A:* A point *reflection* that interchanges *A* and *B*

### 4.13.5.1 Similarities and Sliders

Similarities are sometimes very useful tools for implementing a slider control in a geometric scenario. The following example illustrates this technique. Suppose that you want to create an object that can be smoothly transformed from one shape into another. The transition is to be controlled by a point on a segment that acts as a slider control. Let us suppose that you want to create a figure that "morphs" from the square on the left smoothly to the triangle on the right as the green point is moved.

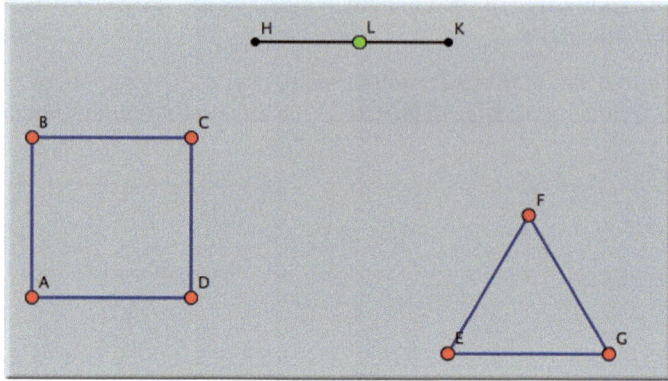

*A morphing task*

One can achieve such an effect by defining many similarity configurations, one for each preimage/image pair of points. For each such pair one defines a similarity that maps the endpoints of the slider to the two points of the pair. One then constructs the image of the sliding point under this map. When the slider is moved, the mapped point moves from the morphing start position smoothly to the end position. The picture below shows a construction of this process.

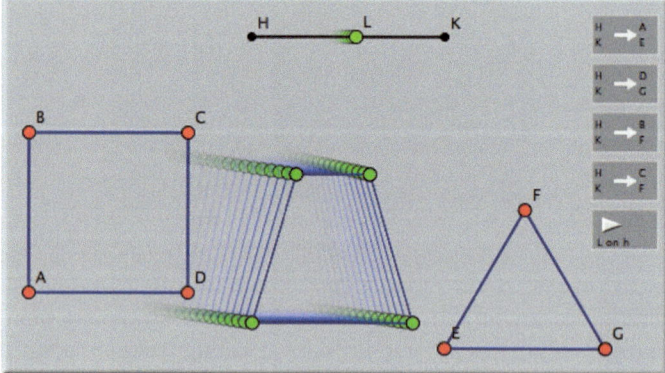

*Morphing made easy*

**Caution**

Similarities are not defined for hyperbolic and elliptic geometry.

**See also**

- Translation (p. 125)

- Affine Transformation (p. 131)

- Projective Transformation (p. 131)

### 4.13.6 Affine Transformation

An affine transformation is something between a similarity (p. 128) and a projective transformation (p. 131). A similarity preserves the angle between two lines, but this is no longer the case for an affine transformation. However, parallelism is preserved by affine transformations, which is not the case for projective transformations.

An affine transformation is defined by three preimage/image pairs of points. The definition is analogous to that of a similarity (p. 128), though with six required points instead of four.

The picture below shows a simple scene in which a blue figure is mapped affinely to a black figure.

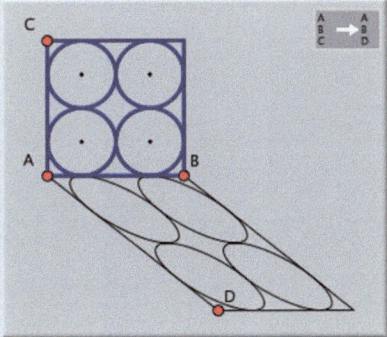

*Application of an affine transformation*

**See also**

- Similarity (p. 128)

- Projective Transformation (p. 131)

### 4.13.7 Projective Transformation

Projective transformations come into play whenever perspective drawings of a scene are required. A projective transformation is defined by four preimage/image pairs of points. The definition is analogous to that of a similarity (p. 128), though with eight required points instead of four.

As an example we consider the situation of producing a projective image of a rectangular grid. We furthermore add a few circles to make visual effect of a projective transformation more visible. For this we simply have to draw a rectangular grid (and the circles) and the corners of the desired image. We define a projective transformation by mapping the corners of the rectangular grid to the corners of the image.

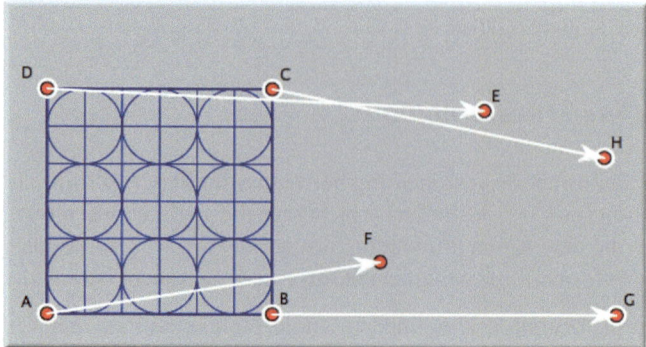

*Defining a projective transformation*

After this we apply the map to the elements to which we want to apply the projective transformation. The image of this operation is the desired projected rectangular grid.

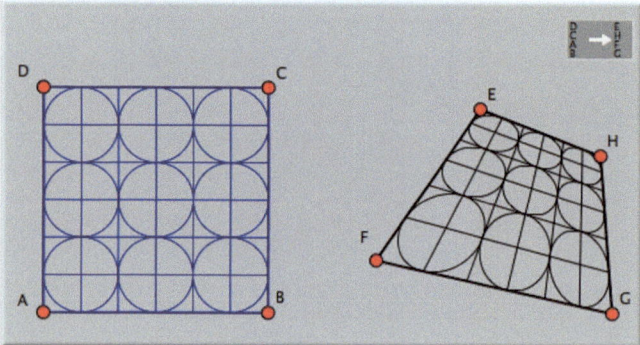

*The image of the rectangular grid*

**See also**

- Similarity (p. 128)
- Affine Transformation (p. 131)

### 4.13.8 Möbius Transformation

Mathematically, perhaps the most interesting transformation available in Cinderella is the Möbius transformation. This is a transformation with the property that circles and lines are again mapped to circles and lines. Any rotation (p. 127), translation (p. 125), or similarity (p. 128) can be considered a special case of a Möbius transformation, whereas neither an affine transformation (p. 131) nor a projective transformation (p. 131) is a Möbius transformation.

A Möbius transformation is defined by three preimage/image pairs of points. Defining a Möbius transformation in Cinderella is analogous to the creation of a similarity (p. 128), though with six required points instead of four.

The following two pictures demonstrate how a circle is mapped under a Möbius transformation. Since the Möbius transformation in the picture maps $A \rightarrow D$, $B \rightarrow E$, $C \rightarrow F$, a circle through $A, B, C$ is mapped to a circle through $D, E, F$.

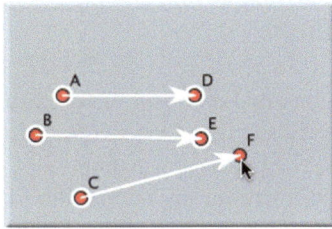
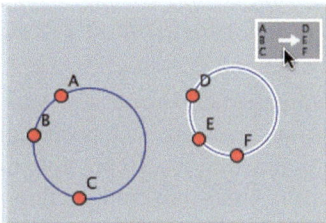

*Defining a Möbius transformation*          *Mapping a circle*

If one iterates a Möbius transformation, the sequence of mapped pictures forms a mathematically and aesthetically very interesting pattern that can be considered a generalization of a logarithmic spiral. The following picture shows the image of an iterated Möbius transformation applied to a single circle and its center.

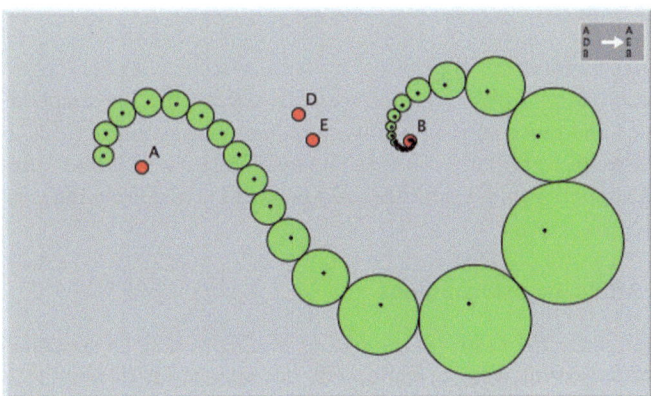

*Iterating a Möbius transformation*

A Möbius transformation usually has two distinct fixed points. In the above drawing these fixed points have been constructed explicitly by making sure that the transformation maps $A$ to itself and $B$ to itself.

**See also**

- Similarity (p. 128)

### 4.13.9 Inversion of Transformations

The mode *inverse* enables one to construct the inverse of any transformation. For this one has simply to choose this mode and click on the transformation to be inverted.

Thus a new transformation that is the inverse of the selected one is generated. This operation is frequently useful in studying the group structure underlying a set of transformations.

**See also**

• Transformation Groups (p. 145)

## 4.13.10 Composition of Transformations

It is also possible to form new transformations by composing two existing ones. For this, one goes into composition mode and then clicks the two transformations to be composed (it is also possible to click the same transformation twice to obtain its *square*). Almost all combinations of transformations are admissible for compositions; the few exceptions are explained below.

### 4.13.10.1 Combining Euclidean Transformations

Euclidean *rotations*, *line reflections*, and *translations* are called Euclidean transformations. They have the property that they leave Euclidean lengths and angles invariant. Composing such transformations results in another Euclidean transformation.

### 4.13.10.2 Similarity

*Translations* and *rotations* are special types of *similarity*. In contrast to a Euclidean transformation, a *similarity* may also scale the objects that are mapped. A similarity leaves angles and ratios of lengths invariant. Combining similarities, rotations, and translations yields a similarity. Combining a similarity with a line reflection leads to a reflecting similarity that is not explicitly present in Cinderella but can be generated in this way.

### 4.13.10.3 Affine Transformations

*Translations*, *line reflections*, *rotations*, and *similarities* are all special kinds of *affine transformation*. Affine transformations do not alter ratios of lengths. However, angles are not invariant. Combining any of these transformations results in another affine transformation. Affine transformations have the particular property that they transform parallel lines to parallel lines.

### 4.13.10.4 Hyperbolic Transformations

If you switch to hyperbolic geometry (as described in the Geometries (p. 114) section), all transformations are interpreted in a hyperbolic fashion. In particular, you get hyperbolic translations and hyperbolic rotations. Combining them generates another hyperbolic transformation.

### 4.13.10.5 Elliptic Transformations

If you switch to elliptic geometry, all transformations are interpreted in an elliptic fashion. In particular, you get elliptic translations and elliptic rotations. Combining them generates another elliptic transformation.

### 4.13.10.6 Projective Transformations

All the previously described transformations are special kinds of projective transformation. Thus all of them can be combined freely, and the result will be a projective transformation.

### 4.13.10.7 Möbius transformations

Möbius transformations are not in general projective transformations. Furthermore, there is no easily representable supergroup of transformations contains both Möbius transformations and projective transformations. Thus in Cinderella it is not permitted to compose Möbius transformations and projective transformations. For similar reasons it is not permitted to compose Möbius transformations and affine transformations.

However, Euclidean transformations are special kinds of Möbius transformations, and therefore it is perfectly permissible to combine them. It is also permitted to combine a Möbius transformation with a reflection or with a circle inversion. The characteristic property of Möbius transformations is that they map circles and lines to circles and lines.

## 4.13.11 Mapping of Elements

As already explained in the section General Use of Transformations (p. 123), it is easy to map an element in move mode. Simply select it and press the desired view button of the transformation.

There is also a special mode, *map an element,* that can be used to map an element. In this mode one has to select an element and the transformation. This selection results in the generation of the mapped element.

**See also**

- General Use of Transformations (p. 123)

## 4.13.12 Transformations by Functions

This transformation mode is quite different from all other transformation modes. While all other transformation modes directly encode geometric transformations, this mode can be freely defined by an arbitrary function. The function must be given as CindyScript code.

To define such a function one has to choose the mode and click somewhere in a view. Then a window pops up in which one can enter the function. The function must be a function with exactly one free variable. The name of this variable can be "#" (the generic run variable in CindyScript), "x", "y", "z", "t", or "v". The selection is kept somewhat restrictive to avoid interference with the names of geometric elements.

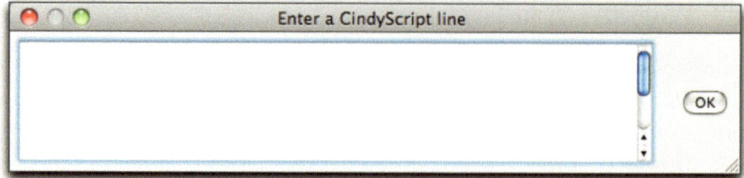

*Input window for function transformations*

In the function, the variable can be either a vector with (x, y)-coordinates or a complex number. If the function requires a vector, it has also to return a vector. If it requires a complex number, it has also to return a complex number.

As usual, defining a transformation creates a view button. In contrast to other transformation modes, the transformations defined by a function can only map points. More complicated objects like lines, circles, and conics cannot be mapped. The function has to describe how the point is mapped to another point. Here two variants are admissible: either the point is treated as a two-dimensional vector or it is treated as a complex number (where the x-coordinate represents to the real part and the y-coordinate represents the imaginary part).

The following entry form shows a simple definition in which a function is defined by a polynomial that interprets a mapped point as a complex number.

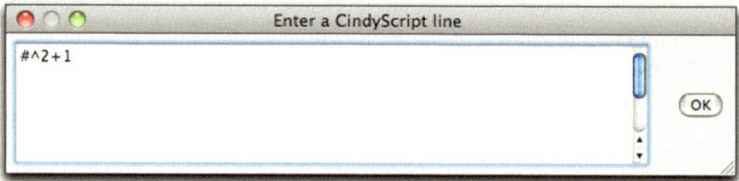

*Defining a complex function*

Applying this transformation to a point generates a new point that represents the result of the function. The picture below shows a point and its image. Furthermore, the trace of a circle is shown in the picture.

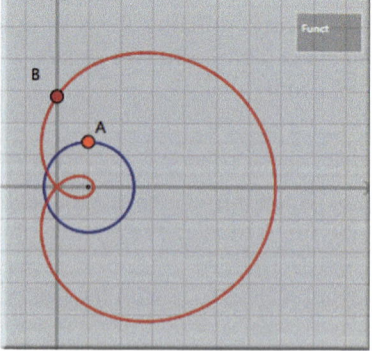

*Applying the function to a point*

### 4.13.12.1 Function Transformations and Iterated Function Systems (IFS)

Function transformations are particularly useful for the definition of iterated function systems. A typical example is the definition of "Julia sets". A Julia set can be defined as the iterated function system generated by the two functions `+sqrt(z-c)` and `-sqrt(z-c)`. Here "c" is an arbitrary constant and "z" is the (complex) input variable. After defining these two functions as transformation functions one can use these functions within iterated function systems (p. 141) mode. In the following example we control the constant "c" by a point $A$ of the construction. The two functions are defined by

```
sqrt(#-complex(A))
```

and

```
-sqrt(#-complex(A))
```

The resulting picture depends on the position of $A$. One instance is shown below:

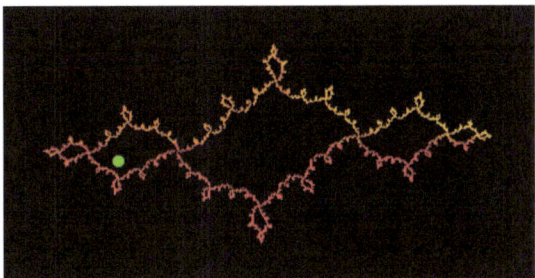

*A Julia set defined by transformation functions*

A more advanced picture can be generated by composing the iterated function system with a color plot programmed in CindyScript. The color plot iterates the function `z2+c` with different start values and counts how long it takes the absolute value of the function to exceed the number 2.

```
m=40;
f(x):=x^2+complex(A);
g(y,n):=if(
  or(|y|>2,n==m),
    n,g(f(y),n+1)
);
colorplot((1-g(#_1+i*#_2,0)/m),B,C,startres->20,pxlres->1)
```

The resulting picture, which is overlaid with the iterated function system, looks as follows.

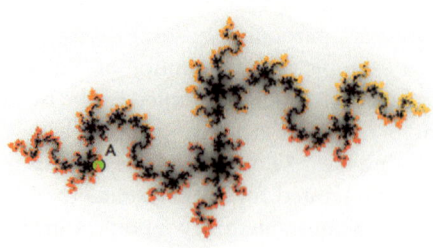

*A Julia set defined by transformation functions and a color plot*

## 4.14  Setting a Basis

Bases are a powerful new experimental concept in *Cinderella.2* for structuring and creating logical dependencies in a drawing. One can set a basis with respect to a frame of up to four points. If a basis is set, then all drawing of free elements is performed with respect to this basis. If the base points move, then everything that is drawn within the basis moves accordingly.

There are four different kinds of bases that can be chosen:

- A *translation basis* requires one base point. If the point is translated, everything that is drawn in the basis is translated as well.

- A *similarity basis* requires two base points. These two points define the origin and the $(1, 0)$ point of a coordinate system. Everything drawn with respect to this basis is scaled and rotated if the points are moved.

- An *affine basis* requires three base points. These three points define the origin, the $(1, 0)$ point, and the $(0, 1)$ point of an affine coordinate system. Everything drawn with respect to this basis is scaled and rotated and sheared if the points are moved.

- A *projective basis* requires four base points. These four points are considered as a basis of a coordinate system in which these points form a square. Everything is drawn accordingly.

After selecting "set basis" mode by pressing the corresponding button in the toolbar, one can select one up to four points. Leaving the mode by selecting a new mode defines a basis and declares it active. If a basis is active, all drawing operations and in particular the location of free objects are performed with respect to this basis. This means that if the points of the basis are moved, then the elements that are drawn with respect to this basis are changed accordingly (even if they were free elements).

### 4.14.1 Example: Creating a Similarity Basis

We will demonstrate the use of a basis with a similarity basis. For this we start by drawing two points to which the basis should be tied. Then we choose "set basis" mode and select these two points one after the other. While we do this, optical hints are shown that indicate which type of basis is currently chosen. At this stage, up to four points can be chosen. Leaving basis mode by selecting another mode triggers the completion of the basis selection, and the basis is then generated and activated.

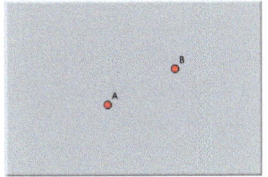
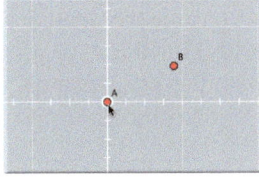
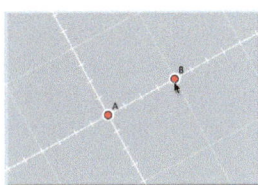

*Two points for defining a similarity basis*      *Selecting the first point*      *Selecting the second point*

After basis mode is exited, a view button is generated that represents the basis. Whenever this button has a yellow border, the basis is active. Everything that is now drawn will have coordinates with respect to this basis. So now you can make an arbitrary geometric construction that may involve free points, circles with radius, and other free elements. In Move Mode (p. 71) you can always change the positions of the free elements of your drawing. Now if you move one of the defining points of the basis, your entire drawing will change and will be adjusted to the current position of your basis.

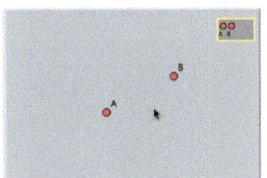
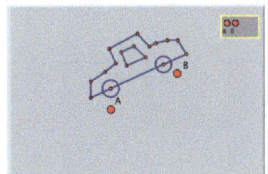
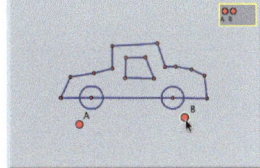

*The basis is generated and selected.*      *Draw something.*      *Move the basis.*

In Move Mode (p. 71) one can activate and deactivate bases by clicking them. Active bases are marked by a yellow frame. Clicking an active basis will deactivate it (and all other bases). Clicking an inactive basis will activate it. In this way it possible to have several bases within one construction that are related to different parts of the drawing. It is even possible to define a basis while another basis is active. Then the bases are cascading.

### 4.14.2 Example: Using a Projective Basis

By selecting four points in set basis mode it is also possible to create a projective basis. If a projective basis is selected, all drawings are performed as though the four points (in the selected order) formed the four vertices of a quadrangle.

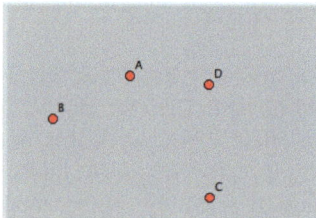
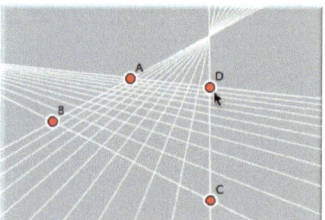

*Four points for a projective basis*            *Clicking them defines the basis.*

This not only affects when the points that define the basis are moved, it affects the generation of several geometric construction operations. In particular, all operations that require (explicitly or implicitly) Euclidean measurements are perspectively distorted. The picture below shows a few constructions within a projective basis. Observe that in particular, midpoints and circles are perspectively adjusted.

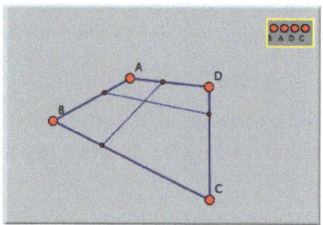
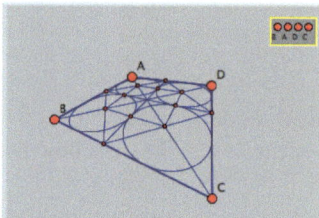

*A simple drawing*                          *A more complex drawing*

The following picture demonstrates the use of a projective basis in a more complicated setting. First a projective basis is constructed. Then in the usual way an iterated function system is defined. The iterated function system (IFS) will automatically appear in a skew perspective that corresponds to the projective basis.

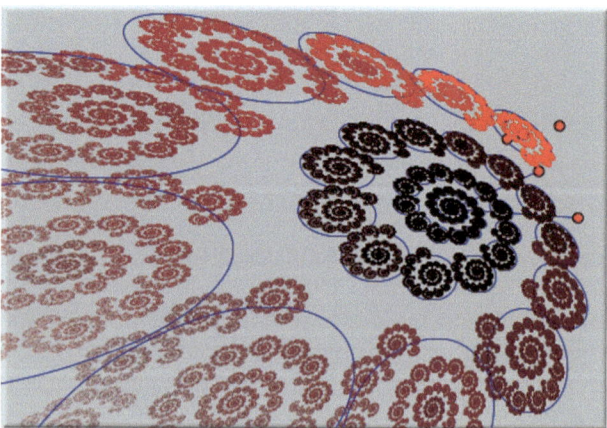

*An IFS in a projective basis*

## 4.15 Iterated Function Systems

Iterated function systems (IFS) can be used to generate fractal images in Cinderella. An iterated function system requires several transformations T1, T2, T3, ... as input. From these transformations a picture is generated as follows:

First one chooses an arbitrary screen point as the start of an iteration. Then one chooses at random one of the transformations and maps the point and draws it. Then one uses the resulting point of the mapping, chooses a transformation again at random, maps the point, and draws it. This operation is repeated many, many, many times. In this way a "cloud of points" on the screen is created.

The amazing fact about iterated function systems is that this cloud of points is simultaneously self-similar with respect all transformations T1, T2, T3, ....

As an example we consider the situation in the following picture. There two different similarity (p. 128) transformations are defined. One maps point *A* to itself and point *B* to Point *C*. Thus this transformation causes a rotation around *A* combined with a small contraction (at least for the positions of *A*, *B*, and *C* in the picture below). The second transformation maps *A* to *D* and *B* to *E*. This also causes a contraction rotation with a different rotation center and contraction rate. The iterated function system defined by these two transformations is shown in the picture below.

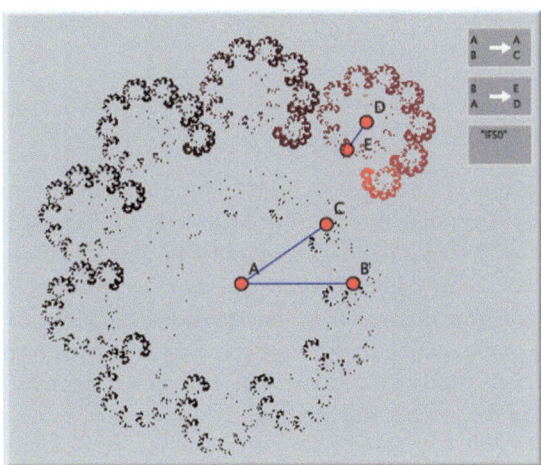

*An IFS by two similarities*

### 4.15.1 Defining an IFS

In order to define an IFS one has first to select several transformations. For doing this one chooses move mode (p. 71), holds the shift key, and clicks on the desired transformations one after another. In this way a set of transformations is selected. Then one chooses in the menu "Modes → Special" the mode "IFS." Then an IFS is automatically added. In our case the cloud of points that is generated is self-similar with respect to both similarities.

### 4.15.2  Enhancing the IFS

Admittedly, the IFS so far does not look very impressive. The reason for this is that during the creation of the cloud of points the two transformations were chosen at random with equal distribution. However, the second transformation is by far more contractive than the first one, which causes the points to accumulate in a region close to the fixed point of the second transformation. One can influence the "importance" of a transformation in the inspector by adjusting its relative probability. For this one selects the IFS by shift-clicking it in move mode (p. 71) and opening the inspector. In the appearance tab one finds a slider and a color selector for each transformation involved.

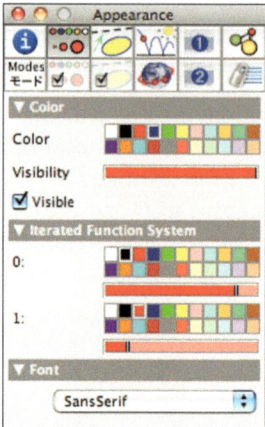

*The IFS inspector*

The slider controls the relative probability with which the transformation is chosen. The color controls the color that is associated with this transformation in the point cloud. Roughly speaking, points that are generated by choosing one transformation relatively often are more likely to have that transformation's color. Lowering the relative probability of the second (more contractive) transformation has a nice effect on our picture. The first similarity (the royation around $A$) becomes "more important," and much more of the inner spiral is filled.

*A different probability distribution*

If one does not need an extremely fast and immediate response to mouse actions, the picture can be further enhanced by altering the visibility of the IFS. By setting the visibility to a small value one causes each individual dot to be printed with a high level of transparency. If one waits long enough, a picture will be produced that has structure even on the subpixel level.

*Rendering with low visibility*

### 4.15.3  Changing Parameters

IFS are remarkably rich structures. Depending on the position of the points involved, the IFS may look qualitatively very different even for identical Cinderella constructions. The picture below shows two different choices of parameters for the IFS of the previous example.

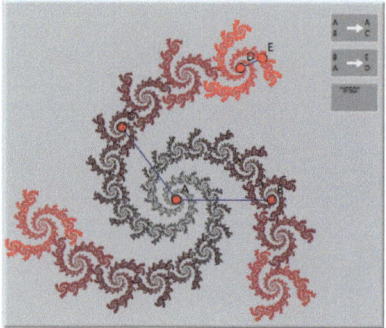 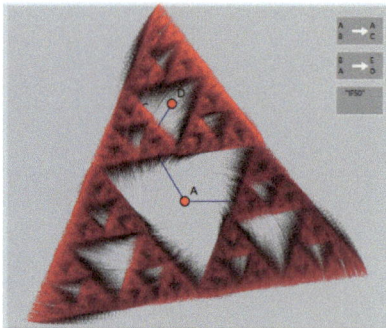

*Different parameters*

Other defining transformations may cause other (sometimes very stunning) visual effects. In the picture below the first example was created by four different affine transformations. The second picture was created by eight different symmetrically chosen circle inversions.

*The Bernsley fern*                           *Hyperbolic fractals*

### 4.15.4 IFS and Transformation Groups

Iterated function systems are very closely related to the Cinderella concept of transformation groups (p. 145). A transformation group is a collection of transformations that when applied to a geometric object causes the iterated mapping of this object under the collection of transformations. An IFS can be considered the limit set of this iteration process (if it exists). Consider the example below, in which again two similarities are defined. In the first picture the image of a point under the iterated application of the transformations is shown (together with green segments that symbolize the traces under the mappings). One now can easily create an associated IFS by simply selecting the transformation group (by shift-clicking it in move mode) and then selecting IFS mode. This operation causes the corresponding IFS to be automatically generated. The resulting picture is shown on the right.

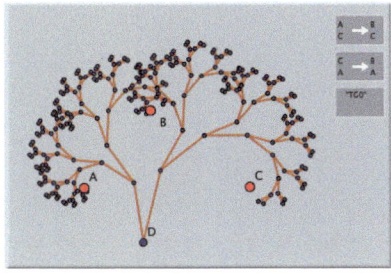
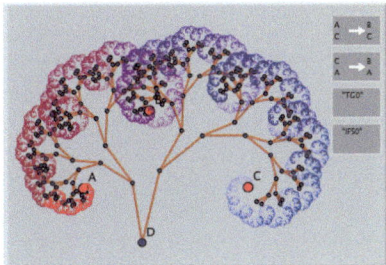

*A transformation group ...*                              *... and its IFS*

**Caution**

The structure of the IFS depends heavily on the properties of the defining transformations. If the transformations are not "contractive," then one usually will not see any reasonable effect of an IFS, since in this case the points of the cloud accumulate very fast at locations close to the "line at infinity."

**See also**

• Transformation Groups (p. 145)

• Transformation Modes (p. 122)

# 4.16  Transformation Groups

Transformation groups are perhaps among the most sophisticated objects currently available in Cinderella.2. A transformation group is constructed from a collection of transformations. A transformation group is can be applied to an arbitrary geometric object. This object is mapped by applying the elements of the transformation group iteratively. The transformation group has several applications and structural features. We will exemplify them by several concrete examples. It should be mentioned that the concept of transformation groups in Cinderella is different from the mathematical notion of a transformation group. The two main differences are that in Cinderella a transformation group does not per se contain the inverses of the transformations and that only a large but finite number of iterations are performed.

## 4.16.1  Creating a Transformation Group

We first will explain how to construct a transformation group that generated a simple plant-like structure on the screen. For this we place three points $A$, $B$ and $C$ as shown in the picture and create two Similarity (p. 128) transformations $A$->$B$; $C$->$C$ and $A$->$A$; $C$->$B$. Selecting the two similarity transformations by shift-clicking them in the move mode and then choosing the mode *Special->Transformation group* creates a view button for the transformation group.

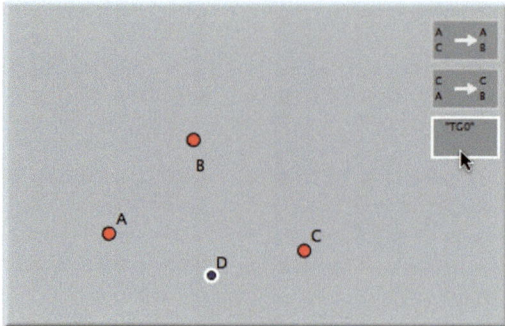

*The starting situation*

If we add a single point *D*, select it and click the button of the transformation group
a picture similar to the following is created:

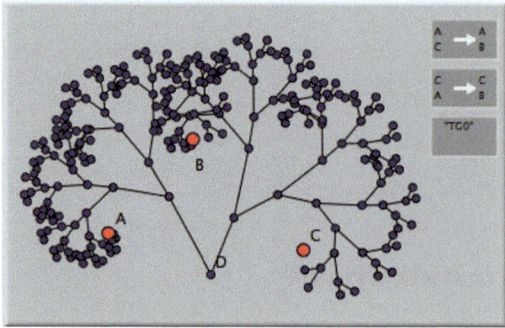

*Applying a simple transformation group to a poin*

This picture is created by applying both transformations to the point D. Then both
transformations are applied to the generated images. This procedure is repeated until
a either certain interation depth is reached or until the number of elements exceeds
a certain limit. These parameters can be controlled in the inspector of the Transformations group.

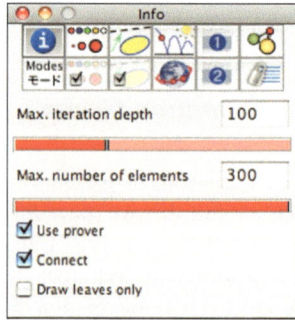

*The transformation group inspector*

The parameters controllable in the inspector are by default applied to all mapped elements. Changing them does not change the parameters of the mapped elements. Parameters of already mapped elements can be changed by selecting the whole group of elements (by clicking any of the generated images) one can change their individual settings in the inspector.

The appearace details of mapped elements can be controlled by the following part of the inspector.

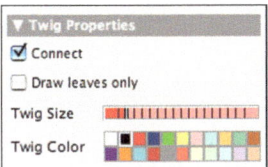

*The twig settings*

One can observe that the points are connected by segments the segments exemplify the tree structure of the derivation process. One can turn off the connecting branches by the corresponding slot in the inspector. It is also possible to draw only the "leaves" of the generated tree. The following table shows all four combinations of drawing the points and conncecting them.

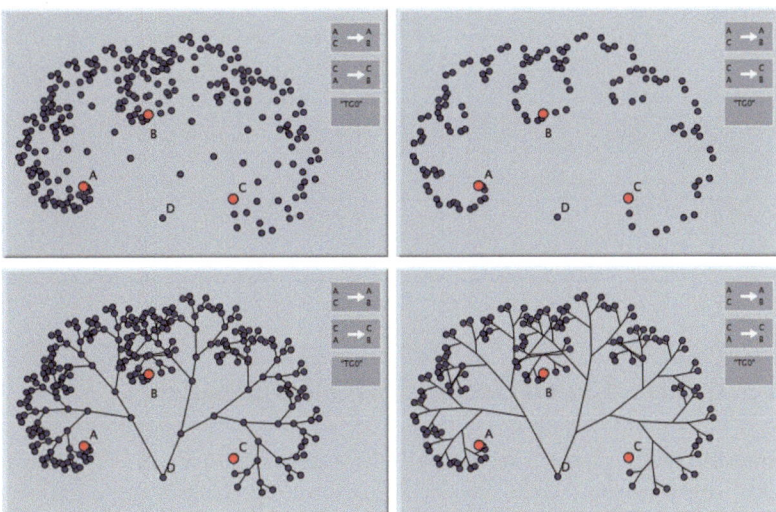

*Drawing all points...*                              *...or drawing only leaves*

Elements are only connected if the mapped element is a point. The following picture adds the mapping of an additional circle.

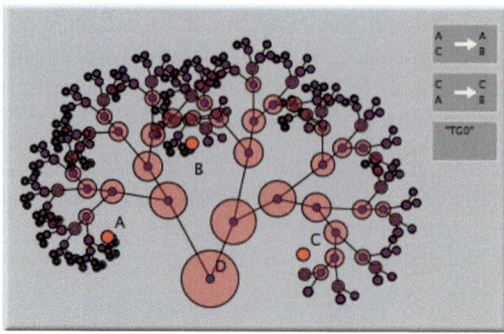

*Mapping a circle*

### 4.16.2 *Transformation Groups and Iterated Function Systems*

Transformation groups are closely related to Iterated Function Systems (p. 141). On can generate the corresponding IFS by selevting a transformation group and choosing the *Special->IFS* mode. The following picture shows the effect of a transformation corresponding to a pythagoras tree. The picture on the left shows the corresponding IFS. The IFS corresponds of all the limit points of the transformation group.

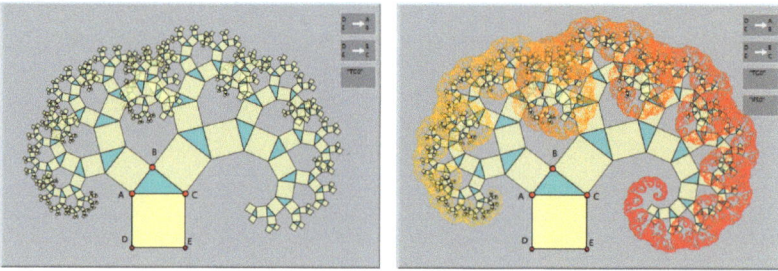

*A Pythagoras tree...*                          *...and its IFS*

### 4.16.3 *Transformation Groups With Dependencies*

The transformation groups considered in the examples so far were "free" in a mathematical sense (if the coordinates of the controlling points were choosen generally enough). This means that in this general case no matter how long we proceed with the iteration process the same composed transformation will never occur twice. However if the parameters of the defining transformations were in a more special position it may happen that there are several ways to generate the same transformation by composition.

For instance take a simple reflection $R$. If $R$ is applied twice the identity map is generated. So for instance $R$ and $R^3$ produce an identical transformation. If as described before the transformationn group is generated by a simple ennumeration of all possible compositions of transformations in the group and $R$ is contained in the group,

then many identical transformations will be constructed. This would be an unec-
cessary waste if computation power and memory. Luckily, Cinderella is equipped
with a geometric proving engine, that allows to neglect transformations if they were
already be generated before. This feature is used when the *Use prover* checkbox is
checked in the inspector.

We exemplify the difference by a very simple example. Consider two different translations *P* and *Q*. Since translations commute with each other applying *P* after *Q*
generates the same effect then applying *Q* after *P*. The pictures below exemplify the
effect of using or not using the prover.

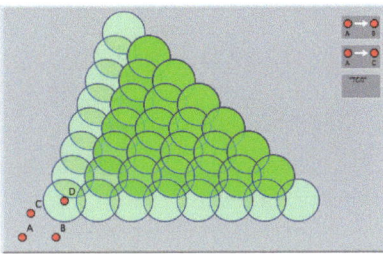

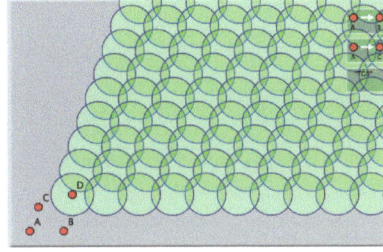

*No prover used*                          *Prover used*

The first situation is generated by not using the prover. A filled circle with opacity 20% is mapped by a translation group generated by two translations. One can
observe that not may different circles seem to be generated. However the situation
is different. Many circles were generated most of them overlap (you recognize this
by the increasing opacity). If instead the prover is used, then the picture on the left
is generated. Considerably more circles are visible no two of which overlap. The
prover has thrown out all superfluous repetitions of transformations.

### 4.16.4 Ornament Groups

It is particularly interesting to study planar ornament groups (or cristallographic
groups) with this feature of Cinderella.2. We will exemplify this by a few simple
examples. First consider a triangle build of three mirrors, that form edge angles of
90°, 45° and 45°. These three reflections generate a reflection group like a kalei-
doskopic prism. Applying this reflection gruop to an arbitrary geometric object creates a nicely symmetric picture in the plane.

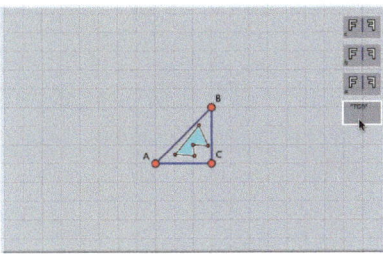

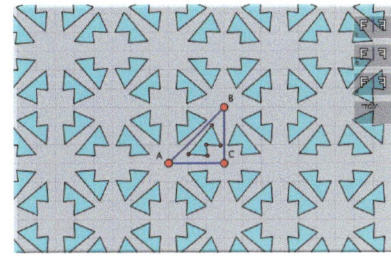

*Generating a reflection group...*           *...and applying it.*

As a second example we consider a cristallographic group that is generated by rotations that come from a regular hexagon. We first draw a regular hexagon by using an appropriate Polygons (p. 112) mode. Then one defines a 60° rotation around the center and two 120° rotations at two consecutive verticies by using the Similarity (p. 128) mode. After generating a transformation group from these transformations one can again map arbitrary objects to obtain an ornamental pattern.

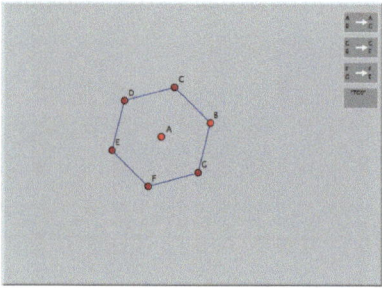

*Generating a rotation group...*                    *...and applying it.*

### 4.16.5  The Generation Order

It is interesting to study which the transformations are processed if the prover is used for a reflection group. In general a tree-like structure is generated that reached each cell of the reflection group exactly once. The following pictures show the situation for two different reflection groups. The default setting of transformation groups are used. This generates approximately 300 elements and uses the prover. If a point is mapped the *Connect* feature that is used by default connects points that differ by exatly one transformation. The tree of points represents the traversion order of the elements of the transformation group. Observe how different positions of the mirrors cause structurally very different trees.

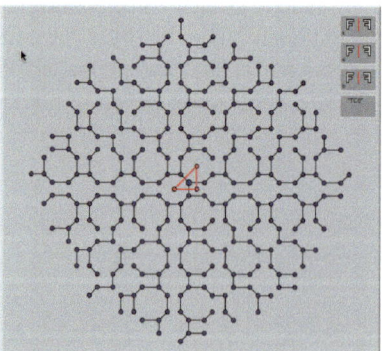
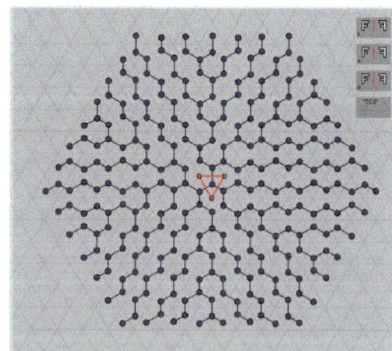

*A (90°,45°,45°) triangle group*                *A (60°,60°,60°) triangle group.*

### 4.16.6 Hyperbolic Transformation Groups

In hyperbolic geometry one can observe particularly interesting transformation groups. They are generalizations of the usual crystallographic groups. The main reason for the existing of those interesting symmetric patterns is the fact, that in hyperbolic geometry there exist regular *n*-gons with arbitrarily small vertex angles. For instance there exist regular polygons with 90° vertex angles. Four of these pentagons fit exactly around a hyperbolic corner. Thus the hyperbolic plane can be seamlessly tiled by regular pentagons with right angles. We will demonstrate how to construct such a tiling.

First we need a hyperbolic pentagon with right angles. For this we switch to hyperpolic geometry (see Geometries (p. 114)) and open a hyperbolic view (see Views (p. 116)). Then one has to construct a right angled pentagon. This can be done by using the *Polygon->Regular Pentagon* mode. If one changes the size of the pentagon, snap values are shown that indicate how many polygons fit around a corner (see Polygons (p. 112)). Stopping at a snap value of *4* (Figure 1) we obtain the desired right angled pentagon. By chosing the mode Rotation (p. 127) LL one can construct two rotations of 90° around two corners of the pentagon (Figures 2 and 3).

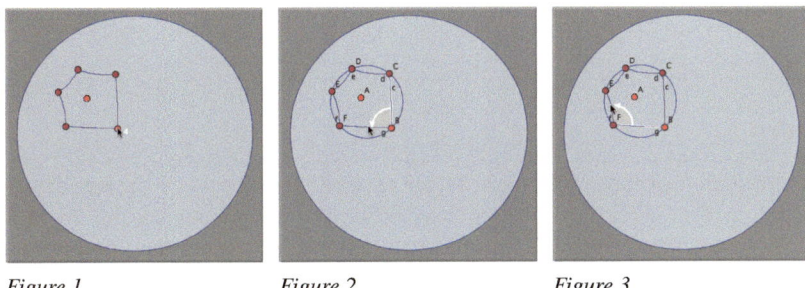

Figure 1                        Figure 2                        Figure 3

Creating the transformation group of these two transformations and mapping all segments creates the desired teselation.

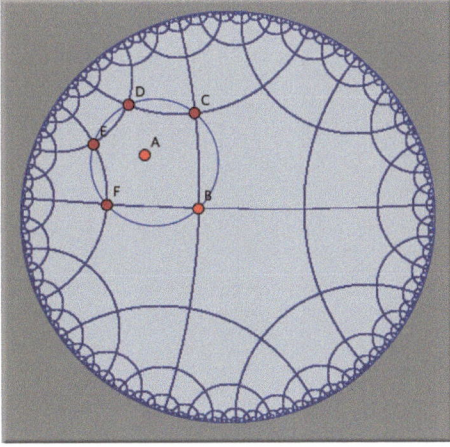

*A hyperbolic tesselation by regular pentagons*

The picture below shows the application of this hyperbolic group to a small triangle. Observe that it shows only rotational symmetry but no mirror symmetry. For the image the iteration depth was considerably raised.

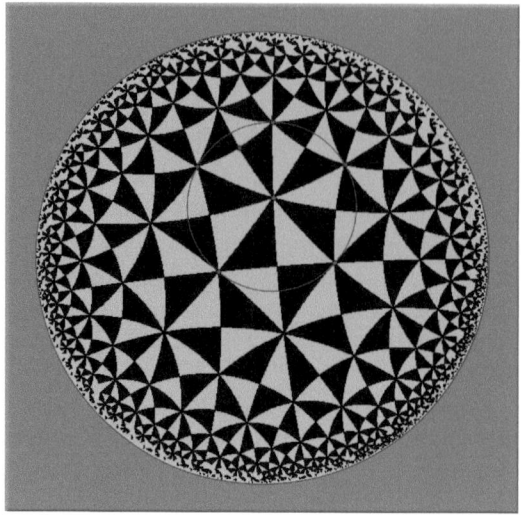

*A hyperbolic tesselation by right triangles*

**Caution**

Transformation groups with high iteration depth may be very time and memory demanding. So interactive drawings may be slowed down considerably by transformation groups with a huge iteration depth. Eventually transformation groups may even cause an *out of memory error*.

**See also**

- Transformation Modes (p. 122)
- Iterated Function Systems (p. 141)
- Polygons (p. 112)
- Geometries (p. 114)

# Chapter 5

# The Inspector

The inspector is the fundamental distributing center for all properties of geometric elements. It is one of the fundamental improvements in *Cinderella.2*, and it completely replaces the appearance editor in previous releases.

Understanding the functionality of the inspector will help you to control and personalize your Cinderella constructions. In this section we will give a general overview on the functionality of the inspector. Specific information on the functionality and inspection parameters of geometric and physical objects can be found in the descriptions of the objects themselves.

## 5.1 A General Overview

You can open the inspector by choosing the menu item *Edit/Information*. Depending on the selection of elements, a window sill pop up that either is almost empty or contains information on the selected elements. In any case, it will contain a button toolbar. With this toolbar you can switch between the different information blocks of the inspector.

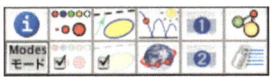

*The inspector buttons*

Each of these buttons represents a choice of properties that can be influenced by the inspector. Clicking one of these buttons switches to the corresponding information block. To provide a rough overview, the following graphic explains the type of parameters that can be inspected within the different blocks:

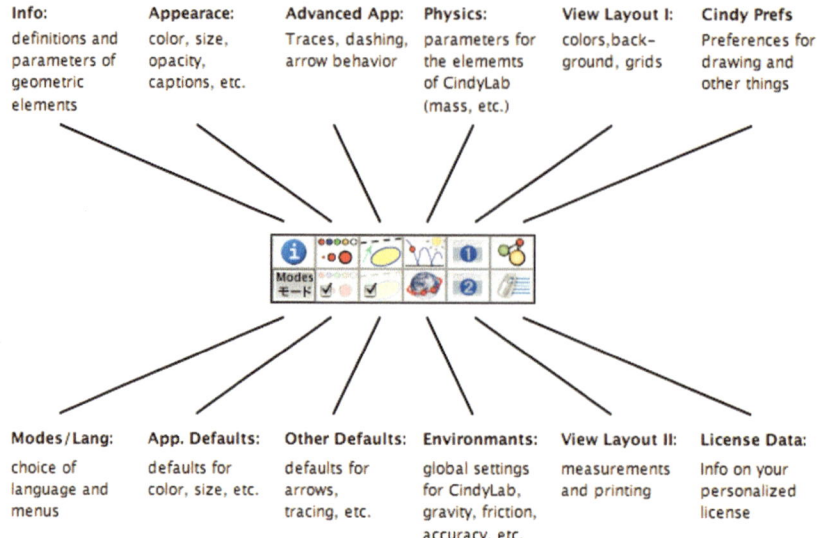

*The information blocks*

Roughly speaking, the first row of buttons represents information on individual elements, while the second row represents Cinderella's default settings. The inspector always refers to all elements that are currently selected. Each element provides several attributes that can be viewed or changed by the inspector. All such attributes that belong to the currently chosen inspector block are shown. In what follows we will briefly demonstrate the possibilities of the inspector. We will organize the sections roughly by the inspector's information blocks:

- *Geometric information:* for definitions, labels, geometric parameters, etc.

- *Basic appearance:* for color, size, visibility, labeling, circle filling, etc.

- *Advanced appearance:* for traces and arrow properties

- *Basic defaults:* defaults for color, size, visibility, labeling, etc.

- *Advanced defaults:* defaults for traces, arrows, rendering, etc.

- *Views I:* for view properties such as mesh, color and background images

- *Views II:* for view properties units, precision etc.

- *Preferences:* for various global settings and various default parameters

- *Menu/language:* for choice of menu and language

- *Physics:* for the physical parameters of all CindyLab elements

- *Environment:* environment for physics, accuracy, friction, many-body setup, actuation, etc.

- *License:* license information

A more detailed description of the most important inspector facilities can be found in the following sections:

## 5.2 The Info Block

The first block presents information on the geometric definition of a geometric object. You can see and edit an object's name, its definition, incidences, and parameters. If you are not already in this block, you can switch to it by pressing the button .

A typical info for a free point will look as follows:

*The info of a free point*

In particular, you can see the name and the position of the point. These two attributes can also be changed by entering text. Furthermore, you see the defining mode of the point and that the point is incident to three lines $a$, $b$, $c$.

There is a particularly nice fact about inspecting free elements. Their position parameters can be changed by typing an arbitrary CindyScript (p. 219) expression. Thus if you type in the position field, for instance [5,4], the point will be set to the exact $xy$-coordinates (5, 6). You can even perform calculations in the position field. If you type in the expression (B+C)/2, the point will be placed exactly in the middle between $B$ and $C$. The point is still freely movable. More information about the places where CindyScript can be used is available in the section on entering CindyScript (p. 395).

### 5.2.1 Information about the Mode

Some modes, e.g., *line by fixed angle* (p. 80) and *circle by fixed radius,* (p. 83) also have information that can be inspected. The fields for entering these are shown below the information area for the currently selected elements. The inspector will open automatically when these modes are selected to allow you to enter those parameters.

There may also be special occasions when the information block can be used to add further functionality. For instance the information block for texts includes a checkbox to turn the text into a button.

## 5.3  Inspecting Appearance

With the inspector you can control the appearance of the geometric objects. There are two levels on which this can be done. The individual appearance inspector is responsible for the currently selected elements. There you can inspect and change the properties of the currently selected items. The default settings are not affected by this. The main setup for appearance issues can be accessed via the button . This section focuses on these basic appearance settings. The following pictures show the parameters for points, lines, and circles that can be inspected.

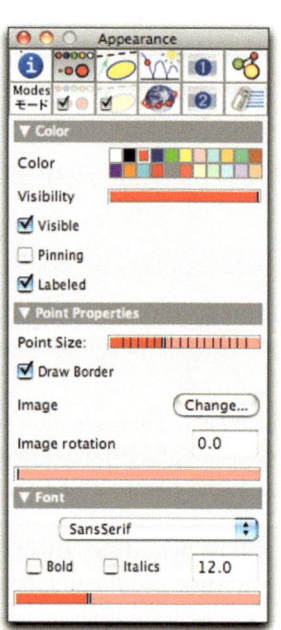
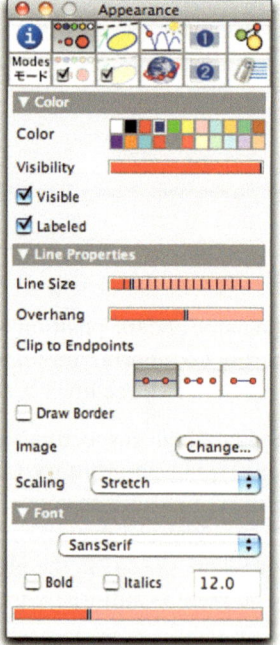
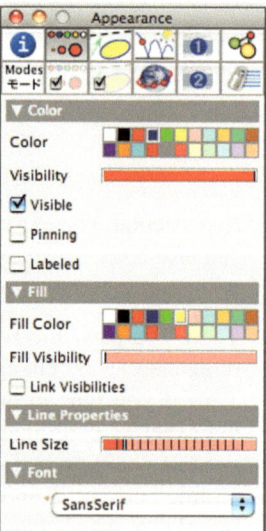

*Point appearance*  *Line appearance*  *Circle/polygon appearance*

Each of these windows has controls for color and visibility. Some of the parameters are specific to the inspected objects.

### 5.3.1 Color and Opacity

For changing the color of the selected elements you simply click on one of the little boxes in the color palette. The color of the selected objects will change immediately.

*The color palette*

It may be the case that you are not satisfied with the color choices offered by Cinderella. You can change a color value in the palette by double-clicking the corresponding box. A color-chooser window pops up in which you can adjust the color of the palette entry. Depending on your operating system, this color chooser may look slightly different. Usually, several different ways of selecting a color are provided. On an Apple Macintosh, two of these ways of selection look as follows:

*Color wheel*

*Extended palette*

It is important to know that by changing the color of an entry in the palette, *you change the color of all geometric elements associated with this color entry.*

Next to the color palette you also have a slider with which you can adjust the opacity of the elements. The opacity can vary from "absolutely invisible" to "solid."

*Grades of opacity*

Absolutely invisible elements are good for auxiliary constructions that should not disturb the rest of the drawing. Slightly visible elements can be used for parts of a construction that are of minor importance. For an element to take active part in the views (being movable and selectable), it needs an opacity of at least 20 percent. You can select invisible elements in the textual view. They cannot, however, be used for clipping lines. Another way to make points or lines invisible is to give them a size of zero.

In *Cinderella.2* it is also possible to fill circles (in contrast to previous releases). Fill colors can be controlled by the fill color properties. The opacity of interior and border can be controlled individually.

### 5.3.2 Size

There are also sliders provided for the control of the size of objects. There is an individual slider for all points, and another slider for lines and circles.

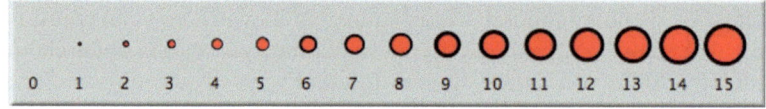

*Possible sizes of points*

Points of size zero are very useful objects. They are invisible, but they are selectable and movable, and they still serve as active clipping points for clipping lines. This makes them ideal elements to use when you want a line that you can move by grabbing its endpoints. This obviates the need to create unnecessary "decorations" with visible points.

### 5.3.3 Clipping

You can control whether a line is "clipped." For this purpose the inspector offers three buttons that turn clipping on and off.

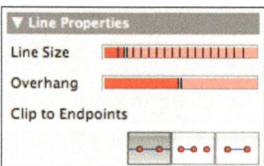

*Controls for clipping, line size and overhang*

An unclipped line will be drawn over the whole extent of a view. A clipped line can be truncated in two different ways. The second button clips a line defined by two points with respect to exactly these points. The third button clips with respect to all points that are incident to the line. The rules that govern this clipping logic are a bit intricate, but they lead to natural behavior:

- To be a "clipping point" of a line the point must be incident to the line. Furthermore, the point must not be completely invisible, and it must not lie at infinity. Hidden clipping points can be created by setting the size of a point to zero.

- All clipping points of a line are consulted for the clipping.

- If the line has at least two clipping points, then the portion of the line that reaches from the first clipping point of the line to the last clipping point is drawn.

- If there are fewer than two clipping points, the line is not clipped.

These rules have the following effect:

- At least a small portion of the line is always shown.

- All visible points incident to the line lie on this portion.

This is what you would expect for geometric drawings. The decision as to whether a point is (always) incident to a line is made by the automatic theorem checking capacity of *Cinderella.2*. In this way, correct and mathematically consistent behavior is ensured.

### 5.3.4 Overhang

It is often undesirable to have a clipped line end directly at certain points. It frequently looks much nicer to have some overhang that suggests that the line continues farther. You can control the size of an overhang using the overhang slider. The slider's position adjusts the overhang on both sides between 0% and 50% of the line's total length.

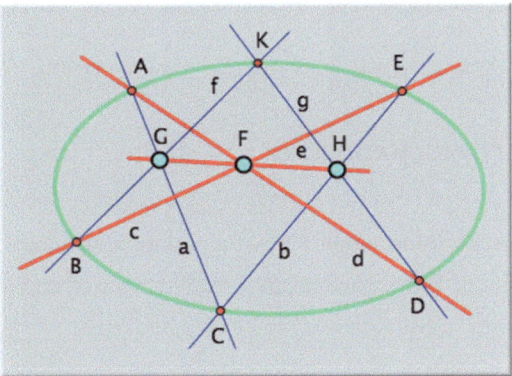

*Pascal's theorem with individual colors, sizes, and overhang*

### 5.3.5 Default Settings

There are two other information blocks in the inspector that are accessible via the buttons ⬜ and ⬜. Changes that are made there affect all selected elements and at the same time define the default settings for elements that will be constructed in the future.

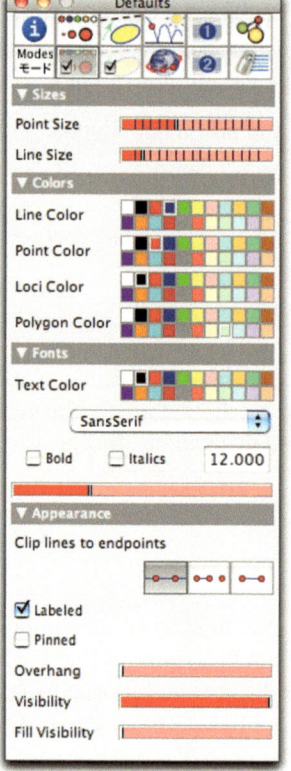

*Basic appearance defaults*

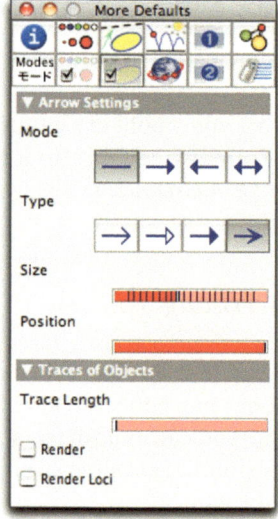

*Arrow and trace defaults*

## 5.4 Traces, Arrows, Rendering

The second important information block for setting appearance can be selected by the button ▭. In this block you can alter the behavior of traces and arrows.

### 5.4.1 Traces

Arrows apply only when segments are selected. So for most selections this inspector block will look as follows:

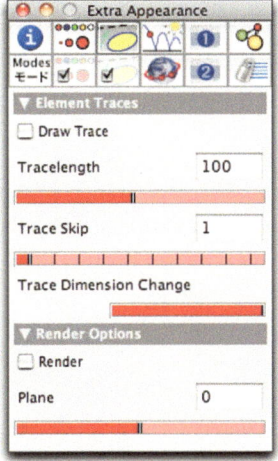

*Trace inspector*

The majority of these controls are related to traces. Traces are perhaps one of the most useful small extensions of *Cinderella.2*. One can draw a trace of *any* object. Traces dim out with time, resulting in the impression of a dynamic process. The following pictures show two applications of traces. In the first picture is shown the trace of six points obtained by reflecting a point with respect to a collection of mirrors. The second picture shows the trace of a line in a configuration during an animation.

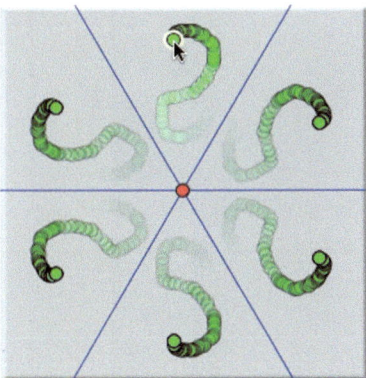

*Trace of reflected points*

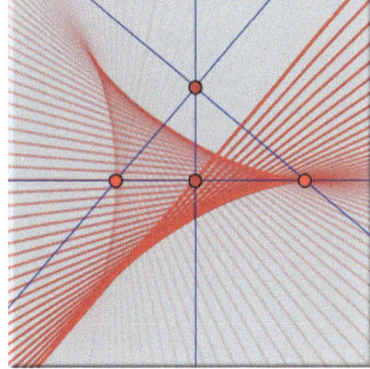

*Animated trace of a line*

For most applications it will suffice to select the element whose trace should be shown and check the "draw trace" checkbox.

The parameter "Tracelength" controls the maximum number of points in the trace. For performance reasons this number should usually not be larger than 1000.

Generally, a trace is generated for every picture. However, the parameter "Trace Skip" makes it possible to draw only every *n*th trace. This is sometimes useful for slow animations involving traces.

Finally, the parameter "Trace Dimension Change" causes the traces of points to become smaller and smaller.

## 5.4.2 Arrows

Segments can support arrows. The appearance of the arrows can be altered by the settings of this inspector block. If a segment is selected, the inspector window contains the following controls in addition to those already mentioned:

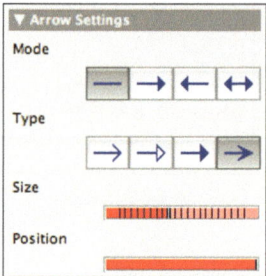

*The arrow controls*

The arrow editor is analogous to the appearance editor. Its choices apply to the currently selected segments. You have the possibility of adding an arrow at either end of the segment. Four types of arrowheads are provided. Two sliders allow you to control the size and the position of the arrow.

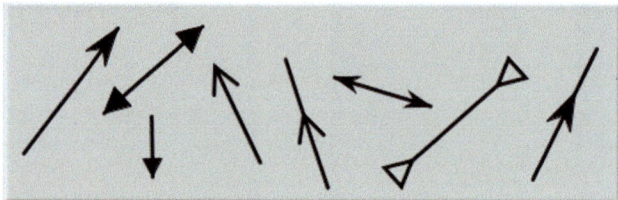

*Different kinds of arrows*

## 5.4.3 Rendering

There is also a single checkbox called "Rendering." if this box is checked, then springs in CindyLab (p. 167) are drawn as wiggly lines. Furthermore, this checkbox determines whether lines and points that have been drawn by hand in scribble mode are to be rendered as hand-drawn elements.

## 5.5 Controlling the Views

With the inspector button  you can access the properties of the currently active view. The corresponding inspector window contains the following controls:

*The view inspector window*

In particular, there are controls that govern the scaling behavior and the opacity of background images. Background images can be loaded with the menu entry "File/Set background image." Background images are extremely useful not only for decorative purposes. By this it is also possible to make geometric analyses of photographs and other pictures. In particular, the Projective Transformation (p. 131) features help in working accurately with perspective distortions.

*Analysis of a Gaudí building*

The inspector window also contains elements for controlling the background color, meshes, and coordinate axes. These controls are also available directly in the ports.

# Part III
# CindyLab Reference

# Chapter 6

# Introduction to CindyLab

CindyLab is the part of Cinderella that for creating simulations of physical experiments with a few mouse clicks. The basic philosophy is similar to the geometry part. You draw the experiments with your mouse, then you just click a start button and you see what happens. While your experiment is running you can always interact with the objects of your construction by picking and dragging them.

The basic model behind the simulations in Cinderella is a particle/forces model. There are particles with certain masses that behave according to Newtons laws of motion. At the same time there is a variety of forces (springs, gravity, magnetic fields, walls, etc.) that influence the movement of the particles. With this simple paradigm a surprisingly rich variety of experiments can be modeled. Just to give one example, you can enter and simulate the motion of a planet around a sun with only four mouse clicks. By adding additional suns or planets you can experience the chaotic motion of a many-body system.

There are many powerful links between CindyLab and the geometry part of Cinderella, as well as to CindyScript, where you can measure forces, speeds, positions and energy, just to name a few. Using this connection you can easily evaluate your experiments and draw quantitative conclusions.

## 6.1 Simulating Masses and Forces

### 6.1.1 Elementary Mechanics

$$\text{acceleration} = \frac{\text{force}}{\text{mass}}$$

This simple law is the key to understanding the movement of mass particles under the influence of forces. Mass particles may be almost anything: electrons, balls, or even planets. Forces may be present for various reasons. They may come from electrostatic attraction, from stretched rubber bands, or from interplanetary gravitation. Independent of the type of mass or type of force, the above law describes how the velocity of a particle will change.

Acceleration is the change of velocity per unit time. Thus we observe a simple but important consequence of the above law: *If no forces are present, then a particle will maintain its velocity and travel forever with the same speed in the same direction.* As soon as forces are present, particles may travel on curved trajectories and they may speed up and slow down. They appear to be attracted to or repelled from the source of the force.

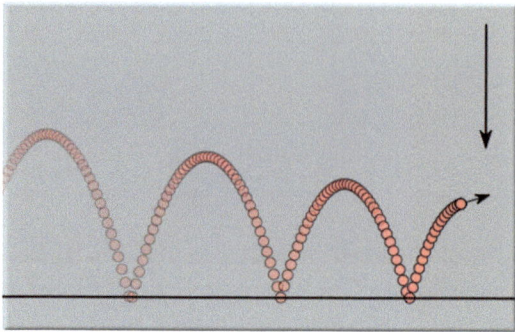

*A ball under the attraction of gravitation*

In principle, if the initial positions and velocities are known as well as all forces that are acting, one can exactly reconstruct the motion of a system of particles. If we apply the correct mathematical formalism, such systems usually lead to differential equations. These differential equations are in a sense the mathematical essence that describes the movement of the system. Unfortunately, even rather small physical systems lead to differential equations that are not solvable explicitly. They may even exhibit chaotic behavior, as shown in the following system (a freely movable planet under the influence of four fixed stars).

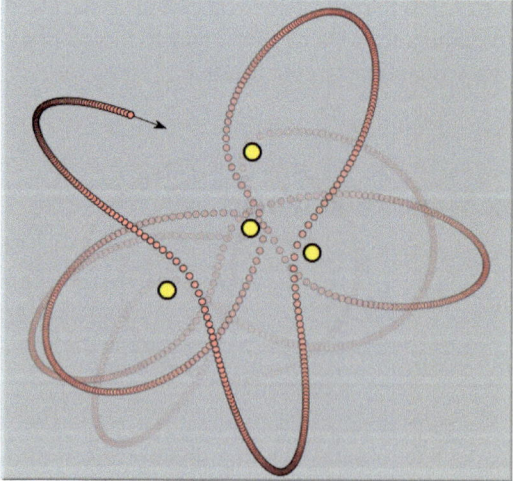

*The chaotic movement of a planet with many suns*

Nevertheless, one can attack such differential equations by numerical methods. These numerical methods model a continuous system by discrete methods. That is, they sample the positions, velocities, and forces at discrete points in time. From the situation at one moment in time (time stamp) (say *t*) such numerical methods try to estimate what the positions and velocities will be after some time interval *delta* has passed. At time stamp *t + delta* the numerical method proceeds with the estimated data to calculate a new estimate for the next time stamp. By this method, a discrete sequence of positions and velocities is generated. Obviously, such a sequence does not match the physical reality precisely. However, if the numerical estimation is good and if the time intervals *delta* are not too big, one can get very reasonable approximations of the physical truth. Furthermore, good numerical solvers have the property that (at least for most cases) the calculated motion becomes a better and better approximation of the physical reality as the time intervals are allowed to tend to zero.

## 6.2 Cinderella and Physics

Cinderella comes with a powerful and easy-to-use physics simulation system. It is based on a numerical method (for experts: a Dormand-Prince-45 integrator with adaptive step width), allowing one simply to draw a physical experiment, press the play button, and observe what happens.

The entire physics simulation CindyLab is based on pointlike particle masses and forces between them. As in the geometry part of Cinderella, there is usually no typed keyboard input necessary. You simply draw the experiment, which may contain masses, charged masses, springs, gravity, fixed planets, walls, etc. After you press the play button, the experiment is simulated by CindyLab and its numerical integrator. An important feature is that you can influence the experiment at any time by grabbing masses with the mouse and moving them. In this way, you obtain a very realistic impression of how the experiment would behave in the real world.

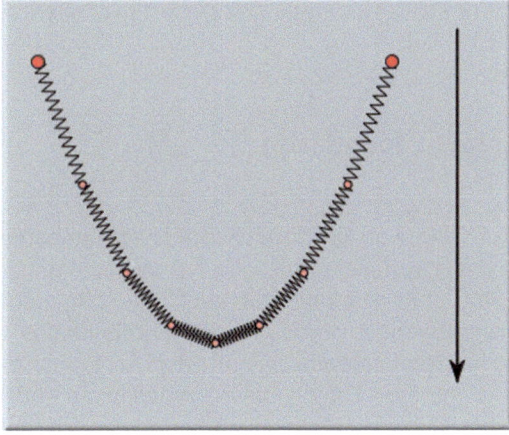

*Masses and springs under the influence of gravity*

Another important feature of CindyLab is that it works seamlessly together with the geometry part of Cinderella and with the programming interface CindyScript. By taking advantage of these connections one can considerably enhance the descriptive power of an experiment. The picture below shows a well-known arrangement of five pendulums with which one can nicely demonstrate the conservation of momentum. The masses are Cinderella points that are treated as more-or-less rigid balls. By constraining the points to be movable only on circles one can simulate the movement of a rigid pendulum.

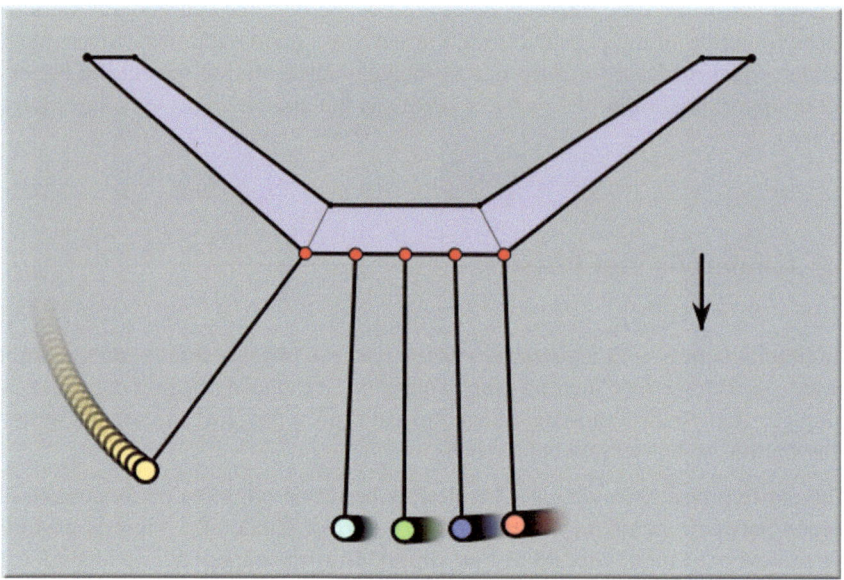

*A conservation-of-momentum system of pendulums*

The physical objects in Cinderella are permitted to exhibit a number of real-world properties. For instance, a particle may have a mass, a charge, and a radius. These properties may be controlled in two different ways: they may be adjusted manually using the inspector or altered programmatically using CindyScript.

## 6.3 CindyLab and CindyScript

The interrelation of CindyLab and CindyScript is very important. CindyScript has full access to all the physical parameters of a system, including positions, velocities, and forces. Using this connection one can use CindyScript either to *influence* or to *analyze* a physical scenario. Analysis of a physics simulation may result either in numerical data or in visual enhacements of the pictures that make the parameters directly apparent to the eye. The picture below shows a bridge construction that is under tension from its own weight. CindyScript was used to highlight the compression or tension in the rods. Red corresponds to strong stress by tension, while blue corresponds to strong compression. The color green indicates the absence of compression and tension. The color effect was generated by adding three lines of CindyScript code to the physical experiment:

```
segs=allsegments();
f(x):=hue(max((min((0.5,x+0.3)),0.0)));
forall(segs,#.color=f(A.y*#.ldiff))
```

The first line defines `segs` to be a list of all line segments in the drawing. The second line defines a function that translates a real number into a reasonable color value. The third line forces the color of each segment to be set in accord with the change in length (from its resting length) of the corresponding spring.

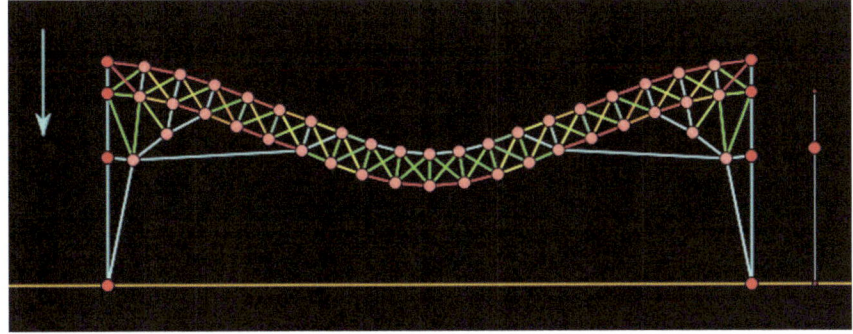

*The tension in a bridge*

CindyScript also provides special operations for analyzing physical experiments. For instance, the special operator `drawcurve(...)` can be used as a kind of curve plotter for parameter values in physics simulations. The picture below shows the energy flow in a coupled pendulum of two harmonic springs. One can observe nicely how the energy is transferred back and forth between the two pendulums.

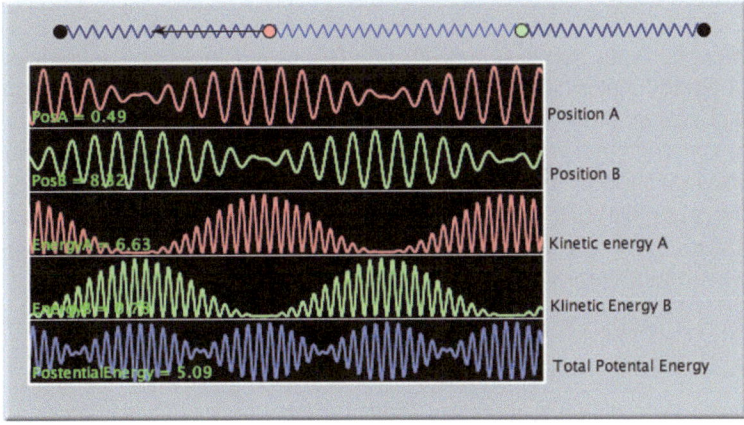

*The movement of a coupled oscillator*

Our last example shows how the `colorplot(...)` operator can be used together with a specialized operator `drawfield` to visualize the flux in an electrostatic field. However, it should be mentioned that in this context the use of colorplot will require a powerful computer in order for fluent movement to be exhibited when points are dragged with the mouse.

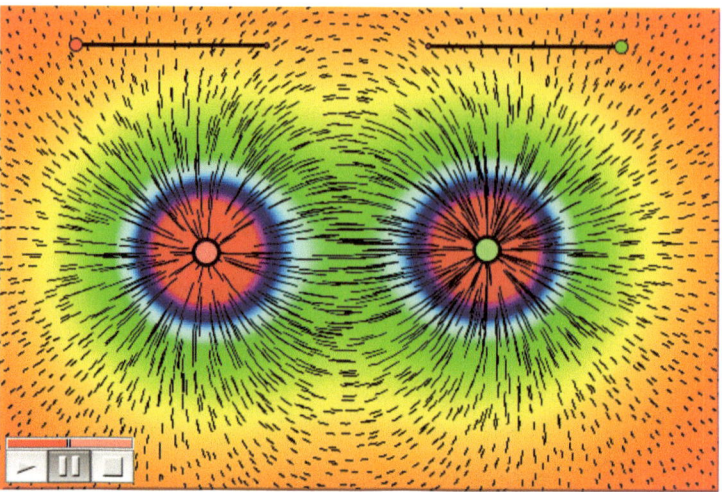

*The flux in an electrostatic field*

CindyScript can be used for more than analyzing physical effects. One can also use CindyScript for controlling the physical parameters of a set of objects. In particular, there is a flag "simulate" for every physical object that can be controlled by CindyScript. This flag controls whether a physical property is to be considered in the simulation. Using these features one can readily use Cinderella to simulate machines or games with many functional dependencies.

## 6.4 The Elements of CindyLab

As mentioned in the previous section, simulations in CindyLab are modeled as systems of pointlike particles and forces that influence their movement. The most fundamental unit in CindyScript is a Free Mass (p. 175). Such a mass is much like a usual geometric point in Cinderella. However, it has a number of additional properties besides its position, including a current velocity, a mass, and a charge. A simulation in CindyLab causes the masses to move according to the parameters of the simulation. Adding a free mass is like adding a free point, and you can use either of these two modes for it:

Free Mass (p. 175)

Velocity (p. 178)

The movements of the masses are influenced by forces. These forces may have various origins.

- First of all, there can be environmental forces such as gravitational forces (constant or otherwise) and magnetic fields. They can be added via the three modes:

Gravity (p. 180)

Sun (p. 183)

Magnetic Field (p. 185)

- Second, there may be forces between a specific pair of masses. Usually, these forces correspond to springs or to an electromagnetic or gravitational interaction. There are several modes for adding such an individual interactions between particles:

Rubber Band (p. 187) (Spring with zero length)

Spring (p. 189)

Coulomb Force (p. 194)

- Finally, the forces may come from settings related to the Environment (p. 197), such as universal gravity or interactions affecting all particles in the system.

A last ingredient to CindyLab is the ability to put walls and floors into a physical experiment. Whenever a particle encounters such a so-called *bouncer*, it bounces off like a tennis ball. One can add bouncers via the following two modes:

Floor (p. 194)

Bouncer (p. 196)

The behavior of any of these objects can be influenced by precise settings, which are entered via the inspector (or CindyScript). The sections on the different elements of CindyLab describe the variety of parameters that are available for each of these objects.

In addition, properties that apply to all physical objects can be controlled by the global setup Environment (p. 197) .

## 6.5 Starting a Physics Simulation

As soon as physical objects are present in a drawing, an animation control device will show up in the geometric views. With this control device one can play/stop/-pause the physics simulation, just as in Animation (p. 107) mode. As in animation mode, it is possible to grab and move free points or free masses at any time.

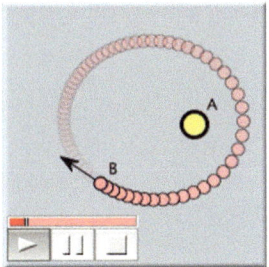

*A simple physics scenario with animation control*

Starting the physics animation will automatically switch to move mode. As soon as you change to another mode, the physics animation will stop. Note that the accuracy and the global frame rate of the simulation can be adjusted by the environment inspector (p. 197).

## 6.6 The Environment

Another extremely important ingredient of CindyLab is the Environment (p. 197). The environment collects properties of physical elements that are considered to be common to all physical objects. There you find such elements as global friction and global gravity. There you can also control whether there is an interaction between arbitrary mass particles by gravity or electrostatic forces. All these properties can be influenced either by the inspector or by suitable CindyScript statements.

# Chapter 7

# CindyLab Objects

## 7.1 Free Mass

Adding a free mass particle is similar to adding a free geometric point. The only difference is that mass particles have to be added with free mass mode, while free points are added with Add a Point (p. 74) mode.

Without the presence of forces or velocities, a free mass appears to be no different from a geometric point. (There is only a very tiny difference between free masses and geometric points, since by default masses are drawn in a slightly brighter color, so that one can easily recognize them in a drawing.)

As soon as a physical object (for instance, a mass) is added, an animation control panel appears in the geometric view. Unless the play button is selected, masses behave like geometric points. Pressing the play button starts the physics simulation. If a Velocity (p. 178) was assigned to the mass, the mass will start moving automatically. If the mass is influenced by forces (say from a Spring (p. 189) or from Gravity (p. 180)), the velocity of the mass will change according to the force while the mass is moving. The change of velocity per time unit is called *acceleration*. The acceleration of a mass in response to a force is governed by the equation

$$\text{acceleration} = \frac{\text{force}}{\text{mass}}$$

Here *force* is the vector sum of all forces that act on the mass.

### 7.1.1 Inspecting Masses

The basic properties of a mass can be set and changed using the physics tab (this is the fourth tab in the top row) of the Inspector (p. 153). The physics inspector for a generic mass looks as follows:

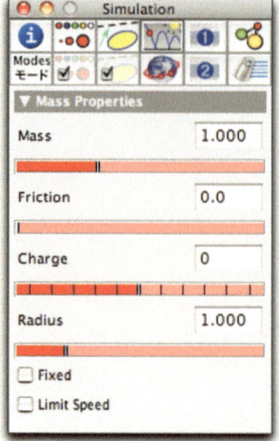

*The mass inspector*

The meaning of the different entries is as follows:

- *Mass:* This is the value of the physical mass of the object. According to the formula acceleration $= \frac{\text{force}}{\text{mass}}$ the acceleration of a particle is larger for smaller masses if the force is constant. This can be considered as the central law of inertia. (Thus with the same force it is much easier to push a tennis ball than a car.) On the other hand, the mass also influences the gravitational force that affects a particle. A gravitational force is proportional to the mass of an object. (Thus heavy objects are more influenced by gravity than light objects. The gravitational force on a tennis ball is much smaller than that on a car.). For instance, in free fall, both of the above effects cancel out exactly, so that in a field of constant gravity a car falls exactly as fast as a tennis ball.

- *Friction:* In a usual environment on Earth, a ball that is rolling over a plane horizontal surface will come to rest at some time. The reason for this is *friction*. In principle, friction can be seen as a force that acts in the opposite direction to that of the motion. This force causes moving objects to slow down over time. For most physical simulations that model situations of everyday life on Earth, it is reasonable to assume that masses are subject to some friction. In contrast, if one models the movement of planets or electrons, it is generally realistic to assume that no friction is present. By default, the friction affecting masses is set to zero. It is also possible to set a global friction value, which is then applied to all masses, in the Environment (p. 197) inspector.

- *Charge:* Electromagnetic forces (Coulomb force and Lorentz force) act only on electrically charged particles. The charge slider allows one to set the charge to a specific value. Charges can be either positive or negative. Positive and negative charges attract each other, while like charges repel each other. The charge of a particle will be relevant only if either the particle is involved in a (Coulomb) interaction or the "charges cause forces" box in the Environment (p. 197) is checked. By default, the charge of a mass is set to zero.

- *Radius:* Usually, CindyLab assumes that particles are pointlike objects. However, if the "masses are balls" box in the Environment (p. 197) is checked, particles will behave like objects with a nonzero diameter, which repel each other when they hit. This slider adjusts the radius of a particle. By default, the radius is set to 1.0.

- *Fixed:* Usually, masses are free to move move around. However, if this box is checked, the mass will not be moved by the physics simulation engine. It is important to know that if this box is checked, the mass may still cause gravitational or electrostatic forces. The following picture shows the movement of three charged particles in a cage of several other similarly charged particles. The outer particles are fixed so that they cannot move.

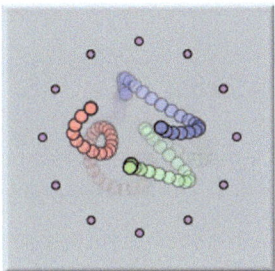

*Three charged particles in a cage of fixed charged particles*

It is important to notice that unlike "pinned" points, a "fixed" point is still movable by a mouse action.

- *Limit speed:* This button puts a limiting value on the speed of a particle. Although this behavior is not physically motivated (aside from the absolute limit of the speed of light), checking this box sometimes helps to avoid numerical problems in simulations.

### 7.1.2 Masses and Geometry

Like any other geometric point, masses can be bound to lines, segments, or circles. In such a case, CindyLab simulates the behavior as if the mass points were attached to frictionless tracks corresponding to the line, segment, or circle. The following picture shows the state of equilibrium of a chain of springs under the influence of gravity, where the endpoints of the springs are attached to two circles:

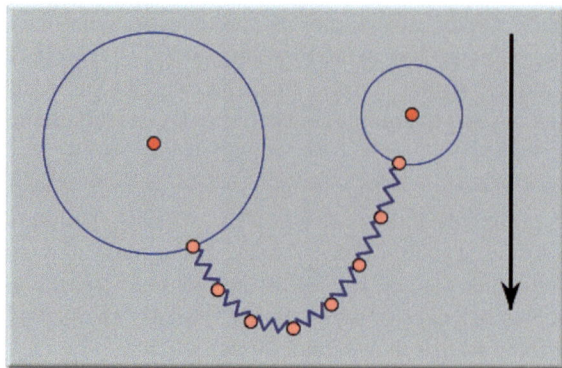

*A chain of springs connected to circles*

### 7.1.3  Masses and CindyScript

Like any CindyLab object, a mass point provides several fields that can be read and generally set by CindyScript. The following list shows the accessible fields for masses:

| Property | Writeable | Type | Purpose |
|---|---|---|---|
| vx | yes | real | $x$-component of velocity |
| vy | yes | real | $y$-component of velocity |
| v | yes | 2-vector | velocity as a vector |
| fx | no | real | $x$-component of force |
| fy | no | real | $y$-component of force |
| f | no | 2-vector | force as a vector |
| kinetic | no | real | the kinetic energy of the point |
| ke | no | real | the kinetic energy of the point |
| mass | yes | real | the mass of the point |
| friction | yes | real | the friction of the point |
| charge | yes | integer | the charge of the point |
| radius | yes | real | the radius of the point |
| simulate | yes | bool | turn on/off simulation for this point |

### See also

- Add a Point (p. 74)

- Velocity (p. 178)

## 7.2  Velocity

The velocity mode requires a press/drag/release sequence. The press generates a free mass, the drag generates a black arrow connected to the mass, and the release freezes the situation. The arrow is an indicator for the velocity (speed plus direction) of the mass point. Its direction indicates the direction of the motion, and its length indicates the absolute value of the velocity, that is, the speed. The longer the arrow, the faster the object. Pressing the mouse button over an existing mass adds a velocity arrow to that mass.

Velocity arrows are used for two purposes:

- *Initialization:* When the simulation is started, a mass with a velocity arrow will be initialized with the corresponding velocity value (this happens only if the "read on init" box of the velocity is checked).

- *Visualization:* During the animation the velocity arrow of a mass always represents the actual velocity of the mass.

A priori it is not clear what length corresponds to what speed of a mass point. Therefore, some convention is necessary. The convention taken in CindyLab is best understood in the case in which the animation-speed slider is dragged to its maximum. In this case, a mass travels exactly the length of its velocity arrow in one simulation time step. The situation is illustrated in the following picture.

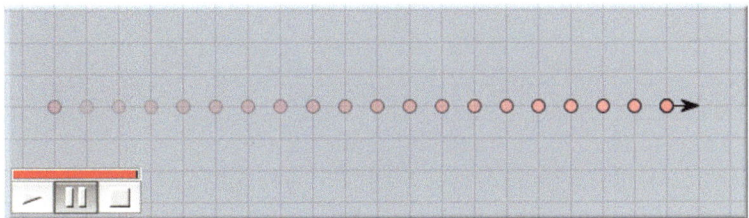

*Trace of a mass point with unit velocity*

For many purposes this behavior may be a bit too fast, but it is a good way of normalizing the speed treatment.

### 7.2.1 Inspecting Velocities

The basic properties of a velocity can be set and changed by inspecting it with the Inspector (p. 153) and opening the physics tab (this is the fourth tab in the top row). The physics inspector for a generic mass looks as follows:

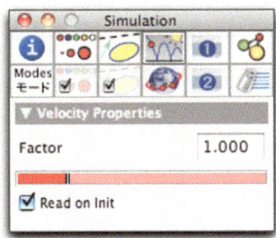

*The velocity inspector*

The precise meaning of the two controls is as follows:

- *Factor:* With this factor one can change the scaling of the velocity arrow. The factor is usually set to 1.0. Doubling the factor means that in the same time interval, twice as much distance is traveled.

- *Read on Init:* This checkbox is checked by default. If it is checked, then at simulation start the length of the drawn arrow is calculated and stored in the point's data as its initial velocity. If the box is not checked, then the initial position does not matter at all. The arrow will be used only to display the actual velocity.

### 7.2.2  Velocity and CindyScript

Like any CindyLab object, a velocity has several fields that can be read and very often set by CindyScript. The following list shows the accessible fields for masses:

| Property | Writeable | Type | Purpose |
|---|---|---|---|
| factor | yes | real | the scaling factor of the velocity |

**See also**

- Add a Point (p. 74)

## 7.3  Gravity

Gravity is a constant force that acts on all masses. The gravity mode in *Cinderella* is designed to model the gravitational force at the earth's surface, that is, a constant force field that acts on an object with nonzero mass. The force on a mass-object caused by gravity is dependent on two parameters: gravity's strength and the value of the mass of the object (you can consider this as the weight of the object, though technically, mass and weight are two different concepts). The exact relation between these three magnitudes is described by the following equation:

$$\text{force} = \text{gravityconstant} \cdot \text{mass}$$

Gravity also has a direction toward which a mass-object is drawn. In *Cinderella* you create a gravitational force by a press/drag/release cycle similar to that for creating a line in Add a Line (p. 76) mode. The gravitational force field is represented by a black arrow that points in the direction of the force. The longer the arrow, the stronger the gravitational force.

The following pictures demonstrate how gravity acts on a mass with nonzero initial Velocity (p. 178). The first picture shows the situation before the simulation is started. The second picture shows a trace of the mass-object during the simulation: a ballistic parabola.

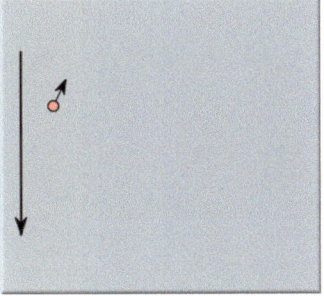

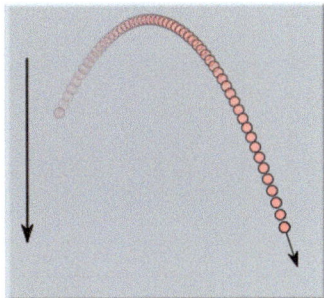

*A mass-object with nonzero*
*initial velocity before start.*

*Trace of the mass-object under*
*the influence of gravity.*

As a second example we consider the action of gravity on a string of masses connected by Rubber Bands (p. 187), whose endpoints are pinned to fixed positions in the drawing. Without a change in the default settings of CindyLab, the system would oscillate wildly. However, if we add some friction to the Environment (p. 197), the system will enter a state of equilibrium after some time. In this equilibrium situation the chain of masses forms a perfect parabola.

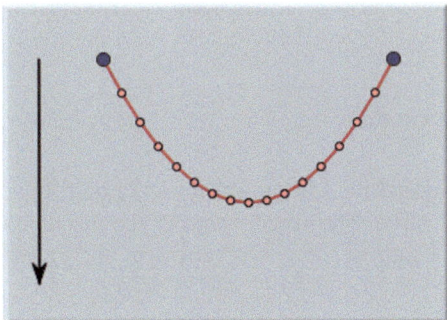

*A chain of mass-objects under the influence of gravity*

The picture below shows an even more sophisticated example. Here a background image of the Golden Gate Bridge has been loaded. The upper left part of the picture shows a physics simulation of a rubber-band chain under the influence of gravity. A Projective Transformation (p. 131) has been used to map the simulated situation to a rectangular frame on the Golden Gate Bridge. Adjusting the gravitational force to the correct value shows that this situation exactly resembles the situation on the supporting cables of the bridge.

*An analysis of the Golden Gate Bridge*

### 7.3.1 Inspecting Gravity

A CindyLab gravitational field is equipped with a built-in scaling factor that is set to a relatively small value. The actual value of the *gravityconstant* is calculated as this factor times the length of the gravity arrow in the drawing.

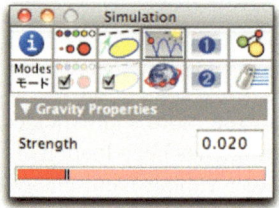

*The gravity inspector*

### 7.3.2 Gravity and CindyScript

Like any CindyLab object, a gravitational force provides several fields that can be read and very often set by CindyScript. The following list shows the accessible fields for gravity:

| Property | Writeable | Type | Purpose |
| --- | --- | --- | --- |
| strength | yes | real | a handle to the scaling factor |
| potential | no | real | the overall potential energy of masses in the gravitational field, defined only up to an additive constant that depends on the choice of coordinates |
| pe | no | real | the overall potential energy of mass objects in the gravitational field, defined only up to an additive constant that depends on the choice of coordinates |
| simulate | yes | bool | turn on/off simulation for the gravitational field |

### See also

- Add a Line (p. 76)
- Sun (p. 183)
- Environment (p. 197)

## 7.4 Sun

While the Gravity (p. 180) mode models a constant gravity field that resembles the situation on Earth's surface, the "sun" mode models a central force field that resembles the situation in space. Its gravity field is modeled to simulate the force field of a pointlike mass. This mass is simply added as a usual geometric point in the Add a Point (p. 74) mode. Unlike a free mass, such a sun will not move during a simulation. It only exerts a force on each free mass. The force is bigger for masses that are closer to the sun. The exact relation between the distance *dist*, the masses involved, and the force is described by the following equation (this is in essence Newton's law of gravity):

$$\text{force} = \frac{\text{sunmass} \cdot \text{mass}}{\text{dist}^2}$$

Free masses with an initial velocity will move in an elliptical, parabolic, or hyperbolic path if they are under the force of a sun. In CindyLab such a situation can be constructed and simulated with a handful of mouse clicks (drawing a sun, drawing a mass with velocity, and starting the animation). The picture below shows the result of such a situation.

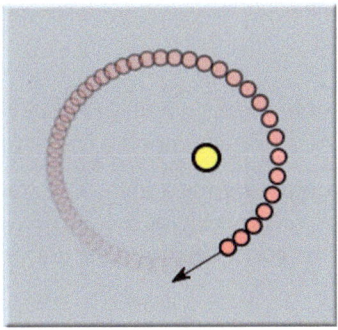

*A planet orbiting a sun*

The situation will become even more interesting if two suns are present. The picture below shows a recorded trace of a planet in the combined force field of two suns.

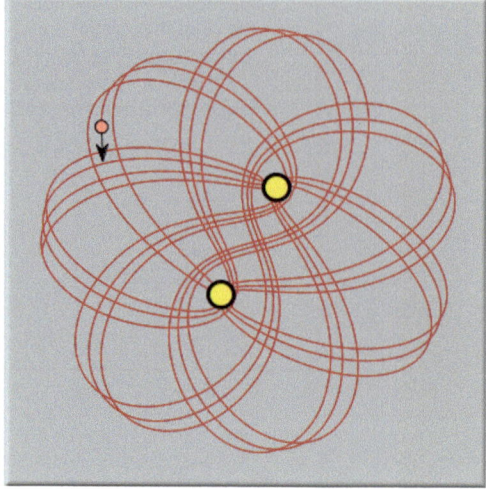

*One planet and two suns*

The following two images give an impression of the force field that is caused by one or by two suns to a generic mass. The pictures were generated by the `drawforces()` statement of CindyScript.

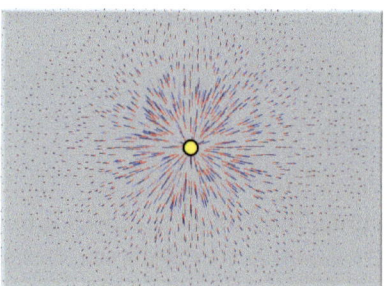

*Force field of one sun*

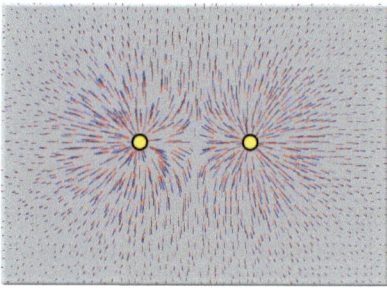

*Force field of two suns*

### 7.4.1 Inspecting Suns

A sun has one item that can be changed by the inspector: its initial mass. By default this mass is set to be 10, which is ten times the mass of a default mass.

A CindyLab gravity is equipped with a built-in scaling factor, which is set to a relatively small value. The actual value of the *gravityconstant* is calculated as this factor times the length of the gravity arrow in the drawing.

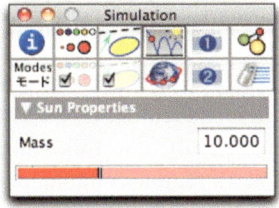

*The sun inspector*

## 7.4.2  Suns and CindyScript

Like any CindyLab object, a sun provides several fields that can be read and very often set by CindyScript. The following list shows the accessible fields for gravity:

| Property | Writeable | Type | Purpose |
| --- | --- | --- | --- |
| strength | yes | real | scaling factor for the force |
| potential | no | real | the overall potential energy of masses in the sun's gravitational field, defined only up to an additional constant |
| pe | no | real | the overall potential energy of masses in the sun's gravitational field, defined only up to an additional constant |
| simulate | yes | bool | turn on/off simulation for this sun |

### See also

- Add a Point (p. 74)

- Gravity (p. 180)

- Environment (p. 197)

## 7.5  Magnetic Field

Perhaps the most complicated force available in CindyLab is the Lorentz force. Whenever a charged particle enters a magnetic field with nonzero velocity, the particle's path is perturbed. The magnetic field causes a force orthogonal (at right angles) to both the direction of movement and the magnetic field. The force is proportional to three quantities: the *charge* of the particle, its *speed*, and its *velocity*. For instance, if an electron moves in a magnetic field without being subject to any additional forces, it will move in a circular path. If friction is present, the path will be a spiral.

In CindyLab one can add magnetic fields that are orthogonal to the drawing surface. Magnetic fields are polygonal regions that are added just as in Polygon (p. 91) mode. The exact relation between the *fieldstrength* of the field, the *velocity* of the particle, its *charge*, and the resulting *force* is given by the following formula:

$$\text{force} = (\text{fieldstrength} \times \text{velocity}) \cdot \text{charge}$$

Since the field is assumed to be orthogonal to the drawing surface, it will result in a force that is orthogonal to the movement of the particle. A charged particle will therefore move along a circular path. The following picture shows the trace of a charged particle that is shot into a region of a magnetic field.

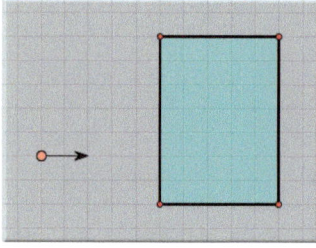
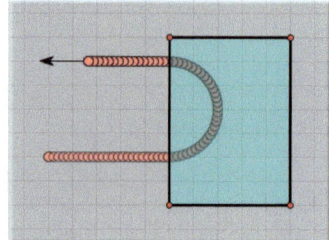

*Shooting a charged particle*
*into a magnetic field...*                       *...causes circular motion*

The magnetic fields in CindyLab admit a second kind of force, one that simulates friction in the field. A particle subjected to a frictional force is slowed down by a force proportional to its speed. The relation of the *friction* and the particle's *velocity* is given by the following equation:

$$\text{force} = -\text{friction} \cdot \text{velocity}$$

Magnetic fields are often used in physics to observe the decay of atomic particles. If, for instance, an electrically neutral particle decays into two smaller particles, one carrying a positive charge, the other a negative charge, the two constituent particles will follow circular paths in a magnetic field if no friction is present, and they will follow spiral paths in the presence of friction. This effect is used in physics research to observe and analyze the constituents of particle decay in a bubble chamber. The following picture compares a real observed decay in a bubble chamber with a simulated decay in CindyLab and CindyScript.

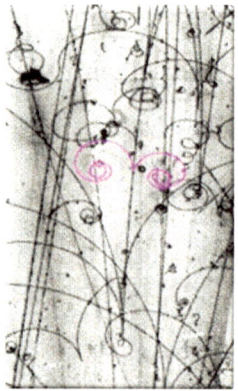
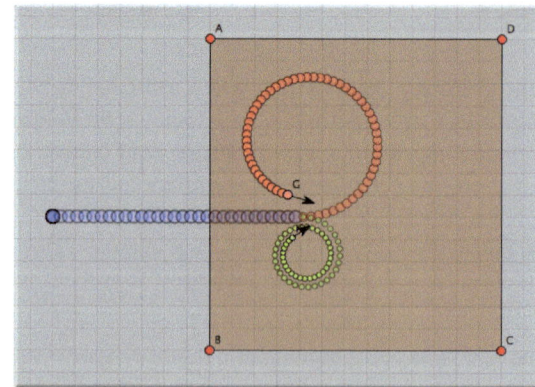

*Real decay*                       *Simulated decay*

### 7.5.1  Inspecting Magnetic Fields

In the inspector one can adjust the *fieldstrength* and the *friction* values of a magnetic field. In particular, it is possible to reverse the direction of the magnetic field by assigning a negative value to it.

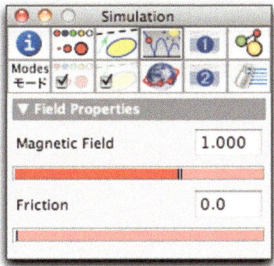

*The magnetic field inspector*

### 7.5.2  Magnetic Fields and CindyScript

Like any CindyLab object, a magnetic field provides several entries that can be read and set by CindyScript. The following list shows the accessible fields:

| Property | Writeable | Type | Purpose |
|----------|-----------|------|---------|
| strength | yes | real | the strength of the field |
| friction | yes | real | the friction in the polygonal region |
| simulate | yes | bool | turn on/off simulation for the field |

**See also**

- Polygon (p. 91)

- Coulomb Force (p. 194)

## 7.6  Rubber Band

In CindyLab a rubber band is a Spring (p. 189) that has a resting length of zero. Rubber bands are constructed by a press/drag/release sequence as in the Add a Line (p. 76) mode. If the endpoints of the spring segment are not already present in the drawing, free masses will be added. A rubber band affects the two masses at the endpoints according to the formula

$$\text{force} = \text{springconstant} \cdot \text{length}$$

Here *springconstant* is a characteristic constant of the rubber band that can be adjusted in the inspector, and *length* is the distance between the two points. Thus if two mass-objects are at a certain distance from each other and connected by a rubber band, they will be attracted to each other. Since the attractive force is proportional to the distance of separation, the farther apart the masses are, the stronger the force.

If no global friction is present in the Environment (p. 197), a system of masses and rubber bands will oscillate wildly. The following picture shows the situation of one free mass-point that is connected to a rubber band whose other end is tied to a regular (fixed) geometric point. The curve on the left shows the resulting oscillation of the point. (The curve was generated by the `drawcurves(...)` operator of CindyScript.)

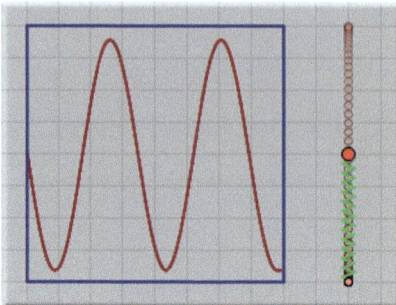

*Trace of a mass-point with unit velocity*

If one uses the Environment (p. 197) to add some global friction, a damping of the oscillation will occur. In such a case, the system will converge to an equilibrium situation in which it comes to rest. The following picture shows a chain of masses connected by rubber bands under the influence of Gravity (p. 180).

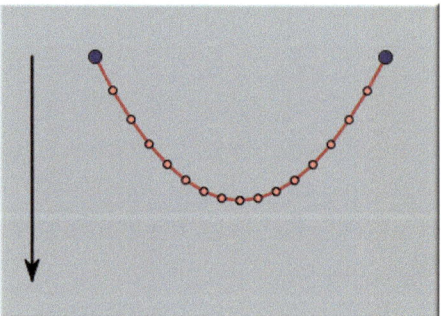

*A chain of mass-objects and rubber bands under the influence of gravity*

Networks of rubber bands exhibit interesting behavior. In a sense, such networks behave like spider webs. The following pictures show a gridlike arrangement of masses and rubber bands with various points pinned to the ground plane in the equilibrium situation.

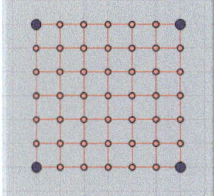  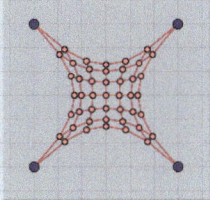  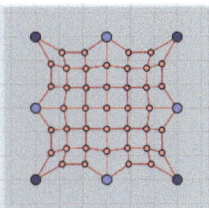

*The net before start*        *Four pinned points*        *ight pinned points*

### 7.6.1 Inspecting and CindyScript

Rubber band, Spring (p. 189), and Coulomb Force (p. 194) are objects that can be transformed one into another. A detailed discussion of the very powerful spring inspector as well as of the CindyScript interface can be found in the documentation to Spring (p. 189) mode.

**See also**

- Add a Line (p. 76)
- Spring (p. 189)
- Coulomb Force (p. 194)
- Environment (p. 197)

### 7.7 Spring

In CindyLab a spring is very similar to a Rubber Band (p. 187). A spring is constructed by a press/drag/release sequence as in Add a Line (p. 76) mode. If the endpoints of the spring segment are not already present in the drawing, free mass-objects will be added. A spring causes forces to act on its two endpoints. Unlike rubber bands, springs have a resting length at which the force is zero. If the distance between the two endpoints is smaller than the resting length, the force will be repulsive. If it is larger than the resting length, the force will be attractive. The force is calculated by the formula

$$\text{force} = \text{spring constant} \cdot |\text{length} - \text{restlength}|$$

Here $|x|$ denotes the absolute value of $x$, *springconstant* is a characteristic constant of the spring that can be adjusted in the inspector, *restlength* is the resting length of the spring, and *length* is the distance between the two points. Springs are very powerful objects that admit many variations via the inspector. So we will first have a look at the inspector before we discuss the various possibilities for defining a spring.

### 7.7.1 Inspecting Springs

The spring inspector comes with a great variety of possibilities for modification. We will discuss these modifications one by one.

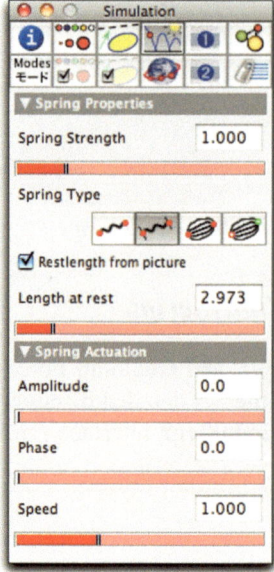

*The spring inspector*

*Spring Strength*

The spring strength slider helps in adjusting the value of *springconstant*. No matter what kind of interaction is modeled by the spring, the *springconstant* acts as a factor by which the final force is multiplied. Thus setting this slider to a value of zero is equivalent to removing the spring from the configuration.

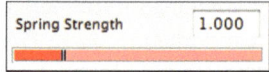

*Spring Type*

In CindyLab all interactions between pairs of points are internally modeled as springs. Thus it is simple to switch between the different types of interaction. This can be done by choosing the *springtype* in the inspector. Currently, there are four different types of interaction implemented. Each of them comes with a specific behavior (and formula) for the spring.

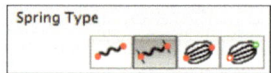

The four types are defined as follows: *rubber band*: For this spring type it is assumed that the spring has a rest length of zero. Thus the force between the two masses is always attractive. Its absolute value is calculated by the formula

$$\text{force} = \text{springconstant} \cdot \text{length}$$

This type is the default spring type when a Rubber Band (p. 187) is added.

*Spring*: For this spring type it is assumed that the spring has a positive resting length. Thus the force between the two masses will be attractive if the two masses are farther apart than the resting length. It will be repulsive if they are closer than the resting length. The absolute value of the force is calculated by the formula

$$\text{force} = \text{springconstant} \cdot (\text{length} - \text{restlength})$$

This type is the default type when a spring is added.

*Newton force*: This spring type models the gravitational attraction between two bodies. This force depends on the masses of the bodies and the distance between them. It is always attractive. The absolute value of the force is calculated by the formula

$$\text{force} = \frac{\text{springconstant} \cdot \text{mass}_1 \cdot \text{mass}_2}{\text{dist}^2}$$

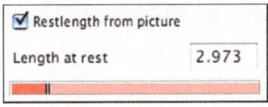

*Coulomb force*: This spring type models the interaction between electrically charged particles. This force depends on the charges of the particles and on the distance between them. It is attractive if the charges have opposite signs and repulsive otherwise. The absolute value of the force is calculated by the formula

$$\text{force} = \frac{\text{springconstant} \cdot \text{charge}_1 \cdot \text{charge}_2}{\text{dist}^2}$$

This type is the default spring type when a Coulomb Force (p. 194) is added.

*Restlength*

The two items *Restlength from picture* and *Length at rest* are relevant only if the spring has a resting length.

| ☑ Restlength from picture | |
|---|---|
| Length at rest | 2.973 |

If *Restlength from picture* is checked, then the resting length of the spring is defined to be the length of the spring in the drawing when the simulation is started. Thus in this case, a single spring with two mass-objects not otherwise connected at the endpoints exerts no force on the masses.

If this box is not checked, the spring's resting length will be defined by the *Length at rest* value in the inspector.

*Spring Actuation* The two items *Amplitude* and *Phase* will be relevant only if the spring has a resting length. In this case, CindyLab provides the possibility to vary the resting length periodically in time. The resting length is then modulated by a sine function, causing it to become periodically longer and shorter.

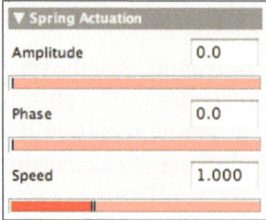

The amplitude of the resting length variation can be adjusted by the *Amplitude* slider. The phase of the vibration with respect to the other springs can be adjusted by the *Phase* slider. The speed of the oscillation is globally adjusted in the environment inspector (p. 197).

## 7.7.2 Examples

Due to their programmatic flexibility, springs can be used in many different circumstances. Here we will present just a few examples and pictures that exemplify the use of springs.

### A Bridge

The first example shows a network of springs that simulates the behavior of a bridge. All springs in this example are used with the default physical setup. However, the appearance of the springs has been slightly altered. Usually, springs are rendered as wiggly objects. This feature can be turned off by the *render* button in the inspector. Furthermore, CindyScript code was added to change the color of the springs according to their interior tension.

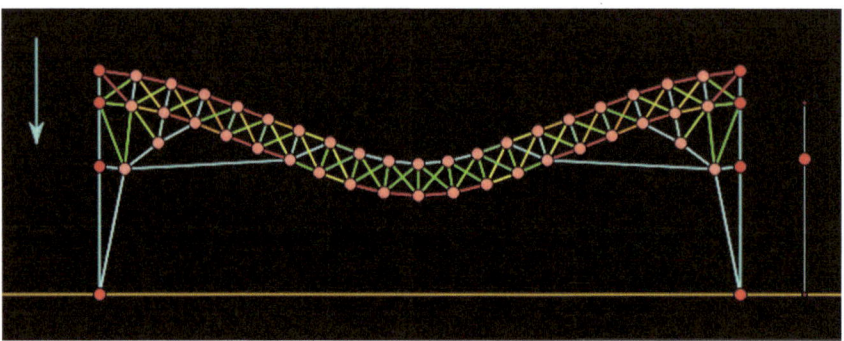

*A bridge constructed with springs*

### Double Pendulum

The next example shows the chaotic movement of a double pendulum. Here two springs are connected and one endpoint is fixed to the ground plane. The *springconstant* is set to a relatively high value so that the springs behave nearly like rigid rods. The picture shows the movement of this double pendulum under the influence of gravity.

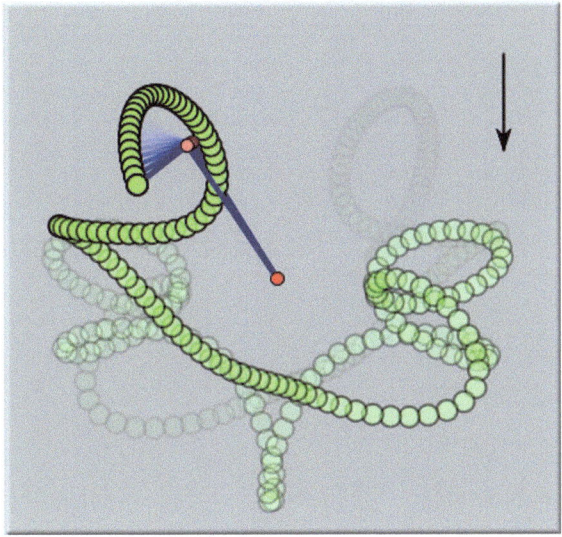

*The chaotic motion of a double pendulum*

**Two-Body Movement**

The next pictures show two mass particles with initial velocities and their behavior under gravitational forces. For this example, the type of the spring was set to *Newton's law*. The two mass particles simulate a system of two stars that orbit each other.

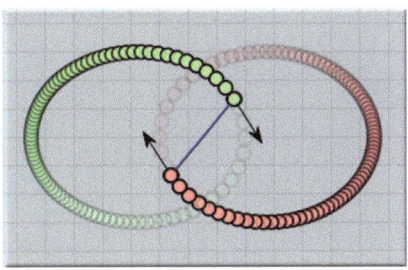 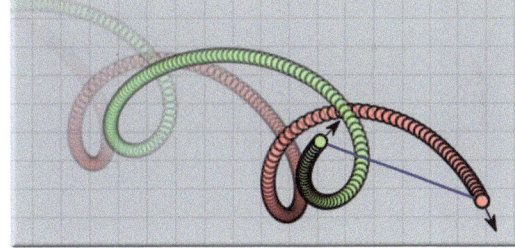

*Two instances of two-body motion*

## 7.7.3 Springs and CindyScript

Like any CindyLab object, a spring provides several fields that can be read and very often set by CindyScript. The following list shows the accessible fields for springs:

| Property | Writeable | Type | Purpose |
|----------|-----------|------|---------|
| l | no | real | current length of the spring |
| lrest | yes | real | resting length of the spring |
| ldiff | no | real | current length difference to the resting length |
| potential | no | real | the potential energy of the spring |
| pe | no | real | the potential energy of the spring |
| strength | yes | real | the spring constant |
| amplitude | yes | real | the amplitude of actuation |
| phase | yes | real | the phase of actuation |
| simulate | yes | bool | turn on/off simulation for this spring |

### See also

- Add a Line (p. 76)

- Environment (p. 197)

- Rubber Band (p. 187)

- Coulomb Force (p. 194)

## 7.8 Coulomb Force

Coulomb forces model the forces between charged particles. In CindyLab a Coulomb force is an interaction that is modeled by a special kind of Spring (p. 189). This force depends on the charges of the particles and on the distance between them. It is attractive if the charges have opposite signs and repulsive otherwise. The absolute value of the force is calculated by the formula

$$\text{force} = \frac{\text{springconstant} \cdot \text{charge}_1 \cdot \text{charge}_2}{\text{dist}^2}$$

For further details on the treatment of springs please consult the section Spring (p. 189).

## 7.9 Floor

A floor is a horizontal line through which no mass particle can pass. Mass particles that hit the line will bounce of like light tennis balls. A floor is added either by a press/drag/release sequence or by a simple click. It may also be attached to an existing point. Floors can either model completely elastic bounces or be equipped with a coefficient of friction that reduces a particle's velocity whenever it hits the floor. The following picture shows the behavior of a point that bounces off a floor (with a small coefficient of friction).

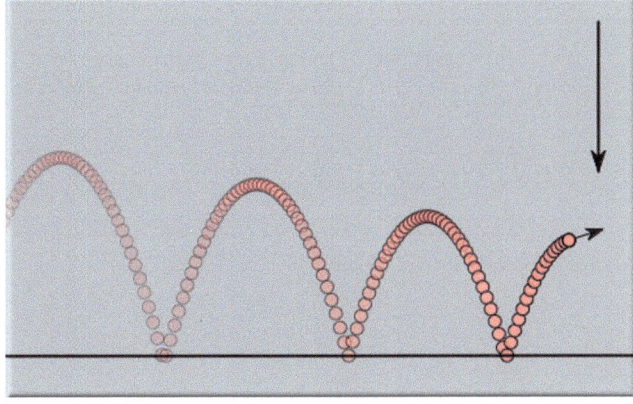

*A ball under gravitational attraction meeting a floor*

A floor is very similar to a Bouncer (p. 196), which behaves like a segment of a floor.

## 7.9.1 Inspecting a Floor

The floor inspector has three values that can be adjusted:

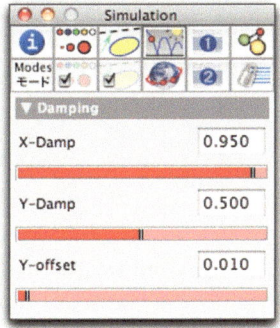

*The floor inspector*

*X-damp* and *Y-damp* are damping factors that can attain values between 0.0 and 1.0. The two numbers $dx = 1 - X\text{-}damp$ and $dy = 1 - Y\text{-}damp$ are factors by which the $x$- and $y$-components of a mass-object's velocity are multiplied whenever the mass-object hits the floor. Thus if both sliders are set to 0.0, the point will bounce off without any damping, while if both sliders are set to 1.0, the particle will stop moving as soon as it hits the floor.

## 7.9.2 Floors and CindyScript

Like other CindyLab object, a floor provides several fields that can be read and set by CindyScript. The following list shows the accessible fields for a floor:

| Property | Writeable | Type | Purpose |
|----------|-----------|------|---------|
| xdamp | yes | real | handle to the *X-damp* factor |
| ydamp | yes | real | handle to the *Y-damp* factor |
| simulate | yes | bool | turn on/off simulation for the floor |

### See also

- Bouncer (p. 196)

## 7.10 Bouncer

A Bouncer is an object that is very similar to a Floor (p. 194). A particle hitting a bouncer will bounce off like a tennis ball. However, unlike a Floor (p. 194), a bouncer is a line segment that is added by a press/drag/release sequence as in the Segment (p. 84) mode. Bouncers are very powerful objects for creating environmental limits to a simulation scenario. The following picture demonstrates a bouncing ball inside a polygonal region. Each edge of the region is an individual bouncer.

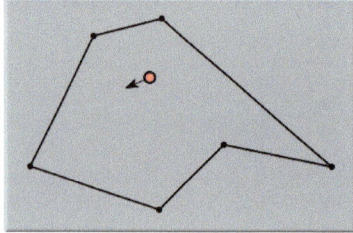
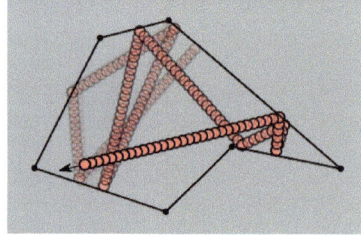

*Starting a mass in a polygonal region*    *The trace of the mass*

Bouncers are also essential for many physically oriented games that one might be interested in implementing in CindyLab, such as billiards and table tennis. The following picture shows a simple scenario in which bouncers and Gravity (p. 180) are combined.

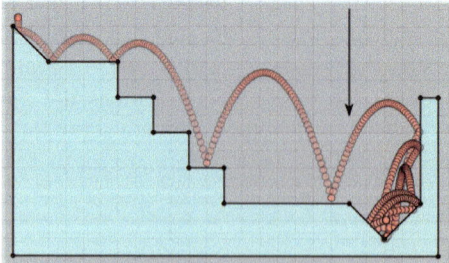

*A ball on a staircase*

### 7.10.1 Inspecting Bouncers

The bouncer inspector is almost identical to the Floor (p. 194) inspector.

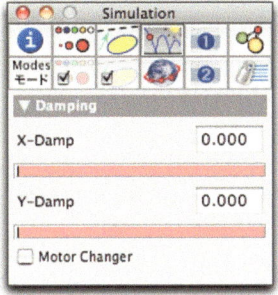

*The bouncer inspector*

In addition, it has a checkbox *Motor Changer*. If this box is checked, the animation direction of the internal spring actuators will be reversed whenever a particle hits the bouncer. For details on the actuation see Spring (p. 189) and Environment (p. 197). This behavior would be very useful in the construction of walking engines that reverse their direction whenever they hit a wall.

### 7.10.2 Bouncers and CindyScript

Like any CindyLab object, a bouncer provides several fields that can be read and set by CindyScript. The following list shows the accessible fields for bouncers:

| Property | Writeable | Type | Purpose |
| --- | --- | --- | --- |
| xdamp | yes | real | handle to the *X-damp* factor |
| ydamp | yes | real | handle to the *Y-damp* factor |
| simulate | yes | bool | turn on/off simulation for the floor |

## 7.11 Environment

The environment plays a very special role in CindyLab. Unlike the other objects in CindyLab there is neither a specific mode nor a specific geometric object associated with it. Since the environment is an omnipresent object, it is more or less *always there*. The behavior of the Environment can be altered or influenced only by the inspector or by CindyScript. The environment allows one to influence the following factors:

- the numerical accuracy of the simulation

- environmental forces such as gravity and friction

- the definition of many-body systems

- the actuation of springs

We will first list all ways that these influences can be made and then present a few examples.

### 7.11.1 Inspecting the Environment

The environment inspector occupies a whole tab in the inspector window. If one presses the tab that is labeled with a small planet Earth, one enters the environment inspector, which contains four major groups.

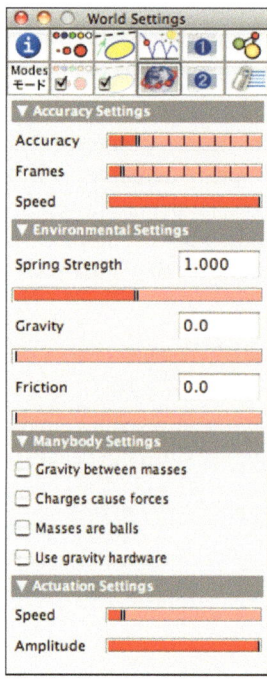

*The environment inspector*

The tasks of the particular groups are listed in detail below:

*Accuracy*

CindyLab is based on a numerical simulation of physical motion. The numerical simulation proceeds in discrete time steps. Between the discrete sample points of the motion there is practically no information about the physical system. This implies that in general, larger time steps imply less accuracy. Using fewer sample points, however, has the advantage of creating faster simulations. So there is a certain trade-off between speed and accuracy. The user of CindyLab must be aware of this effect. Furthermore, it may be useful not to create a picture for every time step that has been created. The default settings of CindyLab are such that the simulation runs relatively fast and every simulated frame is indeed shown. If additional *accuracy* or fewer *frames* are required, one can adjust these parameters with the corresponding sliders in the environment inspector. Moving the accuracy slider to the right increases the accuracy of the simulation. Every tick corresponds roughly to a doubling of the accuracy. Increasing the accuracy without changing the frame rate is like viewing an animation in slow motion. This is sometimes a very useful effect.

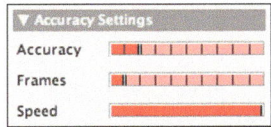

Corresponding to the accuracy slider, every tick of the frame slider corresponds to showing half as many frames. So if the frame slider is in position 4, there are 16 simulation steps between each frame. If the accuracy and the frame slider are exactly aligned, the simulation speed will still be the same as in the initial position (i.e., frames are shown for the identical time steps). The simulation is just much more exact. The additional computational effort that is required may then result in a slowdown of the perceived speed of the animation. We highly encourage the user to play with the two sliders in order to find the right setup for each simulation situation.

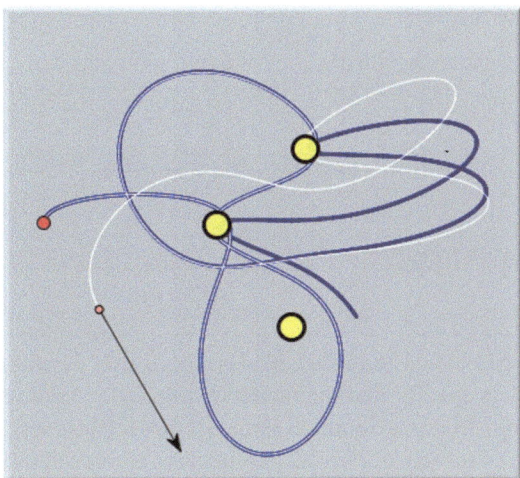

*Traces of various accuracies*

The picture above demonstrates the influence of accuracy on a simulation. The simulation shows a planet moving under the gravitational influence of three suns. This situation is chaotic in the sense that a slight change in the initial parameters will lead to very different passes over time. Both the blue and the white trace are calculated with the same initial parameters. The accuracy for the white trace was four times that for the blue trace. Observe how the traces slightly diverge until they follow qualitatively very different paths.

Finally, the speed slider is synchronized with the animations speed slider in the viewing window.

*Environmental Forces*  In this part of the environment, parameters can be set that simultaneously apply to all simulation objects.

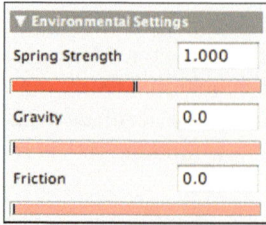

The *Spring Strength* slider is a global parameter that is used to modify all spring constants (see Spring (p. 189)). Thus by moving this slider one can weaken or strengthen all interaction forces simultaneously.

The *Gravity* slider brings into existence a vertical gravitational force that applies to all masses. Sometimes it is more convenient to use this slider then to devise an individual gravity in Gravity (p. 180) mode.

The *Friction* slider is perhaps the most important slider of this group. It produces a frictional force that applies to all moving masses. Whenever one is interested in realistic simulations for everyday scenarios, one should add some friction. In this way, every simulation that is not driven by an external force will eventually come to rest in an equilibrium position. If no friction is present, mass-objects will usually move and oscillate forever.

The picture below shows a ballistic shot of a ball with and without friction.

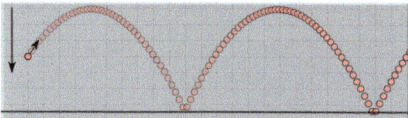

*Without friction*

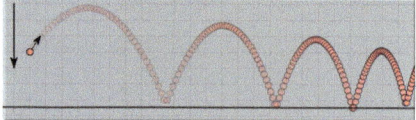

*With friction*

*Many-Body Behavior* With the usual modes interaction between two free masses, one generally has to specify that interaction explicitly by adding a Spring (p. 189), a Rubber Band (p. 187), or a Coulomb Force (p. 194). However, it is often desirable to study an entire ensemble of objects that interact with each other, such as a cloud of electrons, a many-planet system, or a table with a large number of billiard balls. The three checkboxes in this group allow one to define such behavior. Since there is a wide range of applications of such behaviors, we have dedicated an entire section, Many-Particle Systems (p. 201), to this topic.

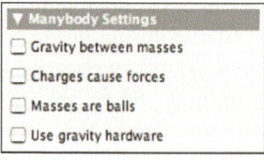

*Spring Actuation*

With the two actuation sliders one can influence the actuation of the resting lengths of a Spring (p. 189). The actuations for all springs are synchronized and in a sense driven by a global motor. With these two sliders one can vary the global speed and strength of this actuation.

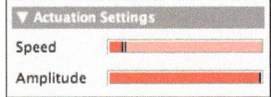

### 7.11.2 The Environment and CindyScript

The environment allows for access to several important magnitudes and settings of the simulation. Since the environment is not directly bound to a geometric object, there is a function `simulation()` that serves as a handle to the environment object. You can access, for example, the total kinetic energy in the simulation using `simulation().ke`. In particular, the values accessible via this object are the following:

| Property | Writeable | Type | Purpose |
|---|---|---|---|
| strength | yes | real | scaling factor of the simulation |
| potential | no | real | the overall potential energy in the simulation; for this, all potential energies of springs, suns, and gravities, as well as interparticle potential energy, are summed up (defined only up to and additive constant that depends on the choice of coordinates) |
| pe | no | real | abbreviation for potential |
| kinetic | no | real | the overall kinetic energy in the simulation; for this value, all kinetic energies of moving particles are summed up (defined only up to an additive constant that depends on the choice of coordinates) |
| ke | no | real | abbreviation for kinetic |
| friction | yes | real | the global friction constant |
| gravity | yes | real | the global gravity constant |

**See also**

- Many-Particle Systems (p. 201)

- Spring (p. 189)

## 7.12 Many-Particle Systems

Sometimes it is desirable to study the interaction of many highly interrelated particles. Imagine, for instance, the balls on a billiard table, a collection of electrons, or a system of several bodies in space. In such a case every particle may exert a force on every other particle. We will briefly discuss these three cases here.

### 7.12.1 Charged Particles

Suppose we want to study the interaction of 20 electrons. There are thus 190 different pairs of electrons, with a force of interaction for each pair. One could in principle construct such a particle system by creating 20 charged free masses and then inserting 190 Coulomb Force (p. 194)s. However, in CindyLab there is a much easier way if doing this. After the masses have been created and equipped with charges (using the Free Mass (p. 175) inspector), one can simply open the Environment (p. 197) inspector and check the box *Charges cause forces*. Then all the particles will interact immediately.

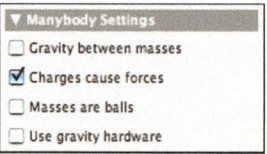

*Charges cause forces*

The picture below shows the situation of several charged particles trapped in a cage of bouncers. CindyScript was used to make fast particles appear blue und slow particles appear red, with the rest of the spectrum for intermediate speeds.

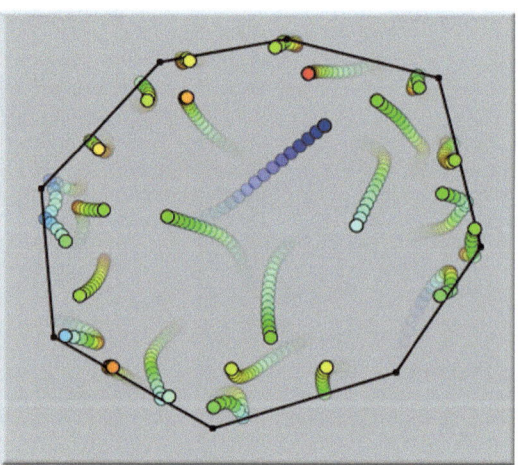

*Many electrically charged particles*

The corresponding fragment of CindyScript used for this drawing is the following:

```
pts=allpoints();
f(x):=hue(max((min((0.5,x)),0.0)));
forall(pts,#.color=f(#.kinetic))
```

### 7.12.2  Gravitational n-Body Systems

In complete analogy to systems of charged particles it is also possible to work with systems of particles that attract each other by gravity, such as systems of celestial bodies. For this one simply has to check the corresponding button in the Environment (p. 197) inspector.

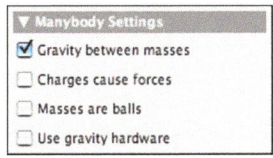

*Gravity between masses*

The following two pictures show two many-body systems subject to gravitational forces. The first one is a very remarkable figure-eight solution to the three-body problem (which actually is a sporadic example that was discovered only recently). The second picture shows a chaotic system of five bodies. For the simulation only the masses together with their initial velocities have to be drawn and the *Gravity between masses* button to be checked.

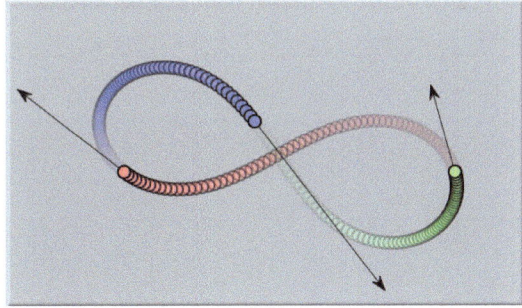

*An island of regularity*

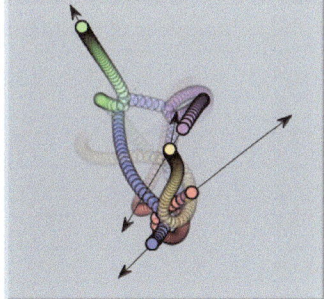

*Chaotic movement*

### 7.12.3  Masses as Balls

Finally, there is also a way in CindyLab to mimic the behavior of balls with nonzero diameter. For this one has to check the *Masses are balls* checkbox. Then each mass-object is equipped with a repulsive potential that is zero at great distances from the object but assumes very high values if the centers of the two mass-objects come within a distance that is close to the sum of their radii. For example, if two tennis balls meet, there will be an interaction between them for an instant of time. Although this approach does not exactly model the usual physical reality, it is still very appropriate for modeling balls and effects involving conservation of momentum.

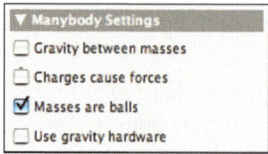

*Masses are balls*

The following picture shows a simulation of a billiard table shortly before a huge collision, shortly after the collision, and a long time after collision.

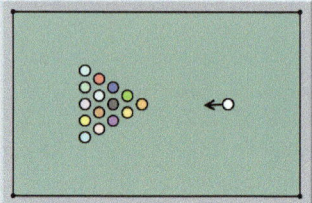 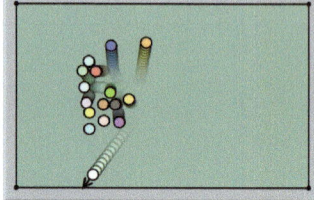 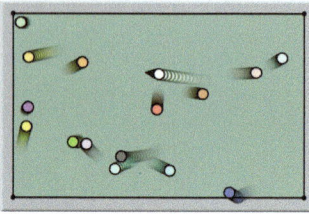

*Before...*                         *...shortly after...*                         *...long after*

The following example shows a standard experiment that exemplifies the conservation of momentum. Here, the masses of the pendulum are bound to circles in order to model the behavior of a stiff pendulum. Observe how the momentum is transferred from the red to the yellow ball. The radius of the force potential can be adjusted by the radius slider of the mass inspector.

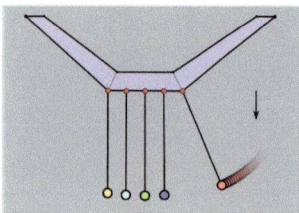 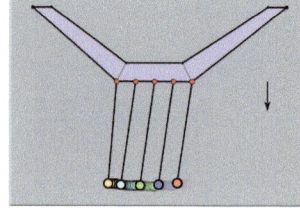 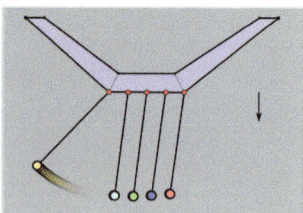

*Before...*                         *...during...*                         *...after*

### 7.12.4 Gravity hardware

It is also possible to control the gravity strength by a build in accelerometer of a laptop computer. This can be achieved by click on the 'Use gravity hardware' button. By this it is possible to control the behavior of the physical objects by actually physically moving the computer. A nice feature for game development.

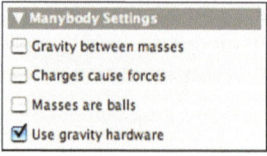

*Gravity hardware*

**Caution**

Many-body simulations may entail expensive (in terms of processing time) computations. In particular, whenever the force is attractive rather then repulsive, there may be many situations in which the automatic step-width adaptation of the CindyLab engine comes into play.

**See also**

- Environment (p. 197)
- Free Mass (p. 175)

## 7.13 Animations and CindyLab

Many devices in the real world are driven by some kind of motor. On an abstract level, a motor pumps energy into the system. It causes movement that goes beyond the pure interaction of free mass-objects. There are three principles in CindyLab that allow for the *external* actuation and control of a simulated environment.

- Actuation of springs
- Use of a geometric Cinderella Animation (p. 107)
- Use of a CindyScript script to change parameters (such as positions and velocities) of the system

In what follows we shall briefly sketch these three possibilities. Since the number of possible interaction scenarios is practically infinite, we shall again focus only on the basic principles and give a few illustrative examples.

### 7.13.1 Spring Actuation

Spring actuation is perhaps the most intrinsic way to drive a CindyLab simulation. It is discussed in the context of Spring (p. 189)s. It can be used to vary the resting length of a spring periodically. If the environment also supplies some friction, then one can use spring actuation to obtain points that exhibit periodic motion. As explained in the Spring (p. 189) section, one can alter the amplitude and the phase of the actuation independently. The picture below shows several springs all actuated with identical amplitudes but with differing phases.

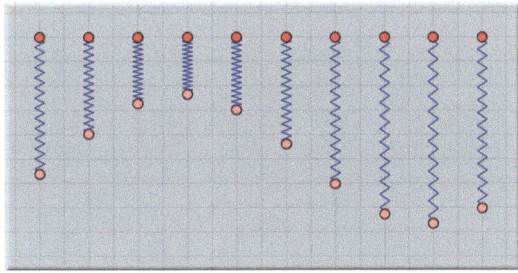

*Many actuated springs: same amplitude, different phases*

A typical application of actuated springs is the generation of legs for small walking devices, whereby one connects a point to several springs that are actuated with a particular phase shift. In the figure below, the two springs of each leg are actuated with a phase shift of 0.25. Each leg has a phase shift with respect to the other. This results in a cyclic walking behavior of the four legs.

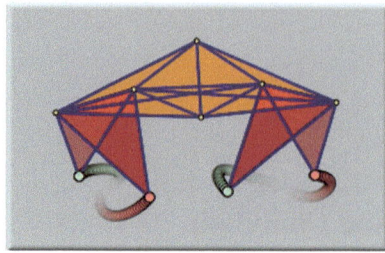

*Walking legs*

If this device is dropped onto a CindyLab Floor (p. 194), it automatically starts to walk like a little animal.

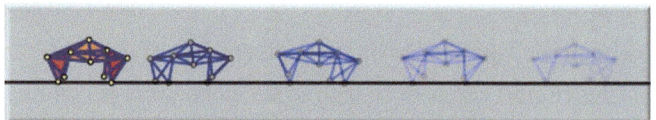

*A walker taking a stroll across the floor*

### 7.13.2  Geometric Animations

In the geometry part of Cinderella there is a standard way of animating points through the use of Animation (p. 107) mode. In this way, one can, for example, create rotating points on a circle or moving points on a segment. The animation mode of Cinderella works seamlessly with CindyLab. One can use animated points and equip them with physical properties or connect them to springs. In the picture below a kind of long elastic rope was created by connecting several rubber bands. One end of the elastic was fixed to the green point at the far left. The other end is driven by a Cinderella animation. The rotating point on the circle is used to move the end of the elastic up and down in periodic motion.

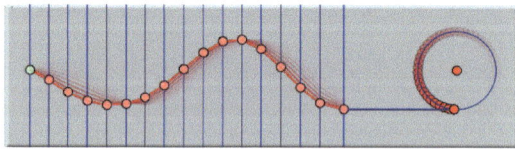

*Animation of an oscillating wave*

Mass-objects can also be used directly in animations. The following example is physically quite unrealistic, but it demonstrates a possible use of mass-objects in animations. Here a Sun (p. 183) is bound to a circle. An animation causes the sun to revolve around the circle. The picture shows the trace of a red planet under the attraction of a rotating yellow sun.

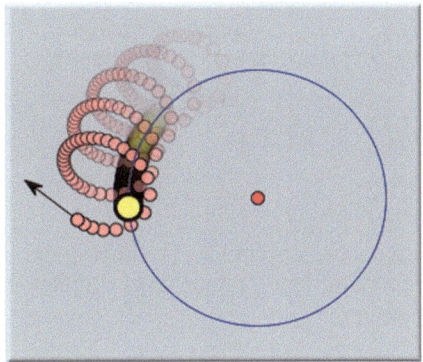

*A planet revolving about a revolving sun*

Animations are synchronized with the physics simulation. This means that if one moves the animation speed slider, then both the physics simulation and the animation itself become slower.

### 7.13.3 Driving Simulations with CindyScript

As usual, CindyScript provides the most flexible and powerful way of influencing a simulation. With CindyScript one can directly influence the parameters of Cindy-Lab objects such as *position*, *velocity*, and *restlength*. In this way, whenever the CindyScript code is executed, one can influence the behavior of the physics objects. As explained in the CindyScript manual, this scripting language allows execution to take place at a specific time. A detailed list can be found in the CindyLab and CindyScript (p. 170) section.

In principle, CindyScript can alter any physics behavior, since it has the "final word" on the parameters of the object. For instance, one can alter the speed of a particle programmatically. The following example shows a simple control loop for balancing a spring by moving one of its points.

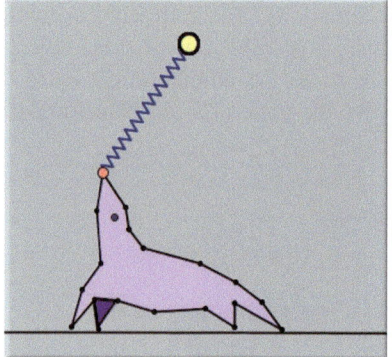

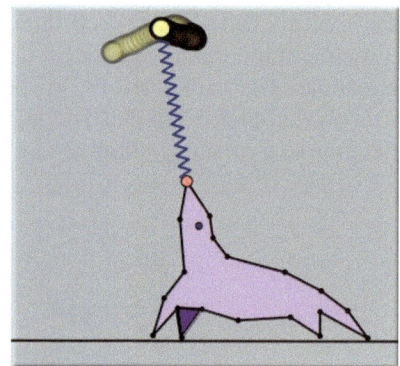

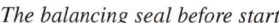

*The balancing seal before start*          *The balancing seal at work*

The tip of the nose of the seal is point C. The other endpoint of the spring is point D. The following lines of CindyScript code perform the balancing by changing the horizontal velocity of point C.

```
if(C.x < D.x,C.vx=C.vx+.05);
if(C.x > D.x,C.vx=C.vx-.05);
C.vx=C.vx+random(0.02);
```

The first line says that whenever the spring tends to fall to the left, the seal has to accelerate in this direction to prevent the spring from falling. The second line says exactly the same for the other direction. Finally, the last line adds a small degree of randomness to the movement of the seal. This makes the animation appear much more realistic.

Sometimes it is useful to perform CindyScript actions at specific moments of time. For this there are two different concepts of time within CindyScript. There is a function `simulationtime()` that is synchronized with the internal time of a simulation. Every time step of the simulation increases this value. Alternatively, one can refer to a timer that is synchronized with real time. For this, the CindyScript functions `time()` and `seconds()` are useful. For details on this concept please refer to the section *Time* on the page Special Operators (p. 381).

## 7.14 Geometry and CindyLab

In a sense this section is not much more than a reminder for the user of Cinderella. One should always be aware of the fact that in CindyLab (p. 167) all masses are still geometric points and all that springs, bouncers and floors are lines and segments. Thus it is always possible to enhance CindyLab (p. 167) experiments by geometric constructions. These geometric constructions can sometimes be extremely helpful in order to analyze the exact meaning of an experiment. We will illustrate this by three small examples.

### 7.14.1 Example 1: Equilibrium Condition for Forces

If a physical situation is in an equilibrium situation then none of the masses is accelerated. This however implies that in such a case at any mass there are no forces present. This in turn implies that if there are several sources of forces affecting at a particle, the all these forces have to sum up to zero. The experiment below exemplifies this situation for a simple network of rubber bands and springs. There a triangle of (blue) springs is constructed. All vertices of the triangle are joined to a central particle by rubber bands. In the start situation the lengths and directions of the rubber bands may be arbitrary. However, if one lets the System come to an equilibrium situation (by starting the simulation and adding some friction) then the forces caused by the rubber bands at the inner vertex have to cancel out. If all rubber bands have identical spring constants, then this implies that the three segments representing the rubber bands can be shifted parallel to form a triangle (sum of forces equals zero).

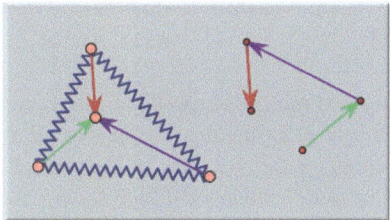
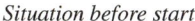
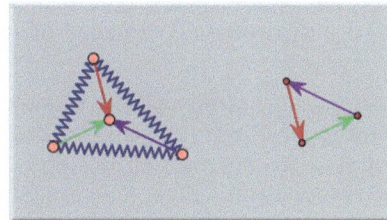

*Situation before start*          *Situation in equilibrium*

This can simply be tested by a small auxiliary construction that concatenates parallel shifts of the three segments (left part of picture). In the start situation one observes that this chain of three arrows does not necessarily have to meet. The second picture shows the equilibrium situation in which the three edges automatically form a perfect triangle.

Sometimes very surprising and interesting relations occur on equilibrium configurations. For instance, the next picture shows the situation of three springs that are surrounded by six rubber bands. The rubber bands may have arbitrary spring constants. In the equilibrium situation the six vertices of the construction will automatically lie on a conic.

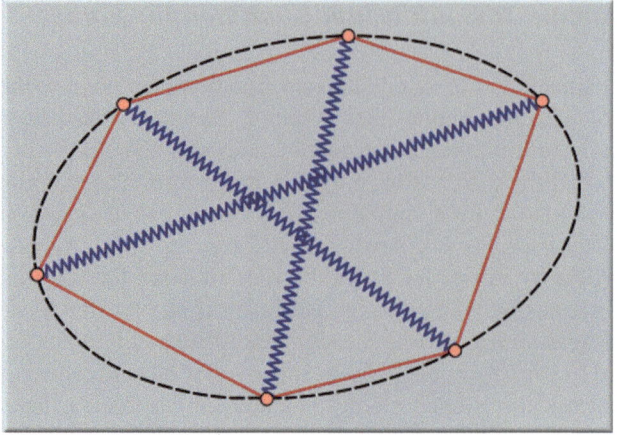

*A surprising equilibrium condition*

## 7.14.2  Example 2: Planet Movement

There is an amazing and not very well known geometric property of planet movement in a two-body system. As already mentioned, a planet orbiting around a sun
will describe the trace of an ellipse. If we consider the velocity of the planet it is
fast whenever the planet is close to the sun and slow whenever it is far away. It also
changes its direction all the time during the movement. In Cinderella it is easy to
explicitly study the path of the velocity vector. For this one first adds a Sun (p. 183)
and a particle with Velocity (p. 178). If one starts the simulation the particle will
move along a Kepler ellipse. Then one adds a free point and defines a Translation
(p. 125) from the planet to this point. Now one can easily translate the velocity
vector to this and obtain a picture of the isolated behavior of the velocity vector.

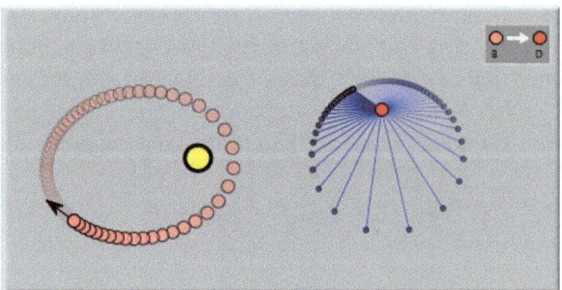

*The trace of the velocity vector*

Viewing the trace of the velocity vector one might conjecture that the trace is circular. It is easy to at least visually verify this conjecture by simply adding a free circle
to the drawing and moving it while the animation is running to a position in which
it matches the trace of the velocity vector. The picture below shows how nicely it
matches up.

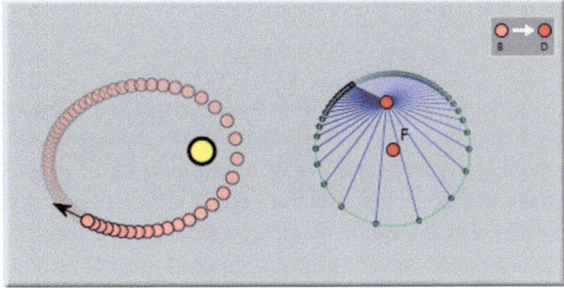

*The trace is circular*

### 7.14.3 Example 3: Forces at the Golden Gate Bridge

The picture below shows an example that was already considered in the section on Gravity (p. 180). Here a background image of the Golden Gate Bridge is loaded. With the use of CindyLab (p. 167) a physics simulation is constructed that models the distribution of forces in the cables of the bridge. Cinderella is used to straighten out the perspective of the photograph. This can be done by using a Projective Transformation (p. 131) that maps the situation in the physics simulation to the situation in the picture. Adjusting the gravity to the correct value shows that this situation exactly resembles the situation on the supporting cables of the bridge.

*An analysis of the Golden Gate Bridge*

## 7.15 Scripting Physical Environments

There are many ways in which CindyScript can be used to enhance the possibilities
of CindyLab. We will present a few of them here. On the one hand, CindyScript provides several special operators that have been implemented for use with CindyScript
(for instance, an oscilloscope for viewing physical magnitudes and a statement for
plotting force fields). On the other hand, CindyScript has direct access to most of
the parameters that are relevant for physics simulations. Moreover, CindyScript can
enable or disable physical functionalities of CindyLab objects. In this way one can
implement machine-like behavior in CindyLab scenarios. We will give several examples here that use CindyScript together with CindyLab.

### 7.15.1 The Oscilloscope

The statement `drawcurves(position, list_of_values)` generates a region in
the geometric view in which the development of magnitudes can be shown. Here
*position* is the position at which the oscilloscope is drawn and *list_of_values* is a
CindyScript list that contains several magnitudes to be represented graphically. The
simplest use of this operator is presented by the following program:

```
drawcurves((0,0),[B.x,B.y])
```

Here the *x* and *y* coordinates of point B are shown. If B is a point moving around a
sun, the picture might look like the one below.

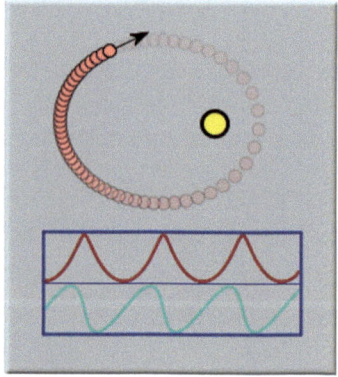

*Curved plots of planetary motion*

By default, the display range of the curve is automatically adapted to the range
of values. The `drawcurves(...)` operator supports many different modifiers that
can be usedto set background colors, curve colors, display texts, as well as many
additional parameters for drawings. You will find a detailed description in the
CindyScript reference manual. The picture below shows the use of the `drawcurves`
operator applied to a coupled harmonic oscillator.

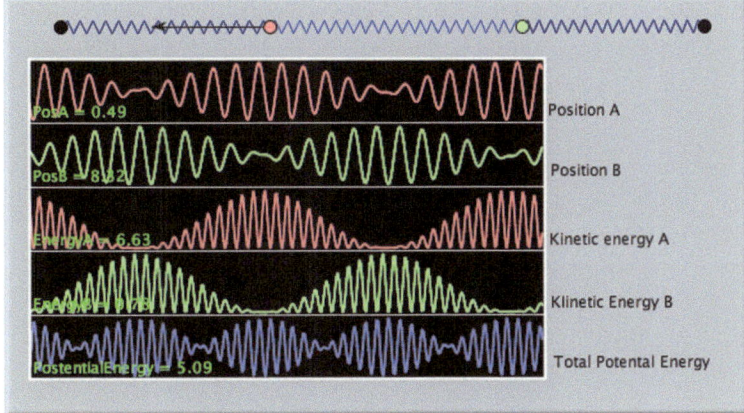

*The movement of a coupled oscillator*

Here many modifiers were used in order to get the desired color behavior and the text display. The example was constructed using the following CindyScript program:

```
linecolor((1,1,1));
drawcurves((-7,-3),
textcolor((1,1,1));    (A.x,B.x,A.ke,B.ke,a.pe+b.pe+c.pe),
  height->50,
  color->(1,1,1),
  back->(0,0,0),
  backalpha->1,showranges->true,
  range->"peek",width->400,
  colors->[
    [1,0.5,0.5],
    [0.5,1,0.5],
    [1,0.5,0.5],
    [0.5,1,0.5],
    [0.5,0.5,1]],
  texts->[
    "PosA = "+ A.x,
    "PosB = "+B.x,
    "EnergyA = "+A.ke,
    "EnergyB = "+B.ke,
    "PotentialEnergy = "+(a.pe+b.pe+c.pe)
]);
```

## 7.15.2 Force Fields

Another CindyScript operator that was designed to support CindyLab is the `drawforces` operator. With this operator one can visualize the direction and strength of forces all over the drawing area. To accomplish this, a virtual probe particle moves across the screen, measuring the forces at different points. If one simply writes in CindyScript the line

```
drawforces()
```

then all gravitational forces, Coulomb forces, ball interactions, etc. are measured and the flux is drawn. The probe particle is standardized to have mass = 1, charge

= 1, and radius = 1. If, for instance, two charged particles are present with opposite charges, then the `drawforces()` operator causes the following flux picture:

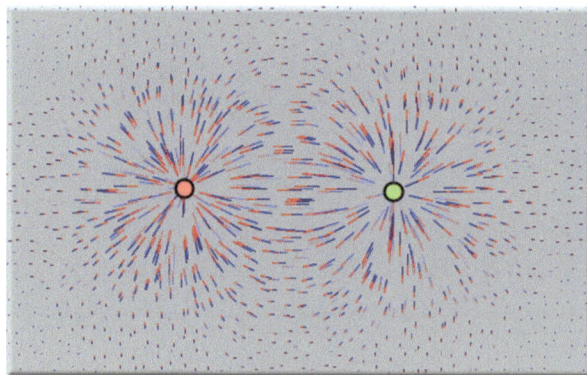

*A simple electrostatic field*

Like the `drawcurves()` operator, the `drawforces()` operator supports many modifiers. They are explained in detail in the CindyScript reference. Here we just present a slightly more advanced example.

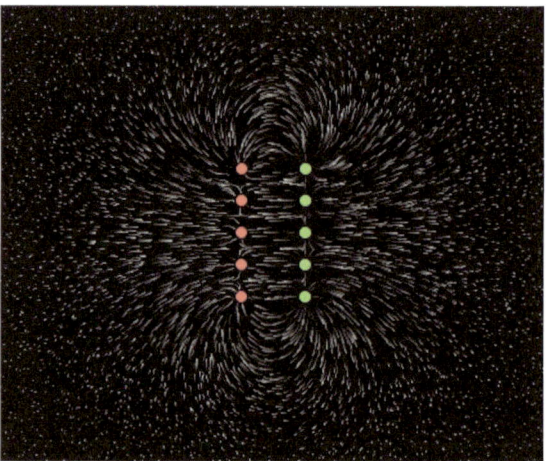

*Another simple electrostatic field*

The code fragment that generated this flux picture uses several modifiers.

```
drawforces(stream->true,
           color->(1,1,1),
           factor->3,
           resolution->5)
```

In particular, the `stream->true` statement replaces the usual display of force fields with small needles with little streamlets that give a more accurate picture of the flux.

### 7.15.3 Programming CindyLab Scenarios

Since CindyScript can influence all the parameters of CindyLab, it is also easy to implement complicated behavior. As an example, we present here an implementation of a simple table tennis game. First, a few bouncers are used to draw the boundary of the table. Another bouncer is constructed geometrically that is bound to a point so that it can be freely moved like a table tennis paddle. Finally, a table tennis ball is added as a particle with velocity. So far, the construction makes use only of CindyLab and Cinderella. The situation is already quite functional. One can indeed play table tennis. However, if the ball leaves the screen, it will be lost forever.

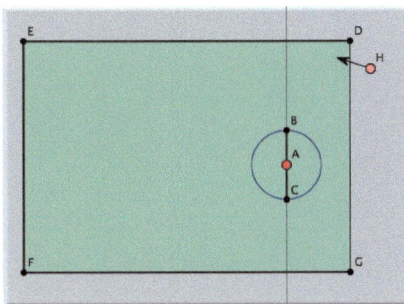
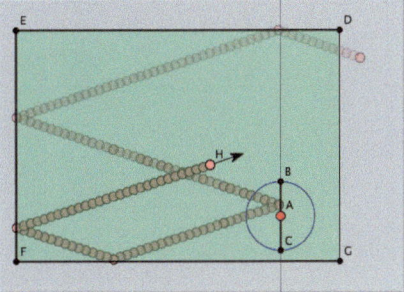

*Construction of a table tennis game*        *A lost ball*

This is the point at which CindyScript comes into play. With CindyScript one can detect whether the ball is leaving the screen and react to such an event. The following program detects the ball leaving the screen with the code line `if(H.x>D.x+9,....`
If this happens, a new point with random position, random speed, and random color is inserted at the end of the table. In addition, a counter is used to count how many balls have been lost.

```
if(H.x>D.x+9,
  H.x=F.x;
  r=random();
  H.y=(r*D.y+(1-r)*G.y);
  w=random()*pi/2+3*pi/4;
  H.vx=-cos(w)*.4;
  H.vy=sin(w)*.4;
  H.color=(hue(random()));
  count=count+1;
);
drawtext((-1,-3),"Missed balls: "+count,size->20);
```

In order to initialize the counter, there must also be one additional line of code in the *simulation start* tag that sets the counter to zero.

```
count=0;
```

The picture below shows a snapshot that was taken during a game. Observe how the yellow ball leaves the game and the blue ball is thrown in.

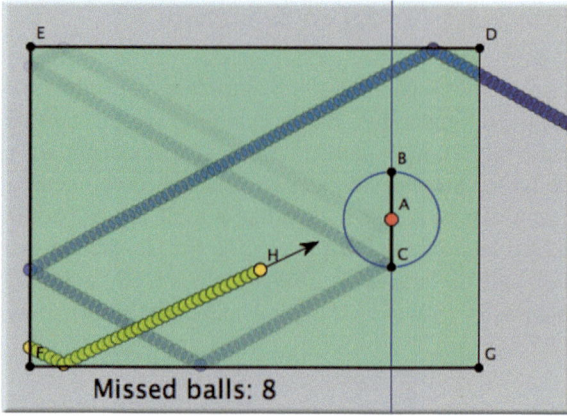

*Scripting the table tennis game*

# Part IV
# CindyScript Reference

# Chapter 8

# CindyScript Fundamentals

Cinderella comes with an easy to learn yet still very powerful functional language. It is designed primarily with the intention to allow for high-level interaction with geometric constructions or physical simulations created in *Cinderella.2*. Nevertheless, it can also be used as a standalone language for performing mathematical calculations.

There are three ways to use the CindyScript language within *Cinderella.2*. First of all, one can write CindyScript expressions as parts of functions (p. 291) in a *Cinderella* construction. Besides, one can enter CindyScript commands in a command shell to, e.g., immediately set properties such as color or size of geometric elements.

However, CindyScript is most commonly used to write programs in the script editor (p. 395). Here, one can specify the occasion on which the program will be executed. For instance, it can be executed whenever the user moves the construction or after any mouse click. Thus one can easily add functional behavior and graphical enhancements to an existing construction.

The following sections will give you an overview of the global design of the programming language CindyScript. The language design follows some guiding principles:

- The language should be easy to learn, write and read

- The language should have only minimal syntactic overhead

- The language should be fast in execution

- The language should interact seemlessly with Cinderella and CindyLab (p. 167)

The following topics will give you an overview over the main language features:

- *General Concepts* (p. 220)*:* Main functionality of the language

- *Entering Program Code* (p. 225)*:* How to write and edit a program

- *Variables and Functions* (p. 229)*:* Declearing, destroying and scope of variables and functions

- *Accessing Geometric Elements* (p. 234)*:* How to interact with Cinderella and CindyLab (p. 167)

- *Control Operators* (p. 243)*:* How to create control structures with `if`, `repeat`, `while`,...

- *Arithmetic Operators* (p. 248)*:* Dealing with numbers (+, -, *, /, sin(), cos(),...)

- *Boolean Operators* (p. 253)*:* Logic statements

- *String Operators* (p. 257)*:* Dealing with strings of characters

For a detailed description of the language it is necessary to consult the documentation on specific parameters. We recommend to browse over the CindyScript manual at least once to get an impression of various possibilities of the language.

## 8.1 General Concepts of CindyScript

This section is a brief introduction to the most fundamental concepts of CindyScript.

### 8.1.1 CindyScript Is a Functional Language

All calculations in CindyScript are performed by executing functions. A function can be considered as a kind of calculation that takes the arguments of the function and produces some kind of output value. Many calculations can already be expressed using only elementary functions. Thus the code fragment

```
sum(1..10)
```

calculates the sum of the first 10 integers. Here `..` is a function that takes two integer numbers, `a` and `b`, and generates as output the list of all integers from `a` to `b`. Thus `1..10` generates a list of 10 integers: `[1,2,3,4,5,6,7,8,9,10]`. The function `sum(_)` is unary (that is, it takes a single argument). It takes as its argument a list of numbers as input and generates as output a number that corresponds to the sum of the list entries. Thus if we type `sum(1..10)` into the command shell, the system will respond with the result `55`.

Moreover, seemingly procedural statements, such as an `if`-statement, are realized as functions. For instance, the expression

```
if(x~60~y,print("Mine"),print("Yours"))
```

demonstrates the function `if`, which takes three arguments. It checks the condition of its first argument `x<y` and depending on the result, evaluates the second or the third argument, that is, either `print("Mine")` or `print("Yours"))`. The result of this evaluation will be the result of the `if(_,_,_)` function. Thus the above expression is equivalent to

```
print(if(x~60~y,"Mine","Yours"))
```

Depending on the evaluation of the condition, the `if` function returns the value of the second argument or the third argument.

## 8.1.2 Side Effects

If a function is evaluated in CindyScript, it may have "side effects." Side effects are important for all kinds of interactions between a CindyScript program and a Cinderella construction. Typical side effects are:

- *Drawing:* A CindyScript statement may cause drawing operations in the construction views.

- *Assignments:* A CindyScript operation may change the position, color, size, etc. of geometric objects.

- *Variable assignments:* A CindyScript statement can create variables and assign values to them.

- *Function creation:* A CindyScript statement can create and define a function that can be used later.

For instance, the statement

```
draw([0,0])
```

produces the side effect of drawing a point at position (0,0). The statement

```
A.color=[1,1,1]
```

sets the color of point *A* to *white*.

## 8.1.3 Control Flow

Most users are probably accustomed to sequential programming languages like C, Java, Pascal, and Basic. In practice, writing sequential code in CindyScript is not so different from writing code in these languages. CindyScript has a ; operator <statement1>;<statement2> that simply first evaluates statement1 and then statement2. The return value of the ; operator is the result of the last statement. Writing a sequential program is relatively simple, and it looks similar to a program written in a sequential language. For instance, the program

```
repeat(9,i,
    j=i*i;
    draw([i,j]);
)
```

creates nine points on a parabola. The function repeat(<number>, <variable>, <program>) creates a loop that performs <number> runs. In each run the variable <variable> is incremented (starting with 1). The body of the loop is the two lines j=i*i; draw([i,j]);.

### 8.1.4 No Explicit Typing

CindyScript is designed to provide a maximum of functionality with a minimum of syntactic overhead. Therefore, CindyScript does not have explicit typing of values. Like many other languages, CindyScript uses the concept of variables. However, in contrast to other languages, the variables do not belong to a specific type. Any value of any type can be assigned to any variable. On the one hand, this gives the programmer a great deal of freedom to generate powerful code. For instance, the following code fragment

```
f(x,y):=x+y;
```

defines a function `f(x,y)` that could be used to add integers or complex numbers as well as vectors or matrices. On the other hand, this freedom requires that the programmer take some responsibility while writing a program in order to produce code that is semantically meaningful. When a function tries to evaluate a meaningless expression, the program will not automatically terminate. Instead, the function will return the value ___, which stands for a meaningless expression. So, in the above example, `f([1,2],[3,4])` will perform a vector addition and evaluate to `[4,5]`, whereas the expression `f(4,[3,4])` is meaningless and evaluates to ___.

### 8.1.5 Local Variables: The # Variable

There are several loop-like constructions in CindyScript. For instance, the operator `select(<list>,<condition<)` traverses all elements of `<list>` and returns a list of objects that satisfy the condition. For this to occur, there must be a way to feed elements that are to be tested to the condition. By default, CindyScript uses a variable #, which serves as a handle for the run variable. For instance, the statement

```
select(1..30,isodd(#))
```

returns a list of all odd numbers between `1` and `30`. Moreover, loops use this run variable, and thus

```
repeat(9,print(#))
```

prints all numbers from `1` to `9`. It is also possible to use an explicit run variable by providing it as the second argument. Thus `select(1..30,i,isodd(i))` and `repeat(9,i,print(i))` are equivalent to the above statements.

### 8.1.6 The Data Types of CindyScript

As already mentioned, CindyScript does not have explicit typing. Nevertheless, any *value* of a variable belongs to an explicit type. The basic types of CindyScript are

- <number>: Any numeric value. Numbers can be integers, real numbers, or complex numbers.

- <list>: A list of arbitrary objects. Such a list may semantically also have the meaning of a vector or matrix.

- <string>: A text expression.

- <geo>: A geometric object that belongs to a construction.

- <boolean>: A value `true` or `false`.

The number type is particularly powerful, since it can contain integers, floating-point numbers, and complex numbers.

### 8.1.7 Variables and Their Scope

Since CindyScript does not have explicit typing for variables, it allows variables to be created "on the fly" as needed. A variable is created when it is assigned for the first time. If x is not already being used, the statement

```
x=7
```

creates the variable x and assigns the value 7 to it. After a variable has been assigned, its value is accessible for the rest of the execution. Values may also be partially overloaded by local variables of a function. Thus in a function defined by

```
f(x,y):=x+y
```

the values of x and y are the local parameters of the function. After the execution of the function is completed, the original value of x is restored. One can also produce additional local variables with the `regional(...)` operator.

### 8.1.8 Access to Geometric Elements and Their Properties

Variables are also used as a kind of handle to geometric objects. They form a major link of CindyScript to Cinderella and CindyLab. If a variable has a name that is identical to the label of a geometric object, it provides a link to that geometric object. The value of the variable can still be overloaded by an explicit assignment of a value to the variable. The different properties of a geometric object (position, color, size, etc.) are accessible via the . operator. Thus if A is a point in a geometric construction, the expression A.size returns an integer that represents the size of the point. The expression A.xy=[3,4] assigns the point to the coordinate [3,4]. Furthermore, properties relevant to physics simulation (mass, velocity, kinetic energy, etc.) are accessible via the . operator.

### 8.1.9 Modifiers

Many operators in CindyScript provide more functionality than one may notice at first glance. Usually these features can be accessed using so-called modifiers. The operators are defined in a way such that their default usage provides a suitable behavior for most situations. However, it may be necessary to modify the default behavior. To that end, one lists corresponding modifiers in the call of the operator. For instance, the statement

```
draw([0,0])
```

draws a point at position (0,0). By default, the point is green and of size 3. The statement

```
draw([0,0],size->15,color->[1,1,0])
```

draws a yellow point of size 15. Modifiers have to be separated by commas. They may occur in any order and at any position of the function call.

### 8.1.10 Lists/Vectors/Matrices

CindyScript offers *lists* as elementary data types. Lists are the fundamental paradigm that is used to define more complex data structures. In addition to the obvious application as enumeration objects, lists can also be used to represent vectors and matrices. A vector is a list of numbers. A list of vectors whose vectors all have the same length will be interpreted as a matrix. CindyScript provides the usual operations for combining vectors, matrices, and numbers. Depending on the content of a and b, the expression a*b may represent a usual multiplication of numbers, a matrix product, or a matrix/vector multiplication.

In CindyScript there is no distinction between row vectors and column vectors on the level of vectors. However, by the use of suitable functions one can convert a vector of length n to an $(n \times 1)$ matrix or to a $(1 \times n)$ matrix.

### 8.1.11 Drawing

CindyScript provides many statements with which one can draw directly on the canvas of the geometric views. Using this feature it is possible to enrich the behavior of Cinderella constructions significantly. It is possible to draw points, lines, segments, polygons, tables, functions, etc. However, it is important not to confuse a script-drawn geometric object with a geometric object that is active in geometry. It is not possible to use such script-drawn elements as definers in Cinderella modes.

If one wants to modify active elements using a script, then it is necessary to first construct them and then alter their positions using CindyScript statements. All free elements can be moved by setting their position parameters.

### 8.1.12 Execution Slots

The script window of Cinderella (p. 395) in which one enters the CindyScript code contains several slots in which the text can be entered. The particular slots are called

- Draw
- Move
- Initialization
- Timer Tick
- Simulation Start
- Simulation Stop

- $\quad\ldots$

Each of these entries corresponds to the occasion that triggers the execution of the script. For instance, scripts in the *Draw* slot is executed directly before a screen refresh in the view. The *Initialization* slot is executed directly after the CindyScript code is parsed. *Simulation Start* is executed before starting an animation when the play button is pressed. Using this mechanism it is possible to write programs that react nicely to user events.

### 8.1.13 Runtime Error handling

CindyScript runs in a runtime environment. In principle, every tiny move in a construction can cause the evaluation of a script. For this to happen, a reasonable design decision in the language had to be made concerning the occurrence of runtime errors. It would be very distractive if the usual user interaction was interrupted by error messages over and over (in particular, if a construction is used as an applet within an HTML page). For this reason, error handling in CindyScript at runtime reports only the first ten errors. However, runtime errors will never interrupt execution. Runtime errors (such as division by zero, or access to a nonexistent array index) are simply ignored in the program flow. Erroneous function evaluations simply produce an undefined result, and the calculation proceeds (perhaps causing more undefined results). This usually guarantees fluent performance of a construction even if errors occur.

This feature may make debugging of programs a little cumbersome, since runtime errors are not reported. For this purpose a special function, `assert(<boolean>,<string>)`, was introduced to check whether a certain assumption about the current data is satisfied. If the assumption in the first argument is not satisfied, the message in the second argument is printed.

## 8.2 Entering Program Code

The editor for CindyScript code can be opened by selecting the menu item *Scripting/Edit Scripts*. When this item is selected, a window for the program code editor is opened. Each construction has an individual program editor window. When first opened, the editor window looks as follows:

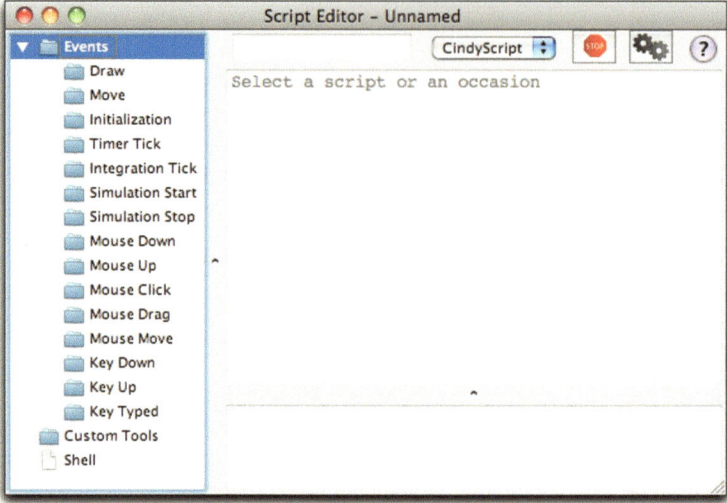

*The script editor*

This window contains several components that require explanation. The main part of the window is occupied by a large white space (the middle part of the right side). This is the text field in which you will enter your script later on. The left side of the window is occupied by a "tree view," in which you can select an occasion upon which a script should be executed. The first step in entering a script consists in choosing a suitable occasion.

The lower part of the right side is reserved for console output, while the top part of the window contains some controls and a field in which you can enter a name for your script.

### 8.2.1 Entering a Small Script

The first step in entering a script consists in choosing the occasion on which the script is to be executed. For this, one has to click on one of the "Events" shown on the left. Let us choose the "Draw" event. A script entered in the Draw event will always be executed when drawing is performed on the screen. After this entry has been selected, the main part of the window is ready for input. Clicking in the main window allows one to enter script code. In the example shown below a small script was entered and given the title "My first script"

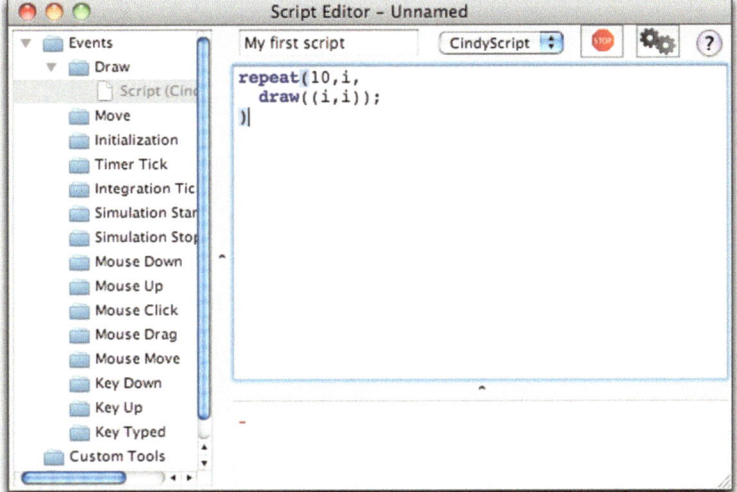

*Entering text*

Pressing the button with the two gears executes the script. In the above example a green dotted line will be drawn consisting of ten simple dots. Alternatively the execution can be started by pressing <shift>+<enter>.

### 8.2.2 Execution Slots

Every script in Cinderella is associated with an occasion on which it will be executed. Since scripts in Cinderella are assumed to run during the runtime of the program, it may well happen that the same script is executed many times during a move. The precise interpretation of the different occasions is described below:

- *Draw:* A script in this slot is executed right before a new screen picture is generated. Scripts entered here will be executed very often. This is the typical place to enter a script that should be automatically updated when the user drags geometric elements.

- *Move:* Scripts in this slot will be executed even more often than those in the "Draw" slot. They are always invoked if the position of free elements has changed. In general, this happens more often than just at screen refresh. Sometimes putting a script in the "Draw" slot produces strange and unexpected results. It might then be a good idea to place it in the "Move" slot.

- *Initialization:* A script in this slot is executed whenever the parse button of the script editor is pressed. This slot is very useful for resetting variables or points to an initial state or position.

- *Timer Tick:* A script in this slot causes an animation (p. 107) controller to be shown in geometric view (if not already present). When the play button of the animation controller is pressed, a script in this slot will be regularly executed every few milliseconds.

- *Integration Tick:* This event is only relevant in context with physics simulations. If a script is contained in this slot it will be executed whenever the internal physics simulation engine requires information on the position and forces of masses. This slot is needed when one wants to implement user specific force potentials.

- *Simulation Start:* This script is executed when the animation controller changes from "Stop" to "Play." It is a very good place to enter initial setups for animations.

- *Simulation Stop:* This script is executed when the animation controller changes from "Play" to "Stop." It is a very good place to evaluate the result of an animation.

- *Mouse Down:* This slot and the following three slots are very useful for programming user interfaces in CindyScript. A script in this slot is executed whenever the mouse button is pressed. With the function `mouse()` the current mouse coordinates can be read within a script.

- *Mouse Up:* Scripts in this slot are executed when the mouse is released.

- *Mouse Click:* Scripts in this slot are executed when the mouse is moved.

- *Mouse Drag:* Scripts in this slot are executed when the mouse is dragged.

- *Key Typed:* Scripts in this slot are executed when the key on the keyboard is typed. This slot is very useful for handling user input via the keyboard. A string that corresponds to the currently pressed key can be accessed via the `key()` and via the `keydownlist()` functions.

- *Key Down:* Scripts in this slot are executed when the key on the keyboard is pressed, exactly at the moment when the key moves down.

- *Key Up:* Scripts in this slot will be executed when the key on the keyboard is released, exactly at the moment when the key moves up.

### 8.2.3 Control Buttons

In the upper part of the script editor window are two buttons to start and stop the program, as well as a text field to provide names to parts of the program and a language chooser for switching between different languages (at present, only CindyScript is documented).

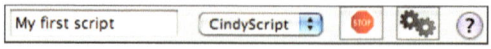

*The script controls*

Perhaps the most important part of these controls is the "Play" button (the one with the two gears). Whenever this button is pressed all scripts are parsed and the execution of the program is started. Every change in a program requires the play button to be pressed to become active. Alternatively, the key combination <shift>+<enter> can be used as an equivalent for pressing the play button.

It may happen that a script gets stuck in an infinite loop or performs a calculation that simply takes too long. As a kind of emergency exit, the "Stop" button can be pressed.

The little text field on the left can be used to give a script an individual name. Scripts with different names can be accessed and edited separately.

Finally, there is a drop-down menu for choosing the language for the script. In this documentation you will find all necessary information for programming in CindyScript. However, it is also possible to program in "Python", "JRuby" or in the core Cinderella geometry language. These options will be documented later and elsewhere.

You find more information about the ways to enter code in the corresponding section of the reference manual (p. 395).

## 8.3 Variables and Functions

In CindyScript, variables and functions are not declared explicitly. They are created on demand and are not explicitly typed. This is in sharp contrast to many other programming languages. In this section you will learn under what circumstances one can create functions and variables. You will also learn how to destroy or clear variables and about their scope.

### 8.3.1 Defining Functions

Defining a function in CindyScript is very easy. One simply has to specify the name of a function, provide a parameter list, and write down the body of the function. No explicit typing of arguments or function values is required. In what follows, we provide some examples of simple functions. For example, function f defined by

```
f(n):=sum(1..n,i,i^2)
```

calculates the sum of the first n squares. For instance, after this definition, f(4) evaluates to 30.

Functions with more than one argument can be defined similarly. The following function assumes that a and b are two-dimensional vectors and draws a square whose edge is defined by these two vectors:

```
sq(a,b):=(
  n=(b-a);
  n2=(-n_2,n_1);
  draw(a,b);
```

```
   draw(a,a-n2);
   draw(b,b-n2);
   draw(a-n2,b-n2);
)
```

In this code a few interesting things happen. First of all, the code is in principle procedural. The body of the function has the form (statement_1;...;statement_k). Furthermore, the function uses the variables n and n1. These variables are created when the function is first called. However, they are (by default) not local. Their values are visible also after the function has been called.

The return value of a function is the value of the last evaluated statement in the function. Thus the following function calculates the arithmetic mean of three entries.

```
mean(a,b,c):=(                     .
   sum=a+b+c;
   sum/3;
)
```

Since functions are not explicitly typed, it is also possible to pass more complex objects as a function's arguments. The function is automatically as polymorphic as possible, restricted only by the generality of the operations used in the function. For instance, mean([3,4],[2,7],[4,7]) evaluates to [3,6].

### 8.3.2 Recursive Functions

Functions may also be defined recursively. Then a new instance of every function parameter is created for each level of recursion. The following code calculates the factorial of a number:

```
fac(n):=if(n==0,1,n*fac(n-1));
```

The following more complicated code calculates the greatest common divisor of two positive numbers:

```
gcd(a,b):=if(b==0,           //End of recursion reached
             a,              //Then return the number a
          if(b>a,            //Perhaps switch parameters
            gcd(b,a),        //switched version
            gcd(b,mod(a,b))  //Recursion
            )
          );
```

### 8.3.3 Defining Variables

Variables in CindyScript are defined on their first occurence in code. Once defined, variables remain accessible throughout the rest of the program. A variable may contain any type of object (numbers, strings, booleans, lists, geometric points, or even programs, ...). The program

```
x=3;
b=[x^2,x^3];
c=2*b;
```

assigns to x the value 4, to b the value [9,27], and to c the value [18,54]. A variable defined in a function remains visible also outside the scope of the function. Exceptions to this rule are the parameters of the function and variables explicitly defined local. The following program exemplifies the scope of variables:

```
f(x):= (
   x=x+x;
   println(x);
   y="User"
);
x="Hello ";
y="World";
println(x+y);
f(x);
println(x+y);
```

It produces the output

```
Hello World
Hello Hello
Hello User
```

Local variables in a function may be defined explicitly using the regional(...) operator. They are automatically removed when the function terminates. In the following code snippet, as a slight variation of the above program, y is defined to be a local variable within the function:

```
f(x):= (
   regional(y);
   x=x+x;
   println(x);
   y="User";
);
x="Hello ";
y="World";
println(x+y);
f(x);
println(x+y);
```

The program produces the output

```
Hello World
Hello Hello
Hello World
```

Run variables in loops are also treated as local variables.

### 8.3.4 Binding Variables to Functions

Variables in a function (unless defined as local variables) remain visible after the execution of the function. Besides, variables used in functions may have initial values that influence the evaluation of the function.

For instance, the following piece of code

```
a=3;
timesa(x):= x*a;
println(timesa(2));
a=5;
println(timesa(2));
```

produces the output

```
6
10
```

The return value of `timesa(2)` depends on the actual value of the (global) variable a at the moment the function is evaluated. So after redefining a the behavior of the function `timesa` changes. Sometimes this effect is intended, sometimes it is not. It may happen that one wants to freeze the behavior of a function to depend on the values of the variables at the moment when the function was defined. This can be achieved by using the operator = to define the function. This operator copies the entire variable assignments and binds them to the function. Therefore, the program

```
a=3;
timesa(x)::= x*a;
println(timesa(2));
a=5;
println(timesa(2));
```

produces the output

```
6
6
```

Every time the function is called, the original value of a is restored. This binding process does not only extend to all variables used in the function itself. It extends to all variables that may be relevant to the execution of the function.

There is one way to intentionally circumvent this binding: The value of a can be set explicitly using a modifier. An example thereof can be seen in the following piece of code:

```
a=3;
timesa(x)::= x*a;
println(timesa(2));
println(timesa(2,a->10));
```

This program fragment produces the following output

```
6
20
```

## 8.3.5 Predefined Constants

In mathematics it is often necessary to use mathematical constants like `pi` or the imaginary unit `i`. These constants are predefined as variables in CindyScript. This allows to write a complex number for instance as `3+i*5`. However, different values can be assigned to those variables. For example, it is still possible to use these variables as run variables in loops. The following program illustrates this feature:

```
println(i);
repeat(4,i,println(i));
println(i);
```

It produces the following output:

```
0 + i*1
1
2
3
4
0 + i*1
```

If, for instance, the complex unit is needed but the variable `i` is overwritten, then it is still possible to access the complex unit using the function `complex([0,1])`. Other predefined variables are `true` and `false` for the logical constants, as well as the empty list, `nil`.

There is another important type of predefined variable. Any geometric element in a construction may be referred to as a predefined variable of the corresponding name. Thus, for instance, a point *A* can be accessed using variable `A`. More detailed information on this topic may be found in the section on Accessing Geometric Elements (p. 234).

### 8.3.6 *User Defined Data*

There is also a possibility to associate user defined data to geometric elements. This can be done by the `:` operator. This is a simple but very powerful feature. After the colon an arbitrary string value can be added as a key to access the data. This key serves as a variable to which arbitrary values may be attached.

The usage of this operator is best explained by a examples. Assume that *A* ad *B* are geometric objects. The following code associates some data to them:

```
A:"age"=17;
B:"age"=34;
A:"haircolor"="brown";
B:"haircolor"="blonde";
```

The data may be accessed by the same key. So the following code

```
  forall(allpoints(),p,
    println(p:"age");
    println(p:"haircolor");
)
```

will produce the output

```
17
brown
34
blonde
```

A list of all keys of a geometric object may be accessed via the `keys(...)` operator. So in the above example the code

```
print(keys(A));
```

will produce the following output:

```
["age","haircolor"];
```

It is also possible to attach key information to lists. By this one can also create custom data that is passed by variables. The following code exemplifies this behavior.

```
a=[];
a:"data"=18
print(a:"data")
```

*Caution:* The functionality of attaching key data is still subject to change. It is planned to support object like data structures. So the currently implemented feature may not be compatible with future releases.

## 8.4  Accessing Geometric Elements

The main communication between CindyScript and the geometry part of Cinderella is accomplished by accessing the geometric objects of a construction. Geometric elements can be accessed in two different ways: either one can access an element by the name of its label or one can access lists of elements by special CindyScript operators. The interaction between Cinderella and CindyScript gives CindyScript the possibility to read essentially all properties of the elements of a geometric construction. Most properties can also be set by CindyScript. The following sections first describe the possible ways to address geometric objects and then provide a detailed list of the supported properties.

### 8.4.1  Accessing Elements by Their Names

Every element in a geometric construction has a unique name, its label. This name can be used as a handle in CindyScript. Formally, in CindyScript the element plays the role of a predefined variable. The different properties can be read and set via the . operator (dot operator). For instance, the line of code

```
A.size=20
```

sets the size of point *A* in a construction to the value 20. If a point or a line is contained in an arithmetic operator without a dot operator, then it is automatically converted to a vector representing its position. Thus a point is converted into an $[x,y]$ vector representing its two-dimensional coordinates. A line is converted to an $[x,y,z]$ vector representing its homogeneous coordinates. However, if one intends to set the coordinates of a point, then one has to use the dot operator explicitly. If a handle to a geometric object is not used in an arithmetic expression, then it is still passed to the calculation as the geometric object. Since these concepts are a bit subtle, we will clarify them with a few examples.

Assume that *A*, *B*, and *C* are points in a Cinderella construction. The line

```
A.xy=(B+C)/2
```

sets the point *A* to be the midpoint of *B* and *C*. These two points are contained in an arithmetic expression, and therefore they are immediately inverted to an `[x,y]` vector. Setting the position of point *A* has to be done by explicitly using the `.xy` property.

The following program sets the color of all three points to green:

```
pts=[A,B,C];
forall(pts,p,
  p.color=[0,1,0];
)
```

In this code the point names are passed as handles to the list `pts`. Traversing the list with the `forall` operator puts this handles one after the other into the variable p, from which their color property is accessed.

### 8.4.2 Lists of Elements

Sometimes it is not necessary to access points individually by their name. In particular, this happens whenever one is interested in performing an operation on all points in a construction. This may happen, for instance, when one wants to calculate the convex hull of a point set. For this, CindyScript provides several operators that return lists of elements. For instance, the operator `allpoints()` returns a list of all points of a construction. We will demonstrate this with a very tiny example. The following program changes the color of the points depending on their position relative to the *y*-axis:

```
pts=allpoints();
forall(pts,p,
  if(p.x<0,
    p.color=[1,1,0],
    p.color=[0,1,0];
  )
)
```

The following picture shows the application of the code to a random collection of points.

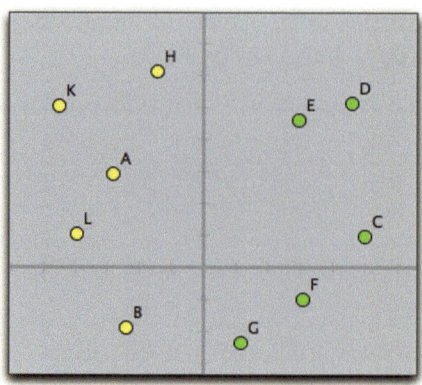

*Working with lists of points*

### 8.4.3 Properties of Geometric Objects

We now start with a complete description of all properties that are currently accessible via CindyScript. Every property is at least readable. For each property we list the type of value expected for the property, whether it is read only or also writeable, and a short description of its purpose. Possible property types are usually as follows:

- *real:* a real number
- *int:* an integer number
- *bool:* either `true` or `false`
- *string:* a sequence of characters
- *2-vector:* a two-dimensional vector
- *3-vector:* a three-dimensional vector
- *3x3-matrix:* a 3 by 3 matrix

Some properties, like the current position, are only writable for free objects. We mark this by the word "free" in the corresponding column.

**Properties Common to All Geometric Objects**

| Property | Writeable | Type | Purpose |
|---|---|---|---|
| color | yes | 3-vector | The (red, green, blue) color vector of the object |
| colorhsb | yes | 3-vector | The (hue, saturation, black) color vector of the object |
| isshowing | yes | bool | Whether the object is shown (this is inherited by all elements that depend on the object) |
| visible | yes | bool | Whether the object is shown (not inherited by dependent objects) |
| alpha | yes | real | The opacity of the object (between 0.0 and 1.0) |
| labelled | yes | bool | Whether the object shows its label |
| name | no | string | The label of the object |
| caption | yes | string | A caption that may replace the name |
| trace | yes | bool | Whether the object leaves a trace |
| tracelength | yes | int | The length of the trace |
| selected | yes | bool | Whether the object is currently selected |

Each geometric element has a unique name. the string that represents this name may be accessed by `.name`. So for instance `A.name` returns the string `"A"`. The name may not be identical with the caption of the element shown in the construction. If `A.caption` is the empty string the name is shown, otherwise the caption.

### Properties of Points

| Property | Writeable | Type | Purpose |
|---|---|---|---|
| x | free | real | The $x$-coordinate of the point |
| y | free | real | The $y$-coordinate of the point |
| xy | free | 2-vector | The $xy$-coordinates of the point |
| coord | free | 2-vector | The $xy$-coordinates of the point |
| homog | free | 3-vector | The homogeneous coordinates of the point |
| angle | free | real | Applies only to PointOnCircle objects. The angle of the point on the circle |
| size | yes | int | The size of the point (0..40) |
| imagerot | yes | real | A rotation angle if the point is equipped with an image |

### Properties of Lines

| Property | Writeable | Type | Purpose |
|---|---|---|---|
| homog | free | 3-vector | The homogeneous coordinates of the line |
| angle | free | real | The angle of the line |
| slope | free | real | The slope of the line |
| size | yes | int | The size of the line (0..10) |

### Properties of Circles and Conics

| Property | Writeable | Type | Purpose |
|---|---|---|---|
| center | no | real | The center of the circle |
| radius | free | real | The radius of the circle |
| matrix | no | real | The matrix describing the quadratic form of the circle or conic |
| size | yes | int | The size of the border line (0..10) |

### Properties of Texts

| Property | Writeable | Type | Purpose |
|---|---|---|---|
| text | yes | string | The content of the text |
| pressed | yes | boolean | The state of this text if it is a button |
| xy | yes | 2-vector | The position of the text |

### Properties of Animations

| Property | Writeable | Type | Purpose |
|---|---|---|---|
| run | yes | bool | Whether the animation is running |
| speed | yes | real | The relative animation speed |

**Properties of Transformations**

| Property | Writeable | Type | Purpose |
|---|---|---|---|
| matrix | no | 3x3 matrix | The homogeneous matrix of the transformation |
| inverse | no | 3x3 matrix | The homogeneous matrix of the inverse transformation |

### 8.4.3.1  Properties of CindyLab Objects

It is not only geometric properties that can be accessed by CindyScript. The simulation parameters of CindyLab constructions can also be read and sometimes set via CindyScript.

**Properties of All CindyLab Elements**

| Property | Writeable | Type | Purpose |
|---|---|---|---|
| simulate | yes | bool | Whether the object takes part in the physics simulation or is neglected |

**Properties of Masses**

| Property | Writeable | Type | Purpose |
|---|---|---|---|
| mass | yes | real | The mass of the object |
| charge | yes | int | The charge of the object |
| friction | yes | real | The individual friction of the object |
| radius | yes | real | The radius if the mass is treated as a ball |
| posx | yes | real | The $x$-component of the mass's position |
| posy | yes | real | The $y$-component of the mass's position |
| pos | yes | 2-vector | The mass's position vector |
| vx | yes | real | The $x$-component of the velocity |
| vy | yes | real | The $y$-component of the velocity |
| v | yes | 2-vector | The velocity vector |
| fx | no | real | The $x$-component of the force acting on the particle |
| fy | no | real | The $y$-component of the force acting on the particle |
| f | no | 2-vector | The force vector acting on the particle |
| kinetic | no | real | The kinetic energy of the particle |
| ke | no | real | The kinetic energy of the particle |

Sometimes one is interested to add a user defined force potential between masses. This can be done by scripting a suitable piece of code in the Integeration Tick event. Since internally the position of masses has a finer time scale than usual geometric movements it is necessary to access their position via the pos, posx and posy accessors.

## Properties of Springs and Coulomb Forces

| Property | Writeable | Type | Purpose |
|---|---|---|---|
| l | no | real | The current length of the spring |
| lrest | no | real | The rest length of the spring |
| ldiff | no | real | The distance to the rest length of the spring |
| strength | yes | real | The spring constant |
| f | no | real | The force vector caused by the spring |
| amplitude | yes | real | The amplitude for actuation |
| speed | yes | real | The speed for actuation |
| phase | yes | real | The phase for actuation (between 0.0 and 1.0) |
| potential | no | real | The potential energy in the spring |
| pe | no | real | The potential energy in the spring |

## Property for Velocities

| Property | Writeable | Type | Purpose |
|---|---|---|---|
| factor | yes | real | The multiplication factor between graphical representation and actual velocity |

## Properties of Gravity

| Property | Writeable | Type | Purpose |
|---|---|---|---|
| strength | yes | real | The strength of the gravity field |
| potential | no | real | The potential energy of all masses in the gravity field |
| pe | no | real | The potential energy of all masses in the gravity field |

## Properties of Suns

| Property | Writeable | Type | Purpose |
|---|---|---|---|
| mass | yes | real | The mass of the sun |
| potential | no | real | The potential energy of all masses in the sun field |
| pe | no | real | The potential energy of all masses in the sun field |

## Properties of Magnetic Areas

| Property | Writeable | Type | Purpose |
|---|---|---|---|
| strength | yes | real | The strength of the magnetic field |
| friction | yes | real | The friction in the magnetic area |

## Properties of Bouncers and Floors

| Property | Writeable | Type | Purpose |
|---|---|---|---|
| xdamp | yes | real | Damping in the $x$-direction |
| ydamp | yes | real | Damping in the $y$-direction |

**Properties of the Environment**

The environment can be accessed by the built-in operator `simulation()`. The following slots of the environment can be accessed:

| Property | Writeable | Type | Purpose |
|----------|-----------|------|---------|
| gravity | yes | real | The global gravity |
| friction | yes | real | The global friction |
| kinetic | no | real | The overall kinetic energy |
| ke | no | real | The overall kinetic energy |
| potential | no | real | The overall potential energy |
| pe | no | real | The overall potential energy |

## 8.4.4 Inspecting Elements

You can also use the generic CindyScript function `inspect(<element>)` to access all the attributes that are available in the Inspector (p. 153). For example, if a point *A* exists in the construction, the function

```
inspect(A)
```

will return the array of strings

```
[name,definition,color,visibility,drawtrace,tracelength,
traceskip,tracedim,render,isvisible,text.fontfamily,
pinning,incidences,labeled,textsize,textbold,textitalics,
ptsize,pointborder,printname,point.image,
point.image.rotation,freept.pos]
```

Using the two-parameter form `inspect(<element>,<string>)` you can read all the attributes of *A* that are listed in the above array:

```
inspect(A,"text.fontfamily")
```

returns

```
SansSerif
```

With the three-parameter form `inspect(<element>,<string>,<expr>)` you can also set the attributes that are not read-only (for example, you cannot change the list of incidences or the definition of an element). The following function will set the font of *A* to a Serif font:

```
inspect(A,"text.fontfamily","Serif")
```

The inspect command is very powerful, as you can automate all actions you normally would have to do in the Inspector using the mouse. Also, it gives you fine grained control over all properties.

**attribute**          **Set a user attribute: `attribute(<geo>,<string1>,<string2>)`**

*Description:* Sets the user attribute of <geo> identified by <string1> to the value <string2>.

**Read a user attribute: `attribute(<geo>,<string>)`**                    `attribute`

*Description:* Returns the user attribute identified by <string> of the geometric element <geo> .

Both versions of the attribute function are mainly used for interaction with the Visage Extension (p. 443).

### 8.4.5 Creating and Destroying Elements

Starting with Cinderella version 2.1 you can also create points on the fly from CindyScript. The function

```
p = createpoint("A",[4,6])
```

creates a point labelled *A* at coordinates [4,6], unless there is already an element *A*. If it exists, it will be moved to the position given as second argument. The value of the function is the point or the already existing element *A*. This means that repeated executions of the function are not harmful to your code - if you need a free point at "*A*" you can ensure that it exists using the `createpoint`-command.

Using the `removeelement` function you can also remove elements from your construction. Be aware that all dependent elements will be removed as well. The function expects an element as argument, so you can use either

```
removeelement(A)
```

or

```
removeelement(element("A"))
```

to remove the element named "*A*".

More functions to create arbitrary elements are also available and discussed in the section section on special operators (p. 382).

# Chapter 9

# Programming

## 9.1 Control Operators

### 9.1.1 Program Flow

We first describe those operators that allow to generate conditional branching and various kinds of loops. The user should be aware that there are also other kinds of loop-like structures that arise from the different ways of traversing lists. They are described in the section Elementary List Operations (p. 265).

**The conditional operator: `if(<bool>,<expr>)`**      `if`

*Description:* The expression <expr> is evaluated, if the Boolean condition <bool> evaluates to `true`. In this case the return value of the `if`-function is <expr>. Otherwise, ____ is returned. A typical use of the `if`-operator is the conditional evaluation of side effects.

*Example:* This code fragment prints a message on the console, if x has a negative value.

```
if(x<0,println("x is now negative"))
```

**The conditional branch operator: `if(<bool>,<expr1>,<expr2>)`**      `if`

*Description:* The expression <expr1> is evaluated, if the Boolean condition <bool> evaluates to `true`. If <bool> evaluates to `false`, then <expr2> is evaluated. In any case, the value of the evaluated expression is returned. Thus this ternary version of the `if`-operator encodes an if/then/else functionality. There are two typical uses of this version of the `if`-operator: Firstly, the `if`-operator is used to force the conditional evaluation of program parts (which usually causes side effects).

*Example:* This code fragment prints a message on the console that shows whether x is positive, negative, or zero.

```
if(x<0,
   println("x is now negative"),
   if (x>0,
     println("x is now positive"),
```

```
        println("x is zero")
    )
  )
```

A second use of the if-operator is to return a certain value depending on the condition encoded by <bool>. This is particularly useful in the definition of functions.

*Example:* This code fragment defines the function f(x) to be the absolute value function (for real values of x).

```
f(x):=if(x>0,x,-x)
```

*Example:* This code fragment takes a geometric element A (most probably a point) and sets its color to red or blue depending on the value of its x-coordinate.

```
A.color=if(A.x>0,(1,0,0),(0,0,1))
```

**trigger**

### The trigger operator: trigger(<bool>,<expr>)

*Description:* The trigger operator is very similar to the if operator. In contrast to if, the trigger operator has a dynamic flavor. The expression <expr> is evaluated whenever <bool> changes from false to true. This means that during the dragging of a construction, <expr> is evaluated, if <bool> was false in the previous instance and is now true. The purpose of this operator is to trigger side effects whenever some event occurs while the construction is being dragged. The following code fragment demonstrates this behavior.

*Example:* This code fragment will print a message whenever point A crosses the y-axis.

```
trigger(A.x<0,println("A now entered the x-negative half-plane"))
trigger(A.x>0,println("A now entered the x-positive half-plane"))
```

**while**

### The while loop: while(<bool>,<expr>)

*Description:* The while operator evaluates the expression <expr> as long as the condition <bool> is true. The result of the very last evaluation is returned as the function's value.

*Example:*

```
x=0;
sum=0;
erg=while(x<4,
        x=x+1;
        sum=sum+x;
        println(x+"  -->  "+sum);
        sum
     );
println(erg);
```

This code fragment produces the output

```
1  -->  1
2  -->  3
3  -->  6
4  -->  10
10
```

After its evaluation, the value of variable `erg` is 10. A word of caution: one should be aware of the fact that `while` operations may easily create infinite loops, if the conditional is never satisfied.

**The repeat loop: `repeat(<number>,<expr>)`**                                        **repeat**

*Description:* This operator provides the simplest kind of loop in CindyScript: `<expr>` is evaluated `<number>` times. The result of the last evaluation is returned. During the evaluation of `<expr>` the special variable # contains the run variable of the loop.

*Example:* This code produces a list of the first 100 integers together with their squares.

```
repeat(100,
    println(#+" squared is "+#^2)
)
```

*Modifiers:* The `repeat` loop supports a variety of modifiers. These modifiers can be used to control the start value, stop value, and step size of the loop. The modifier `start` sets the start value of the loop. The modifier `stop` sets the end value of the loop. The modifier `step` sets the step size. Arbitrary combinations of modifiers are possible. As long as not all three modifiers are set, the loop will always be executed `<number>` times. Only real values are allowed for the modifiers. The table below demonstrates different uses of the modifiers.

| Expression | Result |
|---|---|
| `repeat(6, println(#+" "))` | 1 2 3 4 5 6 |
| `repeat(6, start->4,`<br>`println(#+" "))` | 4 5 6 7 8 9 |
| `repeat(6, stop->2,`<br>`println(#+" "))` | -3 -2 -1 0 1 2 |
| `repeat(6, step->3,`<br>`println(#+" "))` | 1 4 7 10 13 16 |
| `repeat(6, stop->12, step->4,`<br>`println(#+" "))` | -8 -4 0 4 8 12 |
| `repeat(6, start->3, step->2,`<br>`println(#+" "))` | 3 5 7 9 11 13 |
| `repeat(6, start->3, stop->4,`<br>`println(#+" "))` | 3 3.2 3.4 3.6 3.8 4 |
| `repeat(6, start->0, stop->-3,`<br>`println(#+" "))` | 0 -0.6 -1.2 -1.8 -2.4 -3 |
| `repeat(6, start->3, stop->4,`<br>`step->0.4,println(#+" "))` | 3 3.4 3.8 4.2 |

**The repeat loop: `repeat(<number>,<var>,<expr>)`**                                  **repeat**

*Description:* This operator is identical to the operator `repeat(<number>,<expr>)`, except for one difference: the run variable is now assigned to `<var>`. This allows for the use of nested loops with different run variables.

*Example:* This code fragment will draw a $10 \times 10$ array of points.

```
repeat(10,i,
    repeat(10,j,
        draw((i,j))
    )
)
```

**forall**                      **The forall loop: `forall(<list>,<expr>)`**

*Description:* This operator takes a `<list>` as its first argument. It produces a loop in which `<expr>` is evaluated for each entry of the list. In every iteration, the run variable # takes the value of the corresponding list entry.

*Example:*

```
a=["this","is","a","list"];
forall(a,println(#))
```

This code fragment produces the output

```
this
is
a
list
```

**forall**                      **The forall loop: `forall(<list>,<var>,<expr>)`**

*Description:* Similar to `forall(<list>,<expr>)`, but the run variable is now named `<var>`.

**eval**                        **Forcing evaluation: `eval(<expr>,<modif1>,<modif2>,...)`**

*Description:* This operator forces the evaluation of the expression <expr>. Free variables of the expression can be substituted using a list of modifiers. The variables for the substitution are assigned only locally. Afterwards, the variables are set to the values they had before the evaluation.

*Example:* This code fragment evaluates to 7.

```
eval(x+y,x->2,y->5)
```

## 9.1.2  Variable Management

The following descriptions explain how to intentionally create or destroy local variables. The user should be aware of the fact that for most purposes it is completely sufficient to create variables on the fly by simply assigning values to them. Such variables will by default be global. The use of local variables may become necessary or recommended if recursive functions are generated or if one wants to create a library of functions.

**createvar**                   **Creating variables: `createvar(<varname>)`**

and

**Destroying variables: `removevar(<varname>)`**

*Description:* These operators help in the manual administration of the creation of local variables. `createvar(x)` creates a new variable named `x`, while the old value is put on a stack. `removevar(x)` removes the local variable and restores the value from the stack. Notice that usually, variables do not have to be created explicitly. They are automatically generated when they are first used. The `createvar` and `removevar` operators should only be used if one wants to reserve a variable name for a certain local region of the code.

*Example:*

```
x=10;
println("x is now "+x);
createvar(x);
x=5;
println("x is now "+x);
removevar(x);
println("x is now "+x);
```

This code fragment produces the output

```
x is now 10
x is now 5
x is now 10
```

**Creating many local variables for a function: `regional(name1,name2,...)`**

*Description:* This statement can be used at the beginning of a function. It has almost the same effect as the `local` statement and creates several local variables. However, unlike with the `local` statement, the variables are removed automatically, when the function terminates. Therefore, an explicit call of `release` is not necessary. Most often it is much more convenient to use `regional` than to use `local`.

Variables have some kind of persistence within CindyScript. If the value of a variable is set in a statement, it remains set, until it is changed. One can explicitly clear variables using the following operators. Often it is useful to put a `clear()` statement under the `init` event of the program.

**Clear all variables: `clear()`**

*Description:* This operator clears all variables.

**Clear a specific variable: `clear(<var>)`**

*Description:* This operator clears variable `<var>`.

**Handles to key variables of objects: `keys(<var>)`**

*Description:* Gives a list of all keys associated to an object via a `<object>:<key>=<something>` declaration.

*Example:* It is possible to associate a value under a freely chosen key to an object. This can be done by code similar to the following one:

```
A:"age"=34;
A:"haircolor"="brown";
```

These assignments may be accessed also by a similar syntax like:

```
println(A:"age");
println(A:"haircolor");
```

The operator `keys` returns a list of all associated keys of an object. So in this example

```
println(keys(A));
```

will return the list `["age","haircolor"]`.

## 9.2 Arithmetic Operators

The following section summarizes all functions and operators that can be applied to numbers. There are also many other mathematical operations, and these can be found in the sections Vectors and Matrices (p. 271), Geometric Operators (p. 349), and Function Plotting (p. 291).

### 9.2.1 Infix Operators

The elementary mathematical operators +, -, *, /, ^ are accessible in a straightforward manner. They can be applied to numbers and lists. Their particular meaning depends on the type of objects to which they are applied. For example, `5+7` evaluates to `12`, while `[2,3,4]+[3,-1,5]` evaluates to `[5,2,9]`. Usually all these operators apply to real numbers as well as to complex numbers.

**The addition operator: `<expr>+<expr>`**

*Description:* Numbers (integers, real, complex) can be added with the + operator. Lists having the same structure can also be added; then the addition is carried out component wise.

| Expression | Result |
|---|---|
| 7 + 8 | 15 |
| 2.3 + 5.9 | 8.2 |
| [2,3,4] + [3,4,6] | [5,7,10] |
| [2,3,[1,2]] + [3,4,[1,3]] | [5,7,[2,4]] |

*See also:* String Operators (p. 257)

### The subtraction operator: `<expr>-<expr>`

*Description:* Numbers (integers, real, complex) can be subtracted with the - operator. Lists of the same shape can also be subtracted. The subtraction is then performed componentwise. Furthermore, the - operator can be used as a unary minus.

| Expression | Result |
|---|---|
| 7 - 8 | -1 |
| 8.3 - 5.9 | 2.4 |
| [2,6,4] - [3,4,6] | [-1,2,-2] |
| [5,3,[1,2]] - [3,4,[1,3]] | [2,-1,[0,-1]] |

*See also:* String Operators (p. 257)

### The multiplication operator: `<expr>*<expr>`

*Description:* Numbers (integers, real, complex) can be multiplied with the * operator. Lists that represent numerical vectors or numerical matrices can also be multiplied if the dimensions admit a reasonable mathematical operation. See the examples for further description.

| Expression | Result |
|---|---|
| 7 * 8 | 56 (integer multiplication) |
| (1+i) * (2+i) | 1+3*i (multiplication of complex numbers) |
| 2 * [5,3,2] | [10,2,4] (scalar multiplication of number and vector) |
| [5,3,2] * 2 | [10,2,4] (scalar multiplication of number and vector) |
| [2,2,3] * [3,4,6] | 32 (scalar product of two vectors) $(x_1,x_2,\ldots,x_n) \cdot (y_1,y_2,\ldots,y_n) = (x_1y_1,x_2y_2,\ldots,x_ny_n)$ |
| [[1,2],[3,4]] * [1,2] | [5,11] (matrix times vector) |
| [1,2] * [[1,2],[3,4]] | [7,10] (vector times matrix) |
| [[1,2],[3,4]] * [[1,2],[3,4]] | [[7,10],[15,22]] (product of two matrices) |

*See also:* Vectors and Matrices (p. 271)

### The division operator: `<expr>/<number>`

*Description:* Numbers (integers, real, complex) can be divided with the / operator. Also, a vector can be divided by a number.

| Expression | Result |
|---|---|
| 56 / 8 | 7 |
| [6,8,4] / 2 | [3,4,2] |

### The power operator: `<expr>^<expr>`

*Description:* A number (integer, real, complex) can be taken to the power of another number (integer, real, complex). Note that not only integer powers are allowed. In a^b the exponent b can be an arbitrary real or complex number. Formally, the expression exp(b*ln(a)) is calculated. Since ln(...) is defined only up to a period of 2*pi, the expression a^b is in general multivalued. For noninteger values of b only one principal value of a^b will be returned.

| Expression | Result |
|---|---|
| 5^2 | 25 |
| 5^(-1) | 0.2 |
| 2^(1/2) | 1.4142... |

**The degree operator: `<number>°`**

This operator multiplies any number by the constant `pi/180`. This makes possible angle conversion from degrees to radians.

| Expression | Result |
|---|---|
| `180°` | `3.1415...` |
| `cos(180°)` | `-1` |

**The absolute value operator: `|<number>|`**

*Description:* This operator calculates the absolute value of an object. The object may be a simple number, a complex number, or a vector.

It is not allowed to use the `|...|` operator in a nested way, since such expressions can be syntactically ambiguous.

| Expression | Result |
|---|---|
| `|-5|` | `5` |
| `|(3,4)|` | `5` |
| `|1+i|` | `1.4142...` |

**The distance operator: `|<number>,<number>|`**

*Description:* One can use `|...|` with two arguments, in which case this operator calculates the distance between the two objects. The objects may be simple numbers, complex numbers, or vectors. However, they must be of the same type.

It is not allowed to use the `|...,...|` operator in a nested way, since such expressions can be syntactically ambiguous.

| Expression | Result |
|---|---|
| `|-5,8|` | `3` |
| `|(1,1),(4,5)|` | `5` |

## 9.2.2 Functional Operators

The following operators can be applied to numbers (integer, real complex). Some of them can also be applied to vectors.

## 9.2.3 Arithmetic Functions

add       **Addition: `add(<expr1>,<expr2>)`**

sub       **Subtraction: `sub(<expr1>,<expr2>)`**

mult      **Multiplication: `mult(<expr1>,<expr2>)`**

div       **Division: `div(<expr1>,<expr2>)`**

pow       **Exponentiation: `pow(<expr1>,<expr2>)`**

*Description:* These operators are binary functions equivalent to the operators like +, -, *, /, and ^.

| Expression | Result |
|---|---|
| `add(5,6)` | 11 |
| `pow(6,2)` | 36 |
| `mod(23,4)` | 3 |
| `add([1,2],[3,4])` | (4,6) |
| `mult(2,[3,4])` | (6,8) |
| `mult([4,5],[3,4])` | 32 |

**Modulo: `mod(<expr1>,<expr2>)`**                                                    `mod`

*Description:* The `mod` function calculates the remainder of `<expr1>` if divided by `<expr2>`.

## 9.2.4 Standard Functions

**Square root: `sqrt(<expr>)`**                                                        `sqrt`

**Exponential function: `exp(<expr>)`**                                                `exp`

**Natural logarithm: `log(<expr>)`**                                                   `log`

*Description:* These functions map numbers to numbers. Complex numbers are fully supported.

## 9.2.5 Trigonometric Functions

The standard trigonometric functions are available through the following operators:

**Trigonometric sine function: `sin(<expr1>)`**                                        `sin`

**Trigonometric cosine function: `cos(<expr1>)`**                                      `cos`

**Trigonometric tangent function: `tan(<expr1>)`**                                     `tan`

**Inverse trigonometric sine function: `arcsin(<expr1>)`**                             `arcsin`

**Inverse trigonometric cosine function: `arccos(<expr1>)`**                           `arccos`

**Inverse rigonometric tangent function: `arctan(<expr1>)`**                           `arctan`

**Angle of a vector: `arctan2(<real1,real2>)`**                                        `arctan2`

**Angle of a vector: `arctan2(<vec>)`**                                                `arctan2`

The `arc` operators are in principle multivalued. However, the operator returns only one principal value, for which the real value is between +pi and -pi.

| Expression | Result |
|---|---|
| `sin(pi)` | 0 |
| `arccos(-1)` | 3.1415... |
| `arctan2(1,1)` | 45° |
| `arctan2(-1,-1)` | −135° |

## 9.2.6 Numeric Functions

abs                 **Absolute value: abs (<expr>)**

round               **Rounded value: round(<expr>)**

floor               **Largest integer less than or equal: floor(<expr>)**

ceil                **Smallest integer greater than or equal: ceil(<expr>)**

re                  **Real part of a complex number: re(<expr>)**

im                  **Imaginary part of a complex number: im(<expr>)**

conjugate           **Conjugate of a complex number: conjugate(<expr>)**

*Description:* For complex numbers the operators round, floor, and ceil are applied to the real and imaginary parts separately.

The function abs calculates the norms of numbers, complex numbers, vectors, etc. All other functions can also be applied to lists, in which case they are applied component wise.

| Expression | Result |
|---|---|
| round(4.3) | 4 |
| round([3.2,7.8,3.1+i*6.9]) | [3,8,3+i*7] |
| abs([1,3,1,2,1]) | 4 |
| floor(4.8) | 4 |

## 9.2.7 Random Number Operators

The following operators generate pseudo random numbers.

random              **Uniformly distributed random real number between 0 and 1: random()**

randomnormal        **(0,1)-normally distributed random number: randomnormal()**

randombool          **Random boolean value true or false: randombool()**

random              **Uniformly distributed random real number between 0 and <number>: random(<number>)**

randomint           **Uniformly distributed random integer number between 0 and <number>: randomint(<number>)**

*Description:* The random generators also accept negative and complex numbers as arguments. For example, random(-5) generates a random number between -5 and 0; randomint(6+i*10) generates a random complex number for which the real part is an integer between 0 and 6 and the imaginary part is an integer between 0 and 10.

**Initialize the random generator: `seedrandom(<number>)`**                    `seedrandom`

*Description:* The pseudo random generator will always produce unforeseeable new random numbers. If for some reason one wants the same random numbers to be generated for different runs of a script, one can use the `seedrandom(<number>)` operator. After this function is invoked with a certain integer, the same sequence of random numbers will be deterministically generated. Each seeding integer produces a different sequence of random numbers.

## 9.3 Boolean Operators

Conditional branching may very much depend on the outcome of a boolean query. in this section we describe all such types of queries as well as the different ways to process boolean values.

### 9.3.1 Infix Operators

**Testing equality: `<expr1> == <expr2>`**                                     `==`

*Description:* This operator tests whether two expressions evaluate to the same value. The result of this operator is either `true` or `false`.

**Testing inequality: `<expr1> != <expr2>`**                                   `!=`

*Description:* This operator tests whether two expressions do not evaluate to the same value. The result of this operator is either `true` or `false`. It is the logical negation of `<expr1> == <expr2>`.

**Greater than: `<expr1> > <expr2>`**                                          `>`

*Description:* This operator tests whether the expression `<expr1>` is *greater than* the expression `<expr2>`. It returns a `<bool>` value. The comparison is available only for two situations: If both expressions are *real numbers*, then the order of size is the usual ordering of real numbers. If both expressions are *strings*, then the order is the lexicographic (dictionary) order. In all other cases (if the values are not comparable) the value _____ is returned.

**Less than: `<expr1> < <expr2>`**                                            `<`

*Description:* This operator is similar to > but tests for *less than*.

**Greater than or equal: `<expr1> >= <expr2>`**                               `>=`

*Description:* This operator is similar to > but tests for *greater than or equal to*.

**Less than or equal: `<expr1> <= <expr2>`**                                  `<=`

*Description:* This operator is similar to > but tests for *less than or equal to*.

**Fuzzy comparisons: ~==, ~!=, ~<, ~>, ~>=, ~<=**

*Description:* CindyScript provides a *fuzzy* variant for each comparison operator. This Version tests whether the condition is satisfied up to an epsilon bound. Thus the test `a~==0` tests whether is the variable `a` lies between `+epsilon` and `-epsilon`. The small value epsilon is set to `0.0000000001`. This operator is sometimes very useful to circumvent inaccuracies which are unavoidable in purely numerical calculations.

The exact semantics of the exact and the fuzzy operators can be read off from the following diagram. Here for each operator the picture shows for which region of `b` (marked in red) the operator evaluates to true.

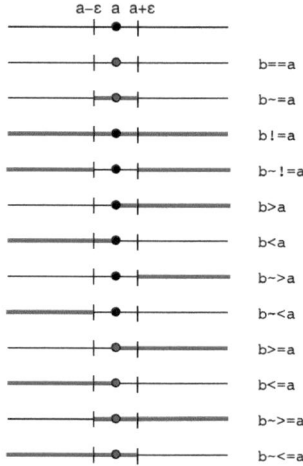

**Logical and: <bool1> & <bool2>**

*Description:* Logical *and* of two Boolean values defined by the following truth table:

```
A     B      A & B
false false false
false true  false
true  false false
true  true  true
```

If one of the two arguments is not a Boolean expression, the operator returns ____.

**Logical or: <bool1> % <bool2>**

*Description:* Logical *or* of two Boolean values defined by the following truth table:

```
A     B      A % B
false false false
false true  true
true  false true
true  true  true
```

If one of the two arguments is not a Boolean expression, the operator returns ____.

**Logical not:** `!<bool>`

*Description:* Logical *not* of one Boolean value defined by the following truth table:

```
A      !A
false true
true  false
```

If the argument is not a Boolean expression, the operator returns _____.

## 9.3.2 Functional Operators

**Logical and:** `and(<bool1>, <bool2>)`

*Description:* `and(x,y)` is equivalent to `x & y`.

**Logical or:** `or(<bool1>, <bool2>)`

*Description:* `or(x,y)` is equivalent to `x % y`.

**Logical not:** `not(<bool>)`

*Description:* `not(x)` is equivalent to `!x`.

**Logical exclusive or:** `xor(<bool1>, <bool2>)`

*Description:* Logical *exclusive or* of two Boolean values defined by the following truth table:

```
A     B      xor(A,B)
false false false
false true  true
true  false true
true  true  false
```

If one of the two arguments is not a Boolean expression, the operator returns ____.

## 9.3.3 Type Predicates

The following predicates test whether the expression `<expr>` belongs to a certain class of objects. The predicates are important in defining functions whose behavior depends on the type of input expressions. Furthermore, these arguments are very useful for debugging, since they can be used to test assertions on the typing of the values in a program.

**Is an integer:** `isinteger(<expr>)`

*Description:* This operator tests whether the expression `<expr>` is an integer.

**Is a real number:** `isreal(<expr>)`

*Description:* This operator tests whether the expression `<expr>` is a real number. Note that integers are also real numbers.

| iscomplex | **Is a complex number:** `iscomplex(<expr>)` |
| --- | --- |
| | *Description:* This operator tests whether the expression `<expr>` is a complex number. Note that real numbers are also complex numbers. |
| iseven | **Is even:** `iseven(<expr>)` |
| | *Description:* This operator tests whether the expression `<expr>` is an even integer. |
| isodd | **Is odd:** `isodd(<expr>)` |
| | *Description:* This operator tests whether the expression `<expr>` is an odd integer. |
| islist | **Is a list:** `islist(<expr>)` |
| | *Description:* This operator tests whether the expression `<expr>` is a list. |
| ismatrix | **Is a matrix:** `ismatrix(<expr>)` |
| | *Description:* This operator tests whether the expression `<expr>` has the shape of a matrix. This means that the entries of the list are themselves lists, all of equal length. If there are $n$ entries each of length $m$ the expression represents an $n \times m$ matrix. |
| isnumbervector | **Is a number vector:** `isnumbervector(<expr>)` |
| | *Description:* This operator tests whether the expression `<expr>` is a list all of whose entries are numbers (integer, real, or complex). |
| isnumbermatrix | **Is a number matrix:** `isnumbermatrix(<expr>)` |
| | *Description:* This operator tests whether the expression `<expr>` is a matrix all of whose entries are numbers (integer, real, or complex). |
| isstring | **Is a string:** `isstring(<expr>)` |
| | *Description:* This operator tests whether the expression `<expr>` is a string. |
| isgeometric | **Is a geometric element:** `isgeometric(<expr>)` |
| | *Description:* This operator tests whether the expression `<expr>` represents a geometric element. |
| isselected | **Is selected:** `isselected(<expr>)` |
| | *Description:* This operator tests whether the expression `<expr>` represents a geometric element and is selected. For a geometric element you can also use the .selected property to check this. |
| ispoint | **Is a point:** `ispoint(<expr>)` |
| | *Description:* This operator tests whether the expression `<expr>` represents a geometric point. |

**Is a line: `isline(<expr>)`**                                                          `isline`

*Description:* This operator tests whether the expression `<expr>` represents a geometric line.

**Is a circle: `iscircle(<expr>)`**                                                      `iscircle`

*Description:* This operator tests whether the expression `<expr>` represents a geometric circle.

**Is a conic: `isconic(<expr>)`**                                                        `isconic`

*Description:* This operator tests whether the expression `<expr>` represents a geometric conic.

**Is a mass: `ismass(<expr>)`**                                                          `ismass`

*Description:* This operator tests whether the expression `<expr>` represents a Cindy-Lab (p. 167) mass.

**Is a sun: `issun(<expr>)`**                                                            `issun`

*Description:* This operator tests whether the expression `<expr>` represents a Cindy-Lab (p. 167) sun.

**Is a spring: `isspring(<expr>)`**                                                      `isspring`

*Description:* This operator tests whether the expression `<expr>` represents a Cindy-Lab (p. 167) spring.

**Is undefined: `isundefined(<expr>)`**                                                  `isundefined`

*Description:* This operator tests whether the expression `<expr>` returns an undefined element (___).


## 9.4 String Operators

This section lists operators that take strings as arguments. Moreover, operators that generate strings from other values are treated as well.


### 9.4.1 Elementary String Operations

**String concatenation: `<string1> + <string2>`**                                        `+`

*Description:* The + operator can also be used to append one string to another. The result of such an operation is again a string. If in an addition operation at least one of the arguments is a string, then the other argument will be automatically converted to a string representation of its value.

| Expression | Result |
|---|---|
| `"Cindy"+"Script"` | `"CindyScript"` |
| `"Four plus three is "+(4+3)` | `"Four plus three is 7"` |
| `""+(4+3)` | `"7".` |

*See also:* The + operator is also used for the addition of usual numerical values. For this see Arithmetic Operators (p. 248).

**text**

### Conversion to string: `text(<expr>)`

*Description:* The operator `text(<expr>)` evaluates the expression `<expr>` and converts the result to a string representation.

**length**

### Length of a string: `length(<string>)`

*Description:* This operator returns the number of characters in a string.

*Example:* `length("CindyScript (p. 219)")` evaluates to the integer 11.

*See Also:* Lists and Linear Algebra (p. 263)

**substring**

### Extracting a substring: `substring(<string>,<int1>,<int2>)`

*Description:* This operator returns the substring of `<string>` that begins after the character indexed by the integer `<int1>` and ends with the character indexed by `<int2>`.

*Example:* `substring("abcdefg",3,6)` evaluates to the string `"def"`.

**indexof**

### Searching for occurrence: `indexof(<string1>,<string2>)`

*Description:* This operator searches for the first occurrence of `<string2>` in `<string1>`. The index of this first occurrence is returned. If `<string2>` is not a substring of `<string1>`, then the value 0 is returned.

| Expression | Result |
|---|---|
| `indexof("CindyScript","i")` | 2 |
| `indexof("CindyScript","y")` | 5 |
| `indexof("CindyScript","z")` | 0 |

**indexof**

### Searching for occurrence: `indexof(<string1>,<string2>,<int>)`

*Description:* This operator searches for the first occurrence of `<string2>` in `<string1>` *after the index* `i`. The index of this first such occurrence is returned. If `<string2>` does not occur in `<string1>` after index i., then the value 0 is returned.

| Expression | Result |
|---|---|
| `indexof("CindyScript","i",1)` | 2 |
| `indexof("CindyScript","i",3)` | 9 |
| `indexof("CindyScript","i",10)` | 0 |

## 9.4.2 *Advanced String Operations*

**Dissecting a string: `tokenize(<string>,<expr>)`**

*Description:* This operator is very useful for parsing input. It creates a list of sub-strings of `<string>`. The second argument `<expr>` must be either a string or a list of strings. If `<expr>` is a string, then the operator searches for occurrences of this string in `<string>`. These occurrences serve as markers for breaking up `<string>` into a list of pieces.

If <expr> is a list of strings, then a hierarchical list is generated that represents the subdivision of `<string>` recursively by the tokens in the list.

| Expression | Result |
|---|---|
| `tokenize(`<br>`"one:two..three:four", ":")` | `["one", "two..three", "four"]` |
| `tokenize(`<br>`"one:two..three:four", ".")` | `["one:two", "", "three:four"]` |
| `tokenize(`<br>`"one:two..three:four", "..")` | `["one:two", "three:four"]` |
| `tokenize(`<br>`"one:two..three:four",`<br>`[".",":"])` | `[["one", "two"], [], ["three", "four"]]` |
| `tokenize(`<br>`"one:two..three:four",`<br>`["..",":"])` | `[["one", "two"], ["three", "four"]]` |

**Replacing in strings: `replace(<string1>,<string2>,<string3>)`**

*Description:* This operator replaces all (!) occurrences of <string2> in <string1> by <string3>.

This operator is extremely useful for creating text replacement systems of the kind they are used in so called Lindenmeyer Systems.

| Expression | Result |
|---|---|
| `replace(`<br>`"one:two..three:four", "o",`<br>`"XXX")` | `"XXXne:twXXX..three:fXXXur"` |
| `replace("F", "F", "F+F")` | `"F+F"` |
| `replace("F+F", "F", "F+F")` | `"F+F+F+F"` |

**Replacing in strings: `replace(<string>,<list>)`**

*Description:* This operator is very similar to the previous one. `<list>` contains a list of replacement pairs, and all such replacements are applied simultaneously to `<string>`.

| Expression | Result |
|---|---|
| `replace("XYX", [["X","one"],`<br>`["Y","two"]])` | `"onetwoone"` |

**parse**

### Parsing a string: `parse(<string>)`

*Description:* This operator parses a string to an expression and evaluates this expression. This operator is particularly useful in processing user input that comes from text fields in a construction.

| Expression | Result |
|---|---|
| `parse("3+7")` | `10` |

The code fragment

```
text="sin(x)+cos(x)";
f(x):=parse(text);
```

defines the function `f(x)` to be `sin(x)+cos(x)`.

**guess**

### Guessing a good representation of a number: `guess(<number>)`

*Description:* This very powerful operator is described in detail in the section Calculus (p. 355). It takes a numerical expression in floating-point representation and attempts to convert it to a mathematical expression that generates that floating-point number with high precision. This expression is then represented as a string.

| Expression | Result |
|---|---|
| `guess(8.125)` | `"65/8"` |
| `guess(0.774596669241483)` | `"sqrt(3/5)"` |

*See also:* Calculus (p. 355)

**format**

### Formating a number to a specified precision: `format(<number>,<int>)`

*Description:* This operator takes a number as first arguments and an integer specifying the number of digits after the decimal point. A string is generated that corresponds to the number up to the specified precision. Up to 14 digits are possible. If the argument of format is a list of objects the format statement is applied to each of the objects recursively.

| Expression | Result |
|---|---|
| `format(sqrt(2),4)` | `"1.4142"` |
| `format(pi,14)` | `"3.14159265358979"` |
| `format([sin(30°),cos(30°)],3)` | `[0.5,0.866]` |

*Warning:* The format statement should only produced to create formatted output elements. The formatted values will always be *strings* and hence usually not valid objects for arithmetic operations. The following example illsustrates this:

| Expression | Result |
|---|---|
| `format(sqrt(2),4)+`<br>`format(sqrt(2),4)` | `"1.41421.4142"` |

## 9.5 String Comparison and Sorting

Like real numbers, strings admit a total ordering. Thus they can be compared using the operators >, <, >=, <=, ==, and !=. Please refer to Boolean Operators (p. 253) for the use of these relations.

The order that is used for strings is lexicographic (dictionary) order. Thus, for example,

$$\texttt{"a"<"abd"<"abe"<"b"<"blue"<"blunt"<"xxx"}$$

**Sorting of lists: `sort(<list>)`**                                                    `sort`

**Sorting of lists: `sort(<list>,<expr>)`**                                            `sort`

**Sorting of lists: `sort(<list>,<var>,<expr>)`**                                      `sort`

*Description:* The various versions of the `sort`-operator can be used to sort lists that contain string values. The sorting order is usually taken to be the lexicographic order of the words. Alternatively, one can specify a user-defined sorting function such as the lengths of the strings.

| Expression | Result |
|---|---|
| `sort(["one", "two", "three", "four", "five"])` | `["five","four","one","three","two"]` |
| `sort(["one", "two", "three", "four", "five"],length(#))` | `["one","two","four","five","three"]` |

*See also:* Lists and Linear Algebra (p. 263)

### 9.5.1 Accessing and Replacing Characters

**Index operator: `<string>_<int>`**                                                   —

*Description:* The infix operator _, which accesses the fields of a list, can be also used to access a character at a specific position in a string. Characters can be returned and set with this operator.

| Expression | Result |
|---|---|
| `"CindyScript"_5` | `"y"` |
| `"CindyScript"_12` | undefined |

After evaluating the code fragment

```
a="CindyScript";
a_5="erella";
```

the variable `a` contains the string `"CinderellaScript"`.

# Chapter 10

# Lists and Linear Algebra

## 10.1 Using Lists in CindyScript

List are among the most fundamental and elementary concepts of CindyScript. In CindyScript lists serve several purposes. They serve as

- enumerative arrays of objects

- structured data arrangements

- vectors

- matrices

This section covers the very general aspects of creating lists and accessing their elements as well as all elementary aspects of array functionality. For information on all other aspects of lists, consult the sections

- Elementary List Operations (p. 265)

- Advanced List Operations (p. 268)

- Lists of Geometric Elements (p. 271)

- Vectors and Matrices (p. 271)

### 10.1.1 Creating Lists

Lists can be created very easily by placing the elements in square brackets, separated by commas. For example,

`[45,12.5,123,2,5.5,5]` is a list of numbers.

`["this", "is", "a", "list", "of", "strings"]` is a list of strings. Objects of different kinds can be mixed in a list:

`["this", 3,"is",5 , "a",654, "mixed",234 , "list"]` Lists can also have lists as elements, and these can be nested arbitrarily.

[[4,6], ["a", "b"], 1, [4, "b", [23, "b"]], [ ]] The last element [ ] of this list does not contain any elements. It is the empty list.

Alternatively lists consisting of at least two elements can also be enclosed by round brackets (...). This is sometimes convenient to have a more mathematical appealing notation in the the code. A three-dimensional vector may then be written as follows:

```
(7.3,9.3,-14.3)
```

**take**

### Accessing Elements of Lists: `take(<list>,<int>)`

*Description:* One can access the individual elements of a list either with the infix operator `<list>_<int>` or the functional operator `take(<list>,<int>)`. The indices start with number 1. If the index that should be accessed is less than 1 or greater than the number of elements in the list, then the value___is returned. Also, a warning message is issued on the console.

| Expression | Result |
|---|---|
| `[2 ,5 ,7 ,3]_3` | 7 |
| `take([2 ,5 ,7 ,3],2)` | 5 |
| `[2 ,5 ,7 ,3]_5` | ____ |

The index can also be an arbitrary calculation. Furthermore, indices can access the nested parts of a nested list.

| Expression | Result |
|---|---|
| `[ [2, [4,5]],1]_1` | `[2,[4,5]]` |
| `[ [2, [4,5]],1]_(7-5)` | 1 |
| `[ [2, [4,5]],1]_1_2` | `[4,5]` |
| `[ [2, [4,5]],1]_1_2_2` | 5 |
| `[ [2, [4,5]],1]_1_2_2_2` | ____ |

If a list is stored in a variable, the individual entries can be set after they are accessed by the _ operator. So for example, after the code fragment

```
a=[[2,[4,5]],1];
a_2="A";
a_1_2_1="B";
```

is evaluated, the value of a is `[[2,["B",5]],"A"]`.

### Advanced usage

The list element accessor has some other powerful options. By using negative numbers as indices one can access the list entries from the end to the beginning. The following examples exemplify this possibility:

| Expression | Result |
|---|---|
| `[2 ,5 ,7 ,3]_(-1)` | 7 |
| `take([2 ,5 ,7 ,3],(-3))` | 5 |
| `[ [2,6] ,5 ,7 ,3]_(-4)_(-1)` | 6 |

It is also possible to use lists of integers as indices. Then a list corresponding to the specified list entries is returned.

| Expression | Result |
|---|---|
| `[2 ,5 ,7 ,3]_[2,3]` | `[5,7]` |
| `[2 ,5 ,7 ,3]_[-1,1,1]` | `[3,2,2]` |

## 10.2 Elementary List Operations

### 10.2.1 Creating and Accessing Lists

**Creating an integer sequence: `<int1>..<int2>`**

`..`

*Description:* The expression `<int1>..<int2>` creates a list of consecutive integers starting with `<int1>` and ending with `<int2>`. If `<int1>` is larger than `<int2>`, then the empty list is returned.

| Expression | Result |
|---|---|
| `4..9` | `[4, 5, 6, 7, 8, 9]` |
| `2..2` | `[-2, -1, 0, 1, 2]` |
| `4..1` | `[]` |

**The length of a list: `length(<list>)`**

`length`

*Description:* This operator returns an integer that is equal to the number of elements in the `<list>`.

| Expression | Result |
|---|---|
| `length([2 ,5 ,7 ,3])` | 4 |
| `length([2 ,[5, 4, 5] ,7 ,3]_2)` | 3 |
| `length(1..1000)` | 1000 |

Combining the `length` and the `repeat` operator allows one to list all elements of a list easily.

```
repeat(length(list),
   println(list_#);
)
```

One word of caution here: CindyScript is designed in such a way that it is seldom useful to traverse all the elements of a list using the `repeat` operator. There are more elegant ways.

**contains**

**Testing for containment: `contains(<list>,<expr>)`**

*Description:* This operator returns either `true` or `false` depending on whether `<list>` contains the element <expr>.

| Expression | Result |
|---|---|
| `contains([1,3,4,5],4)` | `true` |
| `contains([1,3,4,5],7)` | `false` |
| `contains([1,3,4,5],2*2)` | `true` |

### 10.2.2 List Manipulation

**concat**

**Concatenation of lists: `concat(<list1>,<list2>)`**

*Description:* This operator creates a list by concatenation of two other lists. This operator can equivalently be written as `<list1>++<list2>`.

| Expression | Result |
|---|---|
| `concat(["a", "b"], ["c", "d"])` | `["a", "b", "c", "d"]` |

**remove**

**Removing elements from lists: `remove(<list1>,<list2>)`**

*Description:* This operator creates a list by removing all elements that occur in `<list2>` from `<list1>`. This operator can equivalently be written as `<list1> - <list2>`.

| Expression | Result |
|---|---|
| `remove([1,3,4,5,1,5,6], [1,3,7])` | `[4,5,5,6]` |
| `[1,3,4,5,1,5,6]-[1,3,7]` | `[4,5,5,6]` |

**common**

**Intersection of lists: `common(<list1>,<list2>)`**

*Description:* This operator creates a list collecting all elements that are in both `<list1>` and `<list1>`. In the returned list the elements are sorted and each element occurs at most once. This operator can equivalently be written as `<list1>~~<list2>`.

| Expression | Result |
|---|---|
| `common([1,3,4,5,1,5,6], [1,3,7])` | `[1,3]` |
| `[1,3,4,5,1,5,6]~~[1,3,7]` | `[1,3]` |

**append**

**Appending an element: `append(<list>,<expr>)`**

*Description:* This operator returns a list that is created by appending <expr> to the list `<list>` as its last element. This operator can equivalently be written as `<list>:><expr>`.

| Expression | Result |
|---|---|
| `append(["a", "b", "c"], "d")` | `["a", "b", "c","d"]` |
| `["a", "b", "c"]:>"d"` | `["a", "b", "c","d"]` |

**Prepending an element: prepend(<expr>,<list>)**                                          `prepend`

*Description:* This operator returns a list that is created by prepending <expr> to
the list <list> as its first element. This operator can equivalently be written as
<expr><:<list>.

| Expression | Result |
|---|---|
| prepend("d",["a", "b", "c"]) | ["d","a", "b", "c"] |
| "d"<:["a", "b", "c"~34 | ["d","a", "b", "c"] |


## 10.2.3 Traversing Lists

**The forall loop: forall(<list>,<expr>)**                                                `forall`

*Description:* This operator is useful for applying an operation to all elements of a
list. It takes a <list> as first argument. It produces a loop in which <expr> is
evaluated for each entry of the list. For each run, the run variable # takes the value
of the corresponding list entry.

*Example:*

```
a=["this","is","a","list"];
forall(a,println(#))
```

This code fragment produces the output

```
this
is
a
list
```

**The forall loop: forall(<list>,<var>,<expr>)**                                          `forall`

*Description:* Similar to forall(<list>,<expr>), but the run variable is now
named <var>.

**Applying an expression: apply(<list>,<expr>)**                                          `apply`

*Description:* This operator generates a new list by applying the operation <expr>
to all elements of a list and collecting the results. As usual, # is the run variable,
which successively takes the value of each element in the list.

| Expression | Result |
|---|---|
| apply([1, 2, 3, 4, 5],#2) | [1, 4, 9, 16, 25] |
| apply([1, 2, 3, 4, 5],#+5) | [6, 7, 8, 9, 10] |
| apply(1..5, [#,#^2]) | [[1, 1], [2, 4], [3, 9], [4, 16], [5, 25]] |

**Applying an expression: apply(<list>,<var>,<expr>)**                                    `apply`

*Description:* Similar to apply(<list>,<expr>), but the run variable is now
named <var>.

**select**

**Selecting elements of a list: `select(<list>,<boolexpr>)`**

*Description:* This operator selects all elements of a list for which a certain condition is satisfied. The condition is supposed to be encoded by `<boolexpr>`. This expression is assumed to return a `<bool>` value. As usual, `#` is the run variable, which successively take the value of all elements in the list.

| Expression | Result |
|---|---|
| `select(1..10, isodd(#))` | `[1, 3, 5, 7, 9]` |
| `select(0..10, #+# == #^2)` | `[0,2]` |

A high-level application of the `select` operator is given by the following example:

```
divisors(x):=select(1..x,mod(x,#)==0);
primes(n):=select(1..n,length(divisors(#))==2);
println(primes(100))
```

It produces the output

```
[2,3,5,7,11,13,17,19,23,29,31,37,41,43,47,53,59,61,67,71,73,79,83,89,97]
```

In this example, first a function `divisors(x)` is defined by selecting those numbers that divide `x` without any remainder. Then a function `primes(n)` is defined that selects all numbers between `1` and `n` that have exactly two divisors. These numbers are the primes.

**select**

**Selecting elements of a list: `select(<list>,<var>,<boolexpr>)`**

*Description:* Similar to `select(<list>,<boolexpr>)`, but the run variable is now named `<var>`.

## 10.3 Advanced List Operations

There are several operators that take a list as argument and return another list derived from it. This section deals with such operators. These operators form very powerful tools for performing a high-level computation. For examples of how to use and apply these operators in a realistic context, we strongly recommend to read the example section for CindyScript.

### 10.3.1 Pairs and Triples

**pairs**

**Building pairs: `pairs(<list>)`**

*Description:* This operator produces a list that contains all two-element sublists of a list. These are all pairs of elements of `<list>`. This operator is particularly useful for creating all segments determined a set of points.

| Expression | Result |
|---|---|
| `pairs([1, 2, 3, 4])` | `[[1, 2], [1, 3], [1, 4], [2, 3], [2, 4], [3, 4]]` |

### Creating a chain: `consecutive(<list>)`

*Description:* This operator produces a list that contains all pairs of elements of consecutive elements of the argument `<list>`.

| Expression | Result |
|---|---|
| `consecutive([1, 2, 3, 4, 5])` | `[[1, 2], [2, 3], [3, 4], [4, 5]]` |

### Creating a cycle: `cycle(<list>)`

*Description:* This operator produces a list that contains all pairs of consecutive elements of the argument `<list>`. Furthermore, the pair consisting of the last and the first elements is added.

| Expression | Result |
|---|---|
| `cycle([1, 2, 3, 4, 5])` | `[[1, 2], [2, 3], [3, 4], [4, 5], [5, 1]]` |

### Building triples: `triples(<list>)`

*Description:* This operator produces a list that contains all three-element sublists of a list. These are all the triples of elements of `<list>`.

| Expression | Result |
|---|---|
| `triples([1, 2, 3, 4])` | `[[1,2,3], [1,2,4], [1,3,4], [2,3,4]]` |

### Creating the direct product of two lists: `directproduct(<list1>,<list2>)`

*Description:* This operator produces a list that resembles the direct product of two given lists. The direct products consists of all pairs whose first element is taken from `<list1>` and whose second element is taken from `<list2>`.

| Expression | Result |
|---|---|
| `directproduct([1,2,3], ["A", "B"])` | `[[1,"A"], [1,"B"], [2,"A"], [2,"B"], [3,"A"], [3,"B"]]` |

### Flattening a nested list lists: `flatten(<list>)`

*Description:* This operator takes a list that may itself again consist of lists. It returns a single list of elements that results from appending all the second order lists. Using a modifier flattening can be applied recursively. Also the levels of flattening can be controlled.

This operator produces a list that resembles the direct product of two given lists. The direct products consists of all pairs whose first element is taken from `<list1>` and whose second element is taken from `<list2>`.

*Modifiers:* The modifier `levels` can be set either to "all", which results in a complete recursive flattening, or it can be set to an integer, that specifies the maximal recursion level of flattening. The statement `flatten(...,levels->1)` is equivalent to `flatten(...)`.

consecutive

cycle

triples

directproduct

flatten

*Example:* Let us assume that we set

```
list=[[1,2],[3,[4,5],[6,[7,8]]],6];
```

then we get the following responses to various calls of flattening:

| Expression | Result |
|---|---|
| `flatten(list)` | `[1,2,3,[4,5],[6,[7,8]],6]` |
| `flatten(list,levels->0)` | `[[1,2],[3,[4,5],[6,[7,8]]],6]` |
| `flatten(list,levels->1)` | `[1,2,3,[4,5],[6,[7,8]],6]` |
| `flatten(list,levels->2)` | `[1,2,3,4,5,6,[7,8],6]` |
| `flatten(list,levels->3)` | `[1,2,3,4,5,6,7,8,6]` |
| `flatten(list,levels->"all")` | `[1,2,3,4,5,6,7,8,6]` |

## 10.3.2 Order of Elements

The following operators change the order of the elements within a list.

**reverse**

### Reversing a list: `reverse(<list>)`

*Description:* This operator reverses the order of the elements in `<list>`.

| Expression | Result |
|---|---|
| `reverse([1, 2, 3, 4])` | `[4, 3, 2, 1]` |

**sort**

### Sorting a list: `sort(<list>)`

*Description:* Within CindyScript, all elements are in a natural complete order that makes it possible to compare any two elements. Two elements are equal, or one of them is greater than the other. Within the real numbers, the order is the usual numeric order. Within strings, the order is the lexicographic order. Complex numbers are ordered by their real parts first. If two complex numbers have the same real part, then they are compared with respect to their imaginary parts. Two lists are compared by the first entry in which they differ. Furthermore, by convention Cinderella uses the order

$$\text{booleans} < \text{numbers} < \text{strings} < \text{lists}$$

| Expression | Result |
|---|---|
| `sort([4.5, 1.3, 6.7, 0.2])` | `[0.2, 1.3, 4.5, 6.7]` |
| `sort(["one", "two", "three", "four", "five"])` | `["five","four","one","three","two"]` |

**sort**

### Sorting a list: `sort(<list>, <expr>)`

*Description:* This operator takes each element of the list and evaluates a function expressed by `<expr>` applied to it. All elements of the list are sorted with respect to the result of these evaluations.

| Expression | Result |
|---|---|
| `sort([-4.5, 1.3, -6.7, 0.2], abs(#))` | `[0.2, 1.3, -4.5, -6.7]` |
| `sort(["one", "two", "three", "four", "five"],length(#))` | `["one","two","four","five","three"]` |

**Sorting a list: sort(<list>, <var>, <expr>)**                                 `sort`

*Description:* Similar to sort(<list>, <expr>) but with <var> as the run variable.

**Sets from lists: set(<list>)**                                                 `set`

*Description:* This operator sorts all elements of a list and removes occurrences of identical elements. Thus a unique representation of the list is computed if the list is considered as a *set* of objects. Together with the operators concat, remove, and common, this can be used as an implementation of set functionality.

| Expression | Result |
|---|---|
| set([3, 5, 2, 4, 3, 5, 7]) | [2, 3, 4, 5, 7] |
| set([3, 5, 2]++[4, 5, 2]) | [2, 3, 4, 5] |
| set([3, 5, 2]~~[4, 5, 2]) | [2, 5] |

# 10.4  Lists of Geometric Elements

Accessing the geometric elements of a Cinderella construction represents an important interface between CindyScript and a geometric construction. In particular, if for some reason the element names (i.e., labels) are unknown or varying, one needs a way to retrieve the geometric elements. This is done with operators that return lists of elements of a particular kind.

**All points of the construction: allpoints()**                              `allpoints`

**All lines of the construction: alllines()**                                 `alllines`

**All segments of the construction: allsegments()**                         `allsegments`

**All circles of the construction: allcircles()**                           `allcircles`

**All conics of the construction: allconics()**                             `allconics`

**All masses of the construction: allmasses()**                             `allmasses`

**All springs of the construction: allsprings()**                           `allsprings`

**All elements incident to a given one: allsegments(<geo>)**                 `allsegments`

Please refer to the examples section for more on the use of these operators.

# 10.5  Vectors and Matrices

Lists can serve as representation of vectors and matrices. In particular, lists that contain numerical values permit many different arithmetic operations. Furthermore, the coordinate values of geometric elements are usually retrieved as lists of numerical values. For instance, if A is the label of a geometric point, then A.xy returns a list of two numbers, the x and the y coordinates of the point. Similarly, A.homog returns a list of three numbers: the homogeneous coordinates of the point. Several arithmetic operations serve the particular purpose of calculating directly with these coordinate vectors.

### 10.5.1 Definition of Vectors and Matrices

Any list can be considered as a "vector of objects." However, of particular interest are vectors of numbers. Such a vector will be called a "number vector." Whether a certain list is a number vector can be tested with the operator isnumbervector(<expr>).

If the elements of a list are again lists, and if all these lists have the same length, then such a list is called a *matrix*. Whether a list is a matrix can be tested with the operator ismatrix(<expr>). If furthermore all (second-level) elements in the matrix are numbers, this matrix is called a number matrix. Whether a list is a number matrix can be tested with the operator isnumbermatrix(<expr>). The entries of a matrix are vectors of the same length. These vectors are considered as the rows of the matrix. Thus, if a matrix contains $n$ vectors of length $m$, then it is an $n \times m$ matrix.

### 10.5.2 Addition and Multiplication

In the section Arithmetic Operators (p. 248) we explain how the fundamental operations of addition, subtraction, multiplication, and division can be applied to lists of numbers. As a rule of thumb, one can say that on this level, everything that is mathematically reasonable can be performed in CindyScript. So, for instance, if A and B are lists of numbers, then the expression (A+B)/2 calculates the midpoint of these two vectors.

Addition and subtraction of lists is allowed whenever the lists have the same shape. This means that the lists have the same length, and if some of the entries are lists as well, then the corresponding entries of the two summands recursively again have the same shape.

Multiplication with lists is allowed whenever this performs a mathematically meaningful operation. The following table summarizes the different admissible uses of the multiplication operator.

| factor 1 | factor 2 | result | meaning |
|---|---|---|---|
| number | number | number | usual multiplication |
| number | vector of length $r$ | vector of length $r$ | scalar vector multiplication |
| vector of length $r$ | number | vector of length $r$ | scalar vector multiplication |
| vector of length $r$ | vector of length $r$ | number | scalar product of two vectors |
| $n \times r$ matrix | vector of length $r$ | vector of length $n$ | matrix $\times$ vector |
| vector of length $n$ | $n \times r$ matrix | vector of length $r$ | vector $\times$ matrix |
| $n \times r$ matrix | $r \times m$ matrix | $n \times m$ matrix | matrix multiplication |

### 10.5.3 Products, Sums, Max, and Min

**The summation operator: `sum(<list>)`**

*Description:* This operator adds all elements of a list. The elements may be numbers, or themselves lists (or vectors or matrices), or even strings.

| Expression | Result |
|---|---|
| `sum(1..10)` | `55` |
| `sum([4,6,2,6])` | `18` |
| `sum([ [3, 5], [2, 5], [5, 6] ])` | `[10, 16]` |
| `sum(["h","e","ll","o"])` | `"hello"` |

One can, for instance, use the sum operator to define an arithmetic mean function by the following code fragment:

```
average(x) := sum(x)/length(x)
```

This function works for a list of numbers as well as for the average of a list of vectors or matrices.

**The summation operator: `sum(<list>,<expr>)`**

*Description:* This operator is similar to the summation operator, but it takes the sum of results of `<expr>` while a loop traverses all elements of `<list>`. The running variable is as usual `#`.

We can calculate the sum of all squares of the first hundred integers by the following expression:

| Expression | Result |
|---|---|
| `sum(1..100,#^2)` | `338350` |

It is time for a little mathematical mystery:

| Expression | Result |
|---|---|
| `sum(1..10,#^2)` | `385` |
| `sum(1..100,#^2)` | `338350` |
| `sum(1..1000,#^2)` | `333833500` |
| `sum(1..10000,#^2)` | `333383335000` |

**The summation operator: `sum(<list>,<var>,<expr>)`**

*Description:* This operator is similar to the last one, except that the running variable is locally named `<var>`.

**product**

### The product operator: `product(<list>)`

*Description:* This operator multiplies together all elements of a list. The elements are expected to be numbers.

| Expression | Result |
|---|---|
| `product(1..5)` | 120 |

One can, for instance, use the product operator to define the factorial function by the following code fragment:

```
fac(x) := product(1..x)
```

**product**

### The product operator: `product(<list>,<expr>)`

*Description:* This operator is similar to the product operator, but it takes the product of results of `<expr>` while a loop traverses all elements of `<list>`. The running variable is, as usual, #.

**product**

### The product operator: `product(<list>,<var>,<expr>)`

*Description:* This operator is similar to the last one, except that the running variable is locally named `<var>`.

**max**

### The maximum operator: `max(<list>)`

*Description:* This operator finds the maximum value in a list of entries.

| Expression | Result |
|---|---|
| `max([4,2,6,3,5])` | 6 |

**max**

### The maximum operator: `max(<list>,<expr>)`

*Description:* This operator is similar to the max operator `max(<list>)`, but it takes the maximum of results of `<expr>` while a loop traverses all elements of `<list>`. The running variable is, as usual, #.

**max**

### The maximum operator: `max(<list>,<var>,<expr>)`

*Description:* This operator is similar to the last one, except that the running variable is locally named `<var>`.

**min**

### The minimum operator: `min(<list>)`

*Description:* This operator finds the minimum of a list of entries.

| Expression | Result |
|---|---|
| `min([4,2,6,3,5])` | 2 |

**The minimum operator: `min(<list>,<expr>)`**

This operator is similar to the min operator `min(<list>)`, but it takes the minimum of results of `<expr>` while a loop traverses all elements of `<list>`. The running variable is, as usual, `#`.

**The minimum operator: `min(<list>,<var>,<expr>)`**

*Description:* This operator is similar to the last one, except that the running variable is locally named `<var>`.

## 10.5.4 Vector and Matrix Arithmetic

Besides addition and multiplication, as described earlier in this section, there are several operators responsible for vector and matrix administration.

**Dimensions of a matrix: `matrixrowcolum(<matrix>)`**

*Description:* If the argument is a matrix, this operator returns the number of columns and the number of rows of the matrix, encoded as a two-element list.

| Expression | Result |
| --- | --- |
| matrixrowcolumn([[1,2],[3,2], [2,4] | |

**Transposing a matrix: `transpose(<matrix>)`**

*Description:* If the argument is a matrix, this operator returns the transpose of the matrix. In the transpose, the rows and columns are interchanged.

| Expression | Result |
| --- | --- |
| transpose([[1,2],[3,2],[1,3], [5,4]]) | [[1,3,1,5],[2,2,3,4]] |
| transpose([[1],[3],[1],[5]]) | [[1,3,1,5]] |
| transpose([[1,3,1,5]]) | [[1],[3],[1],[5]] |

**Rows of a matrix: `row(<matrix>,<int>)`**

*Description:* If the first argument is a matrix, this operator returns the row with index `<int>` as a vector.

| Expression | Result |
| --- | --- |
| row([[1,2],[3,2],[1,3], [5,4]],2) | [3,2] |

**Columns of a matrix: `column(<matrix>,<int>)`**

*Description:* If the first argument is a matrix, this operator returns the column with index `<int>` as a vector.

| Expression | Result |
| --- | --- |
| column([[1,2],[3,2],[1,3], [5,4]],2) | [2,2,3,4] |

submatrix

**Extracting a submatrix of a matrix: `submatrix(<matrix>,<int1>,<int2>)`**

*Description:* If the first argument is a matrix, this operator returns the submatrix obtained by deleting the column with index `<int1>` and the row with index `<int2>`.

| Expression | Result |
|---|---|
| `submatrix([[1,2,4],[3,2,3],` `[1,3,6],[5,4,7]],2,3)` | `[[1,4],[3,3],[5,7]]` |

rowmatrix

**Converting a vector to a row matrix: `rowmatrix(<vector>)`**

*Description:* If the first argument is a vector, this operator returns the matrix with a single row consisting of this vector.

| Expression | Result |
|---|---|
| `rowmatrix([1,2,3,4])` | `[[1,2,3,4]]` |

columnmatrix

**Converting a vector to a column matrix: `columnmatrix(<vector>)`**

*Description:* If the first argument is a vector, this operator returns the matrix with a single column consisting of this vector.

| Expression | Result |
|---|---|
| `columnmatrix([1,2,3,4])` | `[[1],[2],[3],[4]]` |

zerovector

**Creating a zero vector: `zerovector(<int>)`**

*Description:* Creates a zero vector of length `<int>`.

zeromatrix

**Creating a zero matrix: `zeromatrix(<int1>,<int2>)`**

*Description:* Creates a matrix with `<int1>` rows and `<int2>` columns that contains only zeros.

## *10.5.5 Linear Algebra*

Since lists may be used as vectors or matrices there are also several arithmetic operations from linear algebra that are applicable to lists.

det

**Determinant of a square matrix: `det(<matrix>)`**

*Description:* This operator calculates the determinant of a square matrix, that is, one with the same number of rows and columns. Note that the determinant is an extremely useful function, for many geometric purposes. For instance, the determinant of the $3 \times 3$ matrix formed by the homogeneous coordinates of three points is zero if and only if the three points are collinear. The sign of the determinant carries information on the relative orientation of the three points. In the section Geometric Operators (p. 349) you can find descriptions of the functions `area(<vec1>,<vec2>,<vec3>)` and `det(<vec1>,<vec2>,<vec3>)`. Both are variants of the determinant function that are particularly useful in geometric contexts and exhibit slightly better performance than the general determinant formula.

## Calculating the length of a vector: `|<vec>|`

*Description:* Enclosing a vector between two vertical bars `|<vec>|` can be used to calculate the length of a vector. This operator can also be applied to a real or complex number and returns its absolute value.

## Calculating the distance between two vectors: `|<vec1>,<vec2>|`

*Description:* Enclosing two vectors of equal length within two vertical bars `|<vec1>,<vec2>|` can be used to calculate the distance between the vectors.

## Calculating distances: `dist(<vec1>, <vec2>)`

*Description:* This operator calculates the distance between two vectors and returns it as a number. This operator is also very useful for geometric calculations.

## The Hermitian scalar product: `hermiteanproduct(<vec1>,<vec2>)`

*Description:* This operator returns the Hermitian scalar product of two vectors. It is similar to the dot product `<vec1>*<vec2>`. However, the second vector is complex conjugated before multiplication. In particular, `hermiteanproduct(a,a)` is always nonnegative.

*Example:*

The following code fragment shows the difference between the dot product and the scalar product.

```
a=[2+3*i,1-i];
println(hermiteanproduct(a,a));
println(a*a);
```

produces the output:

```
15
-5 + i*10
```

## Inverse of a square matrix: `inverse(<matrix>)`

*Description:* This operator calculates the inverse of a square matrix, that is, one with the same number of rows and columns. If the matrix is not square or not invertible the operator returns an undefined object. Inverses are sometimes very useful when the same type of linear equations *Ax=b* has to be solved for different right sides *b*. If the matrix *A* does change often it is more preferable to use the `linearsolve` operator.

## Adjunct of a square matrix: `adj(<matrix>)`

*Description:* This operator calculates the adjunct of a square matrix. For invertible matrices the adjunct is the inverse times the determinant. Unlike the inverse the adjunct of a matrix always exists.

**eigenvalues**

**Eigenvalues of a square matrix: `eigenvalues(<matrix>)`**

*Description:* This operator calculates the eigenvalues of a square matrix. The result is returned as a list of values. If an Eigenvalue occurres with algebraic multiplicity 'r' the operator lists this eigenvalue 'r' times. Thus the operator always returns $n$ values for an $n$ by $n$ matrix. In particular the operator assumes the matrix to be embedded over the complex numbers. So also complex eigenvalues are listed.

*Example:*

```
m1=[[1,1,0],[0,1,0],[0,0,.5]];
println(eigenvalues(m1));
m2=[[1,1,0],[-1,1,0],[0,0,.5]];
println(eigenvalues(m2));
```

produces the output:

```
[1,1,0.5]
[1 + i*1,1 - i*1,0.5]
```

**eigenvectors**

**Eigenvectors of a square matrix: `eigenvectors(<matrix>)`**

*Description:* This operator calculates a basis of eigenvectors of a square matrix. The result is returned as a list of vectors. The order of this list corresponds to the order of the eigenvalues in the `eigenvalues` operator.

*Warning:* If the matrix is not diagonalizable the output of this function is meaningless.

**linearsolve**

**Solving a linear equation: `linearsolve(<matrix>,<vector>)`**

**linearsolve**

**Solving a linear equation: `linearsolve(<matrix>,<matrix>)`**

*Description:* The operator `linearsolve(A,b)` calculates a solution $x$ of the system of equations $Ax=b$. The matrix $A$ must be square ($n$ times $n$) and invertible. $b$ can either be an $n$ dimensional vector, it can be a matrix with $n$ rows. If either $A$ is not invertible or the dimension constraints are not met an undefined value is returned.

*Example:*

```
m=[[1,1,0],[0,1,0],[0,1,1]];
x=linearsolve(m,[2,3,4]);
println(x);
println(m*x);
```

produces the output:

```
[-1,3,1]
[2,3,4]
```

### 10.5.6 *Advanced geometric operations*

**Computing a convex hull in 3D: `convexhull3d(<list of vectors>)`**   `convexhull3d`

*Description:* This operator takes a list of 3-dimensional vectors as input and calculates their convex hull. it returns a pair of two lists. The first of these lists contains the vertices of the convex hull. The second list contains a the list of faces of the convex hull. Each facet is given as the indices of the vertices of the first list.

*Example:* The following list of points describes a three dimensional cube with an additional point in its center.

```
[[1,1,1],[1,1,-1],[1,-1,1],[1,-1,-1],
 [-1,1,1],[-1,1,-1],[-1,-1,1],[-1,-1,-1],[0,0,0]]
```

Applying the convex hull operator to this list produces the following output:

```
[
 [[1,1,1],[1,1,-1],[1,-1,1],[1,-1,-1],
  [-1,1,1],[-1,1,-1],[-1,-1,1],[-1,-1,-1]],
 [[6,5,1,2],[3,1,5,7],[3,4,2,1],[8,7,5,6],[8,6,2,4],[8,4,3,7]]
]
```

Observe that the interior point has been properly removed, and that the convex hull operator can nicely handle coplanarities.

The convex hull operator is remarkably robust to degenerate situations. The following image has been computed under usage of the `convexhull3d(...)` operator. It shows the section of a 4-dimensional polytope (a 600-cell) with a 3-dimensional space.

*A section of a 600-cell rendered with CindyScript.*

# Chapter 11

# Drawing

In addition to its computing facilities, one of the most important features of CindyScript is its powerful possibilities for reading position data from a geometric construction and for direct output of graphics to a construction. This section is about the output part. One can easily draw points and lines by invoking a graphics operator within CindyScript. It is important to mention that these drawings of points or lines are not geometric objects of the Cinderella construction. These elements serve purely "decorative" purposes, and it is impossible to apply geometric Cinderella construction steps to them. Nevertheless, they are extremely useful, since often one wants to create complicated-looking output that is not directly constructible by geometric means. Then it is often very easy to write a few lines of CindyScript code that generate the output. In particular, CindyScript is very useful if generation of the output requires repetitive application of constructions. One can also use high-level graphics operations within CindyScript to create a plot of a function directly.

The following topics will be treated:

- *Appearance of Objects* (p. 282)*:* Since the appearance of the elements of a construction is fundamental for all such output operations, we treat the handling of color, size, and transparency first.

- *Elementary Drawing Functions* (p. 285): This section covers the fundamental drawing primitives for points and lines.

- *Function Plotting* (p. 291)*:* Plotting of functions can be very easily done with high-level graphics operations.

- *Texts and Tables* (p. 309)*:* Textual output allows for various ways of creating additional information in a geometric view.

- *TeX Rendering* (p. 313)*:* More details on formula generation inside CindyScript.

- *Script Coordinate System* (p. 340)*:* Finally, it will be explained how the local coordinate system for drawing can be transformed. In this way one can, for instance, create a perspective drawing of a scene.

## 11.1 Appearance of Objects

The drawing engine of Cinderella stores a default appearance for each kind of element (point, line, text) that can be drawn. Whenever a drawing statement is invoked without any modifiers, the default appearance is used to render the object. This is very useful for drawing objects that look graphically identical. The main attributes that are influenced by the appearance are the *color*, *size*, and *opacity* of the object.

gsave           **Push elements on the appearance stack: `gsave()`**

grestore        **Pop elements from the appearance stack: `grestore()`**

greset          **Clear the appearance stack: `greset()`**

*Description:* It is often necessary to switch to temporarily a different default appearance. For this purpose the operators `gsave()` and `grestore()` are provided. The `gsave` operator stores all information about the graphic state (sizes, colors, opacities) in a stack. The `grestore` operator reverses this effect by popping the information from the stack. Finally, the `greset` operator sets the stack back to its initial state. In addition to appearance information, information on the local coordinate system is stored as well.

*See also:* Script Coordinate System (p. 340)

### 11.1.1 Size

pointsize       **Set point size: `pointsize(<number>)`**

linesize        **Set line size: `linesize(<number>)`**

textsize        **Set text size: `textsize(<number>)`**

Default sizes can be set by three operators: `pointsize(<number>)`, `linesize(<number>)`, and `textsize(<number>)`. The size is represented by a real number. For lines and points, the sizes are assumed to be integers between 1 and 20. The sizes encode absolute pixel values. The following code produces the picture below:

```
sizes=1..15;
forall(sizes,
  pointsize(#);
  textsize(#+4);
  draw((#,0));
  drawtext((#,1),#);
)
```

### 11.1.2 Colors

Colors are represented by lists of three real numbers that represent the red/-green/blue component of the color. Each number is assumed to lie between 0 and 1, where 0 means black and 1 means the full color value. With this RGB color scheme the eight fundamental colors are represented by the following vectors:

black: (0,0,0)    red: (1,0,0)    green: (0,1,0)    blue: (0,0,1)

cyan: (0,1,1)    magenta: (1,0,1)    yellow: (1,1,0)    white : (1,1,1)

**Set point color: `pointcolor(<colorvec>)`**                            `pointcolor`

**Set line color: `linecolor(<colorvec>)`**                              `linecolor`

**Set text color: `textcolor(<colorvec>)`**                              `textcolor`

**Set color: `color(<colorvec>)`**                                       `color`

The default colors can be set by three operators: `pointcolor(<colorvec>)`, `linecolor(<colorvec>)`, and `textcolor(<colorvec>)`. Furthermore, the operator `color(<colorvec>)` simultaneously sets the color of all types of objects.

If real values are interpreted as color components, then values below `0` will be replaced by `0` and values above `1` will be replaced by `1`.

*Example:* The following code produces the picture below:

```
n=13;
ind=1..n;
pointsize(9);
forall(ind,i,
  forall(ind,j,
    pointcolor((i/n,j/n,0));
    draw((i,j),noborder->true);
    pointcolor((0,i/n,j/n));
    draw((i+15,j),noborder->true);
    pointcolor((j/n,0,i/n));
    draw((i+30,j),noborder->true);
  )
)
```

### 11.1.3  Opacity

alpha

**Set opacity: `alpha(<number>)`**

*Description:* Opacity is encoded by a real value between 0 and 1. Here 0 stands for completely transparent and 1 for completely opaque. Values that are outside this range are set to either 0 or to 1.

### 11.1.4  Color Functions

To make calculations with colors slightly simpler, a few default functions are declared that return color values.

red

**Red colors: `red(<number>)`**

*Description:* This operator creates an RGB vector whose green and blue values are set to 0. The red value is set to <number>.

green

**Green colors: `green(<number>)`**

*Description:* This operator creates an RGB vector whose red and blue values are set to 0. The green value is set to <number>.

blue

**Blue colors: `blue(<number>)`**

*Description:* This operator creates an RGB vector whose red and green values are set to 0. The blue value is set to <number>.

gray

**Gray colors: `gray(<number>)`**

*Description:* This operator creates an RGB vector whose red, green, and blue values are all set to <number>.

hue

**Rainbow colors: `hue(<number>)`**

*Description:* This operator creates an RGB vector that creates one of the fully saturated rainbow colors. The value of <number> lies between 0 and 1. This range of values represents a full rainbow color cycle. For larger numbers, the cycle repeats periodically.

*Example:* The following code produces the picture below:

```
n=360;
ind=(1..n)/n;
linesize(2);
forall(ind,
  color(hue(#));
  draw((0,0),(sin(#*2*pi),cos(#*2*pi)))
)
```

## 11.2 Elementary Drawing Functions

The operators described in this section are used to draw points, lines, and segments in a construction window.

**Drawing: `draw(<expr>)`**                                                    **draw**

*Description:* The `draw(<expr>)` function is a multifunctional operator. Depending on the meaning of `<expr>`, the corresponding objects will be drawn if possible. Currently, there are three possible inputs that will lead to a drawing action. For the first two we assume that x, y, and z are numbers.

| Expression | Result |
|---|---|
| `draw([x,y])` | Draws a point with $x$-coordinate x and $y$-coordinate y |
| `draw([x,y,z])` | Draws a point with homogenous coordinates given by `[x,y,z]` |

A word on homogeneous coordinates: If `[x,y,z]` are the homogeneous coordinates of a point, then the corresponding point that will be drawn has *xy*-coordinates `[x/z,y/z]`. Points that have homogeneous coordinates with z=0 correspond to "points at infinity." You won't see them in a usual Cinderella Euclidean view. However, they will be drawn in spherical view (or when a local projective basis is set (see Script Coordinate System (p. 340)).

*Drawing a segment:* A segment can be drawn by providing a list of two points. The points can be given in either Euclidean or homogeneous coordinates, which we assume for a and b below.

| Expression | Result |
|---|---|
| `draw([a,b])` | Draws a segment from a to b |

*Example:* The code below produces the following picture. Observe that both Euclidean and homogeneous coordinates are given. Furthermore, the segments appear in front of the points, since their drawing operators are invoked after the drawing operators for the points.

```
A=[0,0];
B=[0,2,2];
C=[1,1,1];
D=[1,0];
E=[0.5,1.5];
linesize(3);
pointsize(10);
draw(A);draw(B);
draw(C);draw(D);draw(E);
draw([A,B]);draw([B,C]);
draw([C,D]);draw([D,A]);
draw([C,E]);draw([B,E]);
```

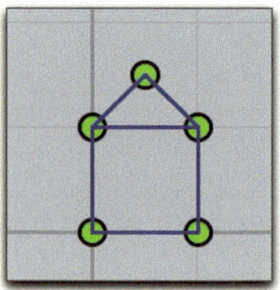

*Drawing a line:* Homogeneous coordinates can also be used to represent lines. One can think of this encoding for lines by three real numbers [a,b,c] as encoding a line with implicit equation a*x+b*y+c=0. A point with Euclidean coordinates [x,y] is on this line if and only if the equation is satisfied. A point with homogeneous coordinates [x,y,z] is on this line if and only if the equation a*x+b*y+c*z=0 is satisfied.

In order to tell CindyScript whether a list of three real numbers [a,b,c] is a point or a line, an internal flag for the list is set by operations that produce lines as output. So for instance, the operation join(A,B) calculates the line through two points A and B and sets the internal line flag. Invoking the draw operator on the result of this operation will draw the line. One can also force the setting of the line flag by applying the operator line(<expr>).

*Examples:* Each of the following two draw operations draws a line:

```
draw(line([1,1,0.5]));
draw(join([1,2],[2,-1]));
```

*Modifiers:* The draw operator can handle the modifiers summarized in the following table:

| point size | *<real>* | sets the point size |
|---|---|---|
| line size | *<real>* | sets the line size |
| size | *<real>* | sets the line size and the point size |
| pointcolor | *[<real1>,<real2>,<real3>]* | sets the point color to an RGB value |
| linecolor | *[<real1>,<real2>,<real3>]* | sets the line color to an RGB value |
| color | *[<real1>,<real2>,<real3>]* | sets the point color and the line color to an RGB value |
| alpha | *<real>* | sets the opacity to the value `<real>` |
| noborder | *<bool>* | `noborder`→`true` turns off the border of points |
| border | *<bool>* | `border`→`true` the opposite of the last modifier |
| dashtype | *<int>* | Specify a certain type of dashing (values 0..4 allowed) |
| dashing | *<real>* | Size of dashing |
| dashpattern | *<list>* | Specify an individual dash pattern |

Modifiers have only a local effect. This means that the default appearance settings are not affected when a modifier is used.

*Example:* The following piece of code illustrates the effect of the dashing modifiers.

```
linesize(3);
draw((0,0),(0,6),dashtype->0);
draw((1,0),(1,6),dashtype->1);
draw((2,0),(2,6),dashtype->2);
draw((3,0),(3,6),dashtype->3);
draw((4,0),(4,6),dashtype->4);
draw((6,0),(6,6),dashing->4);
draw((7,0),(7,6),dashing->6);
draw((8,0),(8,6),dashing->8);
draw((9,0),(9,6),dashing->10);
draw((11,0),(11,6),dashpattern->[0,4,2,4]);
draw((12,0),(12,6),dashpattern->[0,2,2,2,4,2]);
draw((13,0),(13,6),dashpattern->[0,4,2,4]);
draw((14,0),(14,6),dashpattern->[4,2,1,2]);
```

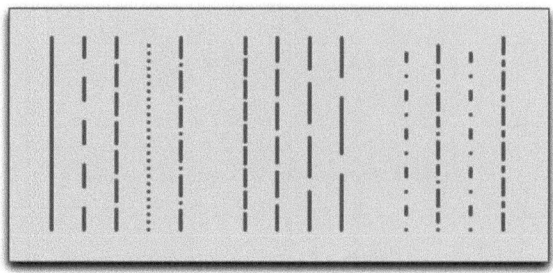

*See also:* Script Coordinate System (p. 340) and Geometric Operators (p. 349)

**Drawing segments: `draw(<expr>,<expr>)`**                          `draw`

*Description:* Invoking the draw operator with two arguments can also be used to draw a segment. Both arguments must represent points either in Euclidean or in homogeneous coordinates. Thus [`draw([0,0],[1,1])`] draws a segment from [`0,0`] to [`1,1`].

*Modifiers:* This operator can handle the same modifiers as the `draw(<expr>)` operator. However, the modifiers that are specific for points have no effect.

**drawall**

**Drawing lists of objects: `drawall(<list>)`**

*Description:* The operator `drawall(<list>)` takes a list as argument. Each element of the list should be such that it could be drawn by the usual `draw(<expr>)` operator. The `drawall` operator will then apply the draw operator to each of the entries of the list. The `drawall` operator is extremely useful for drawing more complicated mathematical pictures that involve structured or highly interrelated mathematical content.

*Example:* The following code produces the picture below. The second line defines a function that maps a number to a point on the unit circle. The line `steps=2*pi*(1..n)/n;` initializes the variable `steps` with a list of 17 angles that correspond to 17 points on the unit circle. These points are assigned to the variable `pts` by the line `pts=apply(steps,f(#))`. The variable `segs` contains all pairs of such points. In the final two lines the `draw` operator is used to draw these lists of objects.

```
n=17;
f(x):=[sin(x),cos(x)];
steps=2*pi*(1..n)/n;
pts=apply(steps,f(#));
segs=pairs(pts);
drawall(segs,alpha->0.9);
drawall(pts,size->4);
```

*Modifiers:* This operator can handle the same modifiers as the `draw(<expr>)` operator.

**connect**

**Connect the dots: `connect(<list>)`**

*Description:* This operator takes a list of points as input and connects them by line segments.

*Example:* The following code together with a collection of points of a construction produces the picture below. The first line assigns to the variable `pts` all points of a construction. The second line assigns to `sortpts` a list of these points that was sorted by the *x*-coordinate of the points. Finally, the `connect` operator connects these points in the given sequential order.

```
pts=allpoints();
sortpts=sort(pts,#.x);
connect(sortpts);
```

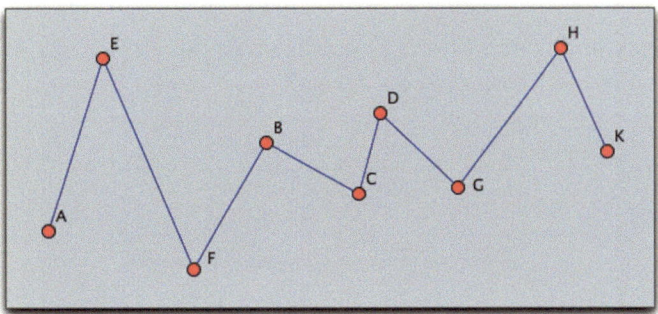

*Modifiers:* This operator can handle the same modifiers as the `draw(<expr>)` operator.

**Draw a polygon: `drawpoly(<list>)`**                                    `drawpoly`

*Description:* This operator takes a list of points as input and draws the border of the polygon described by this list.

*Modifiers:* This operator can handle the following modifiers:

| | | |
|---|---|---|
| **color** | *[<real1>,<real2>,<real3>]* | sets the point color and the line color to an RGB value |
| **alpha** | *<real>* | sets the opacity to the value *<real>* |

**Fill a polygon: `fillpoly(<list>)`**                                    `fillpoly`

*Description:* This operator takes a list of points as input and creates a polygon from them.

*Example:* The following code creates the picture below. In each iteration step of the `repeat` loop the square is drawn, and after this the coordinate system is rotated and scaled.

```
sq=[[-1,-1],[-1,1],[1,1],[1,-1]];
repeat(300,i,
  fillpoly(sq,color->hue(i/10));
  rotate(4°);
  scale(0.95);
)
```

*Modifiers:* This operator can handle the following modifiers:

| | | |
|---|---|---|
| `color` | *[<real1>,<real2>,<real3>]* | sets the point color and the line color to an RGB value |
| `alpha` | *<real>* | sets the opacity to the value `<real>` |

**drawcircle**

**Drawing circles: `drawcircle(<point>,<radius>)`**

*Description:* Draws a circle at `<point>` with radius given by a number `<radius>`. The point may be given either in euclidean or in homogeneous coordinates.

*Modifiers:* This operator can handle the same modifiers as the `draw(<expr>)` operator.

**fillcircle**

**Filling circles: `fillcircle(<point>,<radius>)`**

*Description:* Draws the interior of a circle at `<point>` with radius given by a number `<radius>`. The point may be given either in euclidean or in homogeneous coordinates.

*Modifiers:* This operator can handle the following modifiers:

| | | |
|---|---|---|
| `color` | *[<real1>,<real2>,<real3>]* | sets the fill color to an RGB value |
| `alpha` | *<real>* | sets the opacity to the value `<real>` |

*Example:* The following piece of code shows a combined usage of the `drawcircle` and the `fillcircle` operator.

```
repeat(100,i,
 fillcircle((0,0),1,color->hue(i/70));
 drawcircle((0,0),1,color->(0,0,0));
 translate((1.5,0));rotate(26°);scale(.95);
);
```

## 11.3 Function Plotting

Cinderella provides several operators that allow one to plot information about mathematical functions. Besides simple function plotting, information on extrema, zeros, and inflection points can be shown.

### *11.3.1 Functions*

The following functions allow to plot the graph of a function that maps a real number to a real number.

**Plotting a Function: `plot(<expr>)`**                                          `plot`

*Description:* The `plot` operator can be used to plot a function. The function must be given as an expression `<expr>`. This expression must contain the running variable `#` and calculate either a real value for a real input of `#` or a two-dimensional vector. In the first case, the `plot` operator will simply draw the function. In the latter case, it will draw a parametric plot of a function. The coordinate system is tied to the coordinate system of the geometric views. Instead of `#` also other running variables are detected automatically. If there is only one free variable then this variable is taken as running variable. If there are several free variables the `plot(...)` function searches for typical names in the order `x`, `y`, `t`, `z`.

*Examples:* In its simplest form the `plot` operator can be used directly to plot a function. The line `plot(sin(#))` immediately plots the function $\sin(x)$. The same plot is generated by `plot(sin(x))`.

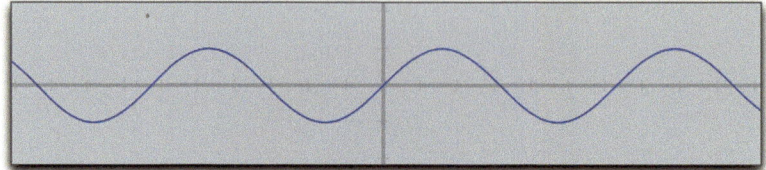

Similarly, one can first define a function whose graph is then displayed by the `plot` operator. The following lines of code produce the picture below:

```
f(x):=1/(x^2+1)*sin(4*x);
plot(f(x));
```

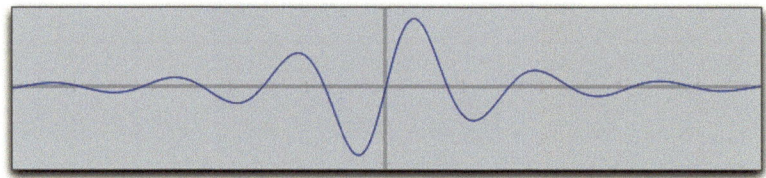

If invoked with an expression `<expr>` that produces a two-dimensional vector as output, the `plot` operator will automatically generate a parametric plot of a curve. Here the value range for the input variable is taken by default to be from 0 to 100. However, these defaults can easily be changed by modifiers. The line `plot([sin(t),cos(t)]*t)` produces the following output:

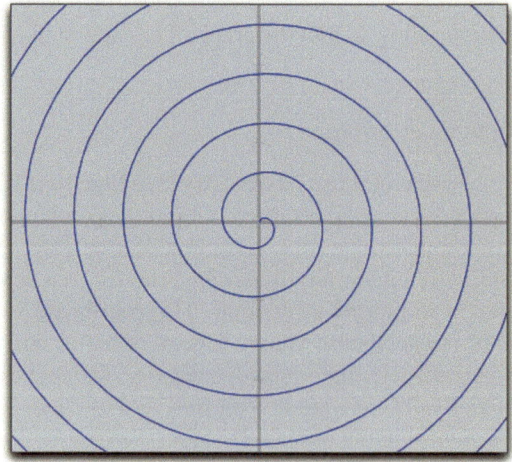

*Modifiers:* The `plot` operator supports many different modifiers. Some of them have to be explained in detail. They can be used to modify the appearance of the curve, to modify the plot range, to modify the position on the screen, and even to display maxima, minima, zeros, and inflection points of the function. An overview of available modifiers is given in the table below. Observe that some of the modifiers may even be invoked with different types of arguments, resulting in a slightly different effect.

*Appearance*

| `color` | *[<real1>,<real2>,<real3>]* | set color to specific value |
|---|---|---|
| `size` | *<real>* | set line size to specific value |
| `alpha` | *<real>* | set opacity |
| `connect` | *<true>* | connect jumps in functions |

*Iteration control*

| `start` | *<real>* | set start value for function drawing |
|---|---|---|
| `stop` | *<real>* | set end value for function drawing |
| `steps` | *<real>* | number of set plot points (for parametric functions only) |
| `pxlres` | *<real>* | pixel resolution of curve plotting (for real functions only) |

*Significant points*

| `extrema` | *<bool>* | mark all extrema |
|---|---|---|
| `extrema` | *[<real1>,<real2>,<real3>]* | mark all extrema in specified color |
| `minima` | *<bool>* | mark all minima |
| `minima` | *[<real1>,<real2>,<real3>]* | mark all minima in specified color |
| `maxima` | *<bool>* | mark all maxima |
| `maxima` | *[<real1>,<real2>,<real3>]* | mark all maxima in specified color |
| `zeros` | *<bool>* | mark all zeros |
| `zeros` | *[<real1>,<real2>,<real3>]* | mark all zeros in specified color |
| `inflections` | *<bool>* | mark all inflection points |
| `inflections` | *[<real1>,<real2>,<real3>]* | mark all inflection points in specified color |

*Line style*

| `dashing` | *<real>* | width of dash patterns (default 5) |
|---|---|---|
| `dashtype` | *<int>* | a specific dash type (values 0...4 are allowed) |
| `dashpattern` | *<list>* | specify an individual dash pattern |

*Examples:* Here are a few examples that demonstrate the use of the modifiers:

One can easily vary the plot range by setting the start and stop values. For instance, `plot(f(#),start->A.x,stop->B.x)` helps to control the plot using by the *x*-coordinates of two free construction points.

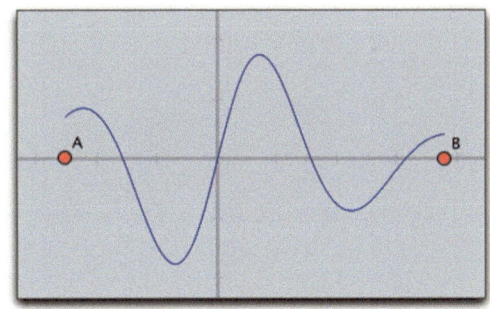

**Plot Appearance**

The resolution of the plot is controlled automatically and adaptively. The `plot(...)` function automatically increases its resolution close to singularities. The following plot shows the output of the call `plot(sin(1/#)*#)`. Observe the quality of the plot close to the origin.

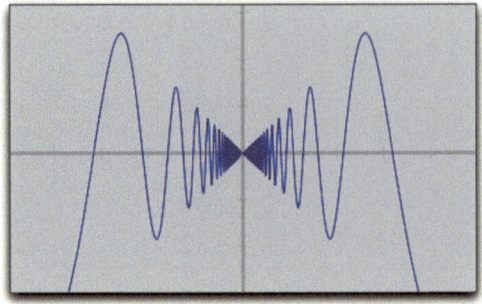

Usually jumps in functions are detected and by default they arenot connected. One can connect them on purpose by setting `connect->true`.

        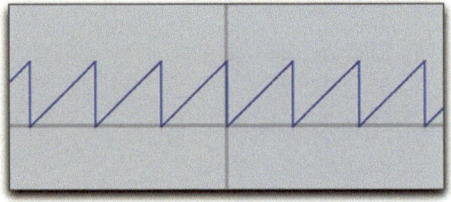

`plot(x-floor(x))`                        `plot(x-floor(x),connect->true)`

**Special Points**

By using the modifiers for a curve's significant points one can display zeros, minima, maxima, and inflection points. The following three pictures demonstrate the use of these operators.

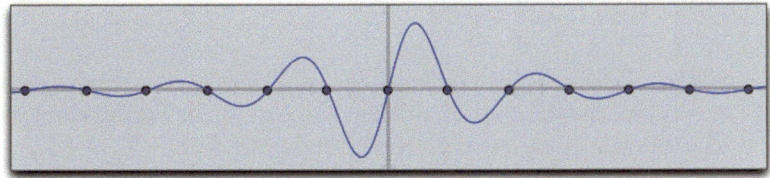

```
zeros->true
```

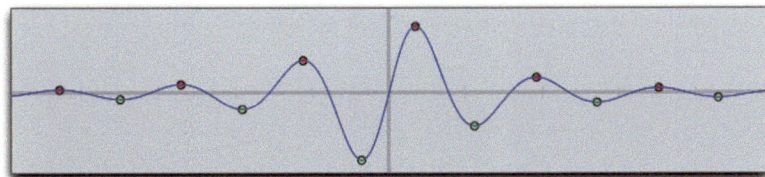

```
extrema->true
```

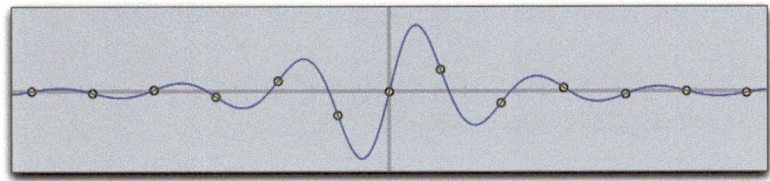

```
inflections->true
```

## Dashing

The Dashing options for a plot statement are identical to those for lines and circles. They can be controlled by the modifiers `dashing`, `dashtype` and `dashpattern`.

- `dashtype` Can be an integer between 0 and 4 and selects one of four predefined dashing patterns. The value 0 creates a solid line.

- `dashing` is a real number specifying the unitsize of the dashes. Simply setting `dashing->5` already creates a standard dashing.

- `dashpattern` Can be set to a list that specifies the length of the successive dashes and empty spaces.

The following picture has been created with `plot(sin(x),dashpattern->[0,3,7,3],size->2,dashing->5)`

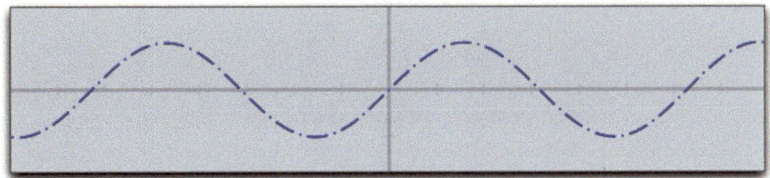

### 11.3.2  Dynamic Color and Alpha

The `color` and the `alpha` can again be functions that depend on the running variable. By the the opacity and the color of the function plot can be varied along with the function. The following plot was generated with the statement

```
plot(sin(x),color->hue(x/(2*pi)),size->3)
```

Here the `hue` function was used that cyclically generates rainbow colors.

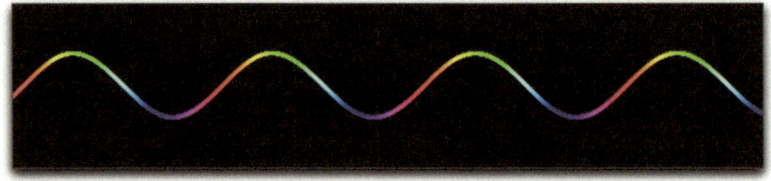

**plot**                    **Plotting a Function: `plot(<expr>,<var>)`**

*Description:* Identical to `plot(<expr>)` but with a specified running variable.

**fillplot**                **Plotting integral-like effects: `fillplot(<expr>)`**

*Description:* Often it is desirable to highlight the area between a function graph and the x-axis of the coordinate system (for instance if one generates an applet for explaining integrals). This can be done using the function `fillplot`. Similarly to `plot`, this operator takes a function as argument (the running variable is determined by the same process as in `plot(...)`). In its simplest form this operator just highlights the area traversed by the function. The function itself is not drawn. This could be done by explicitly calling also the `plot(...)` operator. The following code

```
f(x):=1/(x^2+1)*sin(4*x);
fillplot(f(x));
plot(f(x));
```

produces the following picture:

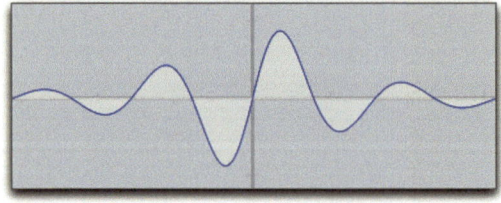

*Warning:* The singularity treatment of the `fillplot(...)` statement is by far less subtle than that of the `plot(...)` statement. So although the modifiers allow to draw functions also with `fillplot` one should use `plot` for function plotting.

*Modifiers:*

*Appearance.*

| **color** | *[<real1>,<real2>,<real3>]* | set color to specific value |
|---|---|---|
| **pluscolor** | *[<real1>,<real2>,<real3>]* | set color for positive function values |
| **minuscolor** | *[<real1>,<real2>,<real3>]* | set color for negative function values |
| **alpha** | *<real>* | set opacity |

*Iteration control*

| **start** | *<real>* | set start value for function drawing |
|---|---|---|
| **stop** | *<real>* | set end value for function drawing |

*Function graph*

| **graph** | *<bool>* | plot also the function graph |
|---|---|---|
| **graph** | *[<real1>,<real2>,<real3>]* | plot also the function graph in specified color |
| **size** | *<real>* | set line size for the function graph |

The `color` and the `alpha` modifier again support the use of functions that control color and opacity (similar to `plot`). The following sampler illustrates different usages of the `fillplot` statement:

```
fillplot(sin(x))
```

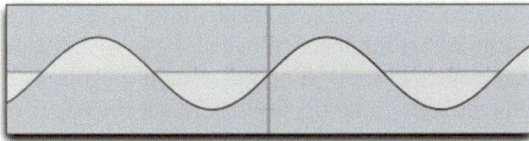

```
fillplot(sin(x),graph->true)
```

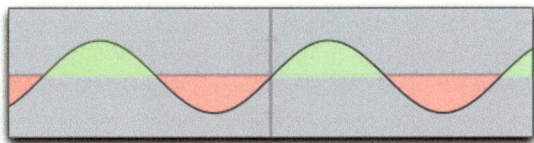

```
fillplot(sin(x),graph->true,pluscolor->(.5,1,.5),minuscolor->(1,.5,.5))
```

```
fillplot(sin(x),graph->true,color->(sin(x),-sin(x),0))
```

**Plotting integral like effects: `fillplot(<expr1>,<expr2>)`**                    `fillplot`

*Description:* This function is very similar to the `fillplot(...)` statement. However, instead of highlighting the are between a function and the x-axis it highlights the area between two functions. The following picture

was created using the statement

```
fillplot(sin(x),cos(x),graph->true,pluscolor->(.5,1,.5),minuscolor
    ->(1,.5,.5))
```

*Modifiers:* This statement supports exactly the same modifiers as `fillplot(...)`.

### 11.3.3 Colorplots

Colorplots are useful to create visual information about functions defined in the entire plane. They can associate a color value to every point in a rectangle.

colorplot

**Creating a colorplot: `colorplot(<expr>,<vec>,<vec>)`**

*Description:* The `colorplot` operator makes it possible to give a visualization of a planar function. To each point of a rectangle a color value can be assigned by a function. In the function `<expr>` the running variable may be chosen as # (for more on running variables, see below). However, it is important to notice that this variable describes now a point in the plane (this is a variable two dimensional coordinates). The return value of `<expr>` should be either a real number (in which case a gray value is assigned) or a vector of three real numbers (in which case an RGB color value is assigned). In any case, the values of the real numbers should lie between 0 and 1. The second and third argument determine the lower left and the upper right corners of the drawing area.

*Example:* The following code and two points A and B produce the picture below. In the first line, a real-valued function is defined that assigns to two points the sine of the distance between them (shifted and scaled to fit into the interval [0, 1]). The first argument of the `colorplot` operator is now a vector of three numbers depending on the run variable # (the red part is a circular wave around A, the green part is zero, and the blue part is a circular wave around B). Finally, C and D mark the corners of the rectangle.

```
f(A,B):=((1+sin(2*dist(A,B)))/2);
colorplot(
    (f(A,#),0,f(B,#)),
    C,D
)
```

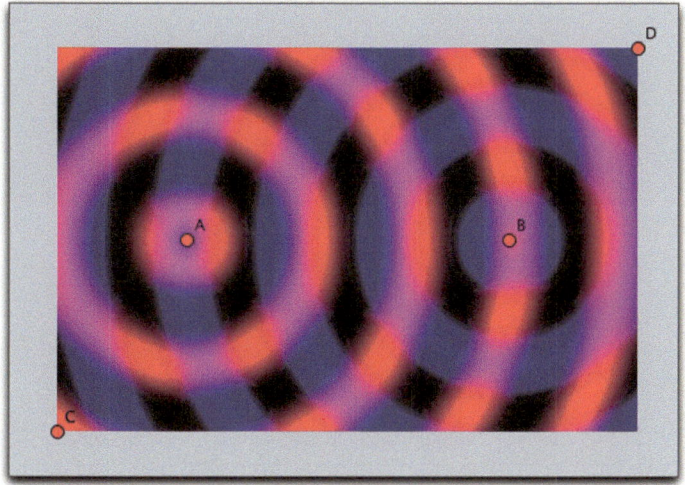

*Running Variables:* Usually # is a good choice for the running variable in the `colorplot` function. However also other choices are possible. The possibilites for running variables are checked in the following order:

- If there is only one free variable in `<expr>` then this variable is taken as running variable and interpreted as a two dimensional vector.

- If `<expr>` contains # the # is taken as running variable (again as a two dimensional vector)

- If `<expr>` contains both x and y as free varaibles the these two variables can be used as running variables the together represent the vector `(x,y)`.

- If exactly one free variable is not assigned yet, then this variable is taken (as vector)

- if none of the above happens also p (for point) and z (for complex number) are checked as running variables.

For instance the following line

```
colorplot((sin(2*x),sin(2*y),0),A,B)
```

produces the following picture:

*Modifiers:* The `colorplot` operator supports three modifiers. The modifier `pxlres` can be set to an integer that determines the size in pixels of the elementary quadrangles of the color plot. The picture above was taken with `pxlres->2` which is the default value. Setting the `pxlres` modifier either to 1 or to 2 produces stunning pictures. However, one should be aware that for each elementary quadrangle of the `colorplot` output the function has to be evaluated once. The computational effort grows quadratically as `pxlres` is linearly reduced. So sometimes it is also good practice to reduce the pxlres modifier in order to gain more preformace. The picture below has been rendered by setting `pxlres->8`.

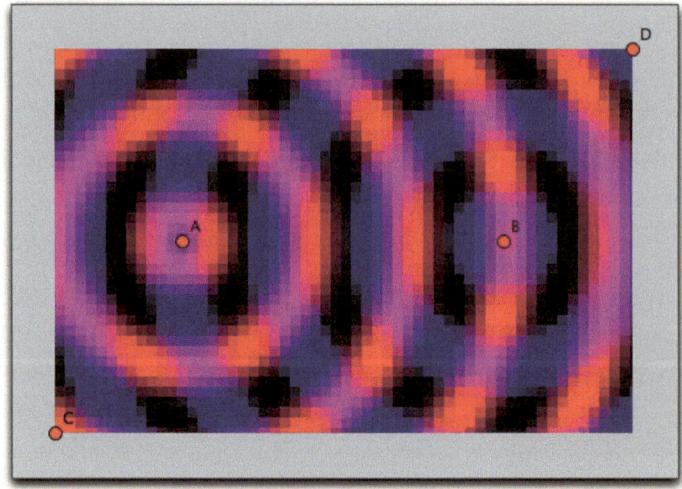

One can also dynamically change the resolution. For this there is another modifier `startres` that can be used to have gradually improving plots, thus combining the best of two worlds. For example, using both `startres->16` and `pxlres->1` you will get a coarse plot during interactive movements, which will be re-calculated automatically with a finer resolution whenever there is enough time.

Furthermore, the `colorplot` operator supports a modifier `alpha` that is used to control the opacity. This modifier can even be set to values that depend parametrically on the running variable.

The picture below was generated by the following code, which is only a slight modification of the code of one of the previous examples:

```
f(A,B):=((1+sin(2*dist(A,B)))/2);
colorplot(
    (f(A,#),0,f(B,#)),
    C,D,
    pxlres->2,
    alpha->-abs(#)+5
)
```

### 11.3.4 Vector Fields

Vector fields can be used to visualize flows and forces. They have many applications in the visualization of systems of differential equations.

**Drawing a vector field: `drawfield(<expr>)`**                           `drawfield`

*Description:* The `drawfield` operator can be used to draw a vector field. The function must be given as an expression `<expr>`. This expression must contain a running variable (usually `#`), which should this time represent a two-dimensional vector (as in color plot). The result should also be a two-dimensional vector. Applying the operator `drawfield` to this expression will result in plotting the corresponding vector field. The field will be animated. This means that it will change slightly with every picture. Therefore, it is often useful to put the `drawfield` operator into the "timer tick" evaluation slot. This creates an animation control in the geometric view. Running the animation will automatically animate the vector field. The running variable policy is identical to the one in the `colorplot(...)` statement. In particular it is possible to use free variables `x` and `y` to represent the two dimensional location `(x,y)`

*Examples:* We consider a vector field defined by the function $f(x,y) = (y, \sin(x))$. The corresponding code with the function definition and the call of the `drawfield` operator is as follows:

```
f(v):=[v.y,sin(v.x)];
drawfield(f(#));
```

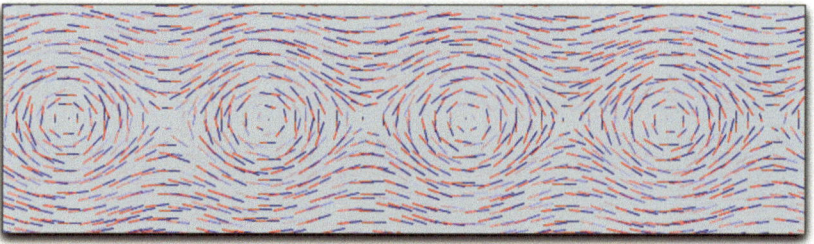

Alternatively the same picture could be generated by

```
drawfield((y,sin(x)));
```

To generate the picture, a collection of needlelike objects is thrown onto the drawing surface. These needles will be oriented according to the vector field. During the animation, the needles move according to the vector field. It is also possible to replace the needles by small snakelike objects that give a more accurate impression of the vector field but take a longer time for calculation. This can be done with a suitable modifier.

```
f(v):=[v.y,sin(v.x)];
drawfield(f(#),stream->true,color->(0,0,0));
```

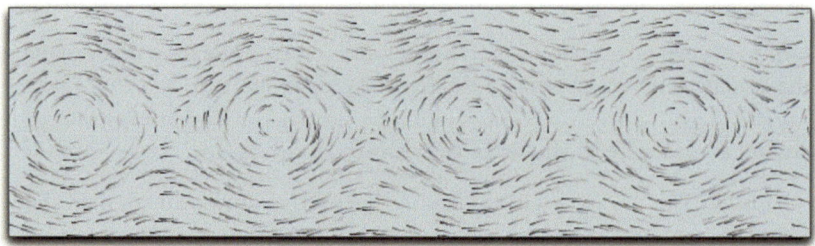

*Modifiers:* The `drawfield` operator supports many modifiers that control the generation process of the vector field. To help in understanding them we first describe in a bit more detail how the pictures are generated.

The pictures are generated by showing the movement of some test objects under the influence of the field. By default, the test objects are needlelike. They are initially positioned on a regular grid. Since this usually creates many visual artifacts, they are randomly distorted within a certain radius around the grid points. During an animation the needles are moved in the direction of the force field. The needles' lengths represent the strength of the field.

*Test objects*

| | | |
|---|---|---|
| **resolution** | *integer* | original grid cell size in pixels |
| **jitter** | *integer* | distortion of the test objects |
| **needlesize** | *<real>* | maximum size of the needles |
| **factor** | *<real>* | scaling factor of the field strength |
| **stream** | *<bool>* | use needles or streamlets |
| **move** | *<real>* | speed of moving objects |

*Appearance*

| | | |
|---|---|---|
| **color** | *[<real1>,<real2>,<real3>]* | set streamlet color or first needle color |
| **color2** | *[<real1>,<real2>,<real3>]* | set second needle color |

The following picture demonstrates the original grid. It has been rendered with `move->0` and `jitter->0`. It shows clear artifacts resulting from unnatural alignment in the horizontal or vertical direction.

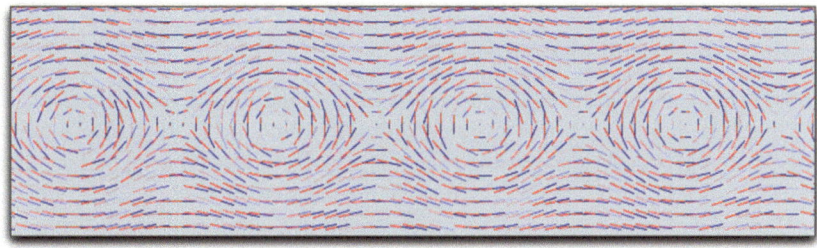

The following picture has been rendered with `resolution->5` and `stream->true`.

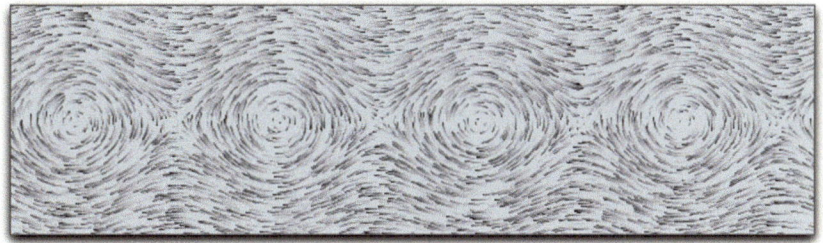

**Drawing a complex vector field: `drawfieldcomplex(<expr>)`**          `drawfieldcomplex`

*Description:* This operator is very similar to the `drawfield` operator. However, it takes as input a one-dimensional complex function. The real and imaginary parts are treated as *x* and *y* components for the vector field. Otherwise, the operator is completely analogous to the previous one.

*Example:* The following example demonstrates the use of the operator with a complex polynomial whose zeros are determined by four points in the drawing:

```
f(x):=(x-complex(A))*(x-complex(B))*(x-complex(C))*(x-complex(D));
drawfieldcomplex(f(#),stream->true,resolution->5,color->(0,0,0))
```

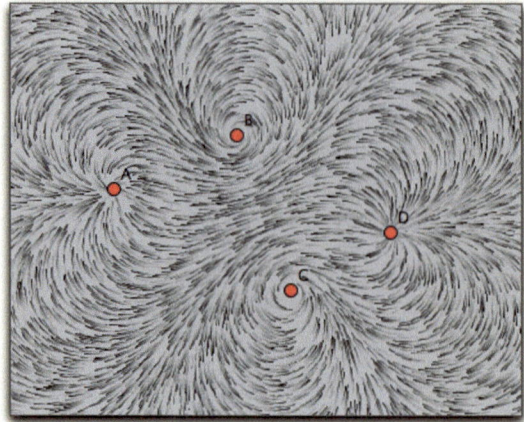

The modifiers are analogous to those for the `drawfield` operators.

`drawforces`   **Drawing a force field: `drawforces()`**

*Description:* This operator is again very similar to the `drawfield` operator. However, this time it is related to a physics simulation in CindyLab (p. 167). No arguments are required, and it shows the forces on a potential test charge that is placed at various locations on the screen. The test charge has mass = 1, charge = 1, and radius = 1. However, no other particle will interact with it. Sometimes it will be necessary to use the `factor` modifier to amplify the force field. The following example shows the interaction among four charged particles.

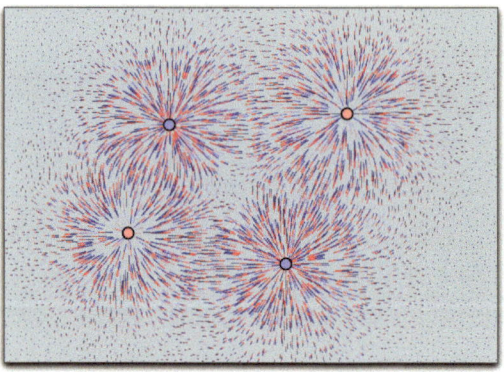

`drawforces`   **Drawing the force field of a point: `drawforces(<mass>)`**

*Description:* There is another operator that draws the force field with respect to a fixed mass particle. The particle itself takes no part in the calculation of the forces. In this way, one can visualize the forces that act on a certain particle.

### 11.3.5 Grids

Grids can be used to visualize transformations that map the plane onto itself. With grids, the deformation induced by such a map can be visualized.

**Mapping a rectangular grid: `mapgrid(<expr>)`** `mapgrid`

*Description:* This operator takes a rectangular grid and deforms it by a function given in `<expr>`. By default the range of the original grid is taken to be the unit rectangle in the plane. The bound of this rectangle may be altered using modifiers. It is also possible to visualize complex maps by using the `complex->true` modifier.

*Modifiers:* There are several modifiers controlling the behavior of this function.

*Appearance*

| `color` | *[<real1>,<real2>,<real3>]* | set color to specific value |
|---|---|---|
| `alpha` | *<real>* | set opacity |
| `size` | *<real>* | set the size of the grid lines |

*Iteration control*

| `xrange` | *[<real>,<real>]* | *x*-range of the source rectangle |
|---|---|---|
| `yrange` | *[<real>,<real>]* | *y*-range of the source rectangle |
| `resolution` | *<int>* | number of grid lines in both directions |
| `resolutionx` | *<int>* | number of grid lines in *x*-direction |
| `resolutiony` | *<int>* | number of grid lines in *y*-direction |
| `step` | *<int>* | refinement in both directions |
| `stepx` | *<int>* | refinement in *x*-direction |
| `stepy` | *<int>* | refinement in *y*-direction |

*Type*

| `complex` | *<boolean>* | use complex functions |
|---|---|---|

*Examples:* The following piece of code exemplifies the usage of the `mapgrid` operator. Is also illustrates the effect of the `xrange` and `yrange` modifiers. It displays the effect of a two-dimensional function that squares the *x* and the *y* coordinate separately.

```
f(v):=(v_1^2,v_2^2);
linesize(1.5);
mapgrid(f(v),color->(0,0,0));
mapgrid(f(v),xrange->[1,2],color->(.6,0,0));
mapgrid(f(v),yrange->[1,2],color->(.6,0,0));
mapgrid(f(v),xrange->[1,2],yrange->[1,2],color->(0,.6,0));
```

In the following example we see that generally the grid lines do not have to stay straight or parallel.

```
f(v):=(v_1*sin(v_2),v_2*sin(v_1));
linesize(1.5);
mapgrid(f(v),xrange->[1,2],yrange->[1,2]);
```

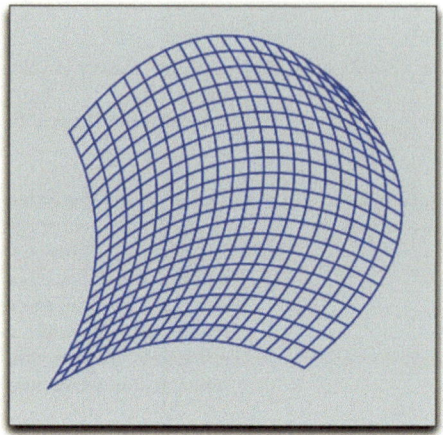

The following simple example illustrates the usage of mapgrid for complex functions.

```
mapgrid(z^2,complex->true);
```

Using the resolution modifier, one can specify the number of mesh lines that are generated.

```
mapgrid(z^2,complex->true,resolution->4);
```

By default, the mapgrid command directly connects the mesh points. This may lead to pictures that do not really map the mathematical truth. Using the step modifier, one can introduce additional steps between the mesh points.

```
mapgrid(z^2,complex->true,resolution->4,step->5);
```

The results of the last three pieces of code are shown below.

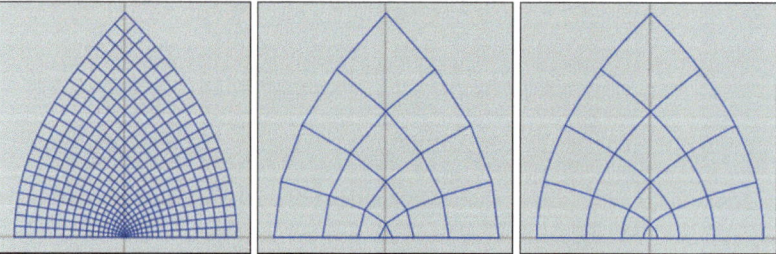

Grids carry very characteristic information about complex functions. The following three pictures show grids for the functions $z^2$, $\sin(z)$, $\frac{1}{z}$, and $\tan(z)$ respectively.

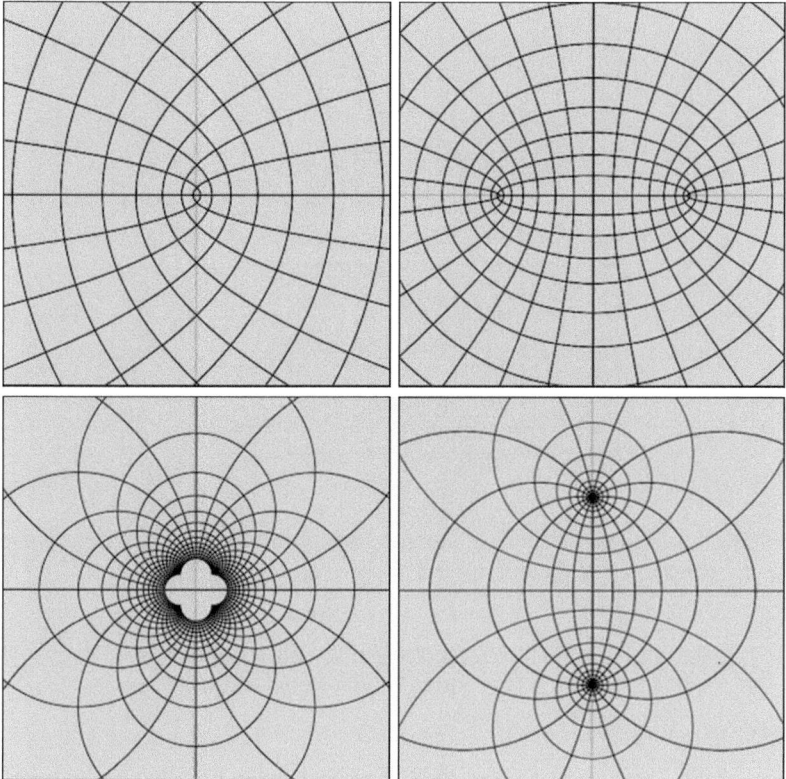

## 11.3.6 Oscillographs

Oscillographs allow to visualize dynamic changes of values in physic simulations and animations.

**Curve drawing of physics magnitudes: `drawcurves(<vec>,<list>)`**     `drawcurves`

*Description:* In real and simulated physical situations one is often interested in plotting curves that show how magnitudes evolve over time. For this, the `drawcurves` operator was created. Here `<vec>` is a two-dimensional vector that refers to the lower left corner of the drawing area, and `<list>` is a list of values that are to be observed. When the animation runs, the values are updated and the corresponding curves are drawn.

*Example:* The next picture shows a very simple application of the `drawcurves` operator. In CindyLab (p. 167), a physical pendulum was constructed. The following code produces a curve plot of the *x* coordinate of the moving point and of its *x* velocity:

```
drawcurves([0,0],[D.x,D.vx])
```

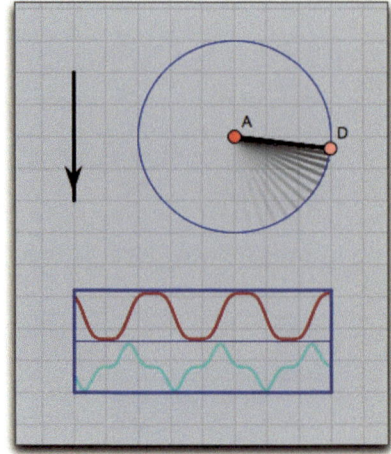

*Modifiers:* The `drawcurves` operator supports many modifiers. Than can be used to change the appearance of the curves and to show additional information.

*Dimension*

| | | |
|---|---|---|
| **width** | *<real>* | pixel width of the plot range |
| **height** | *<real>* | pixel height for each curve |

*Appearance*

| | | |
|---|---|---|
| **border** | *<bool>* | show the borders of the table |
| **back** | *<bool>* | show a background |
| **back** | *[<real1>,<real2>,<real3>]* | show background in specified color |
| **backalpha** | *<real>* | opacity of background |
| **colors** | *[<col1>,<col2>,<col3>,...]* | provide a color for each curve |

*Information*

| | | |
|---|---|---|
| **texts** | *[<text1>,<text2>,...]* | provide a caption for each curve |
| **showrange** | *<bool>* | show the max and min values for each curve |

*Rendering*

| | | |
|---|---|---|
| **range** | *<string>* | "peek" scales to the absolute measured maximum, "auto" scales to the currently shown part of the curve |
| **range** | *[<string1>,<string2>,...]* | individual "peek"/"auto" for each curve |

The following piece of code demonstrates the usage of the modifiers. It shows a weakly coupled pendulum and its energy behavior.

```
linecolor((1,1,1));
textcolor((0,0.8,0));
drawcurves((-7,-3),
  [A.x,B.x,A.ke,B.ke,a.pe+b.pe+c.pe],
  height->50,
  color->(1,1,1),
  back->(0,0,0),
  backalpha->1,
  range->"peek",width->400,
  colors->[
    [1,0.5,0.5],
```

```
    [0.5,1,0.5],
    [1,0.5,0.5],
    [0.5,1,0.5],
    [0.5,0.5,1]],
  texts->[
    "PosA = "+ A.x,
    "PosB = "+B.x,
    "EnergyA = "+A.ke,
    "EnergyB = "+B.ke,
    "PotentialEnergy = "+(a.pe+b.pe+c.pe)
 ]
);
```

The corresponding drawing looks as follows:

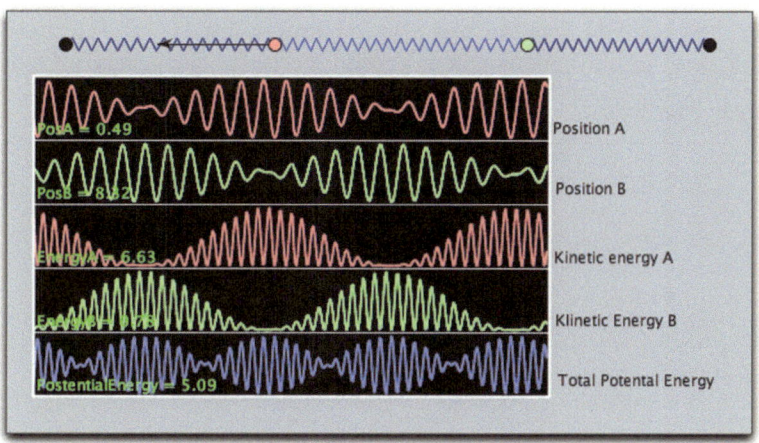

## 11.4  Texts and Tables

With CindyScript you can produce controlled and styled text for a drawing. The essential functionality is covered by the `drawtext` operator. Furthermore, with `drawtable` a table output can be generated. Using texts and tables in scripts is very useful, since one can use the script language to control when and where text is displayed in a construction. This is a very important tool for providing explanatory texts and functional exercises that react to input.

**Drawing Text: `drawtext(<vec>,<expr>)`**                              `drawtext`

*Description:* The `drawtext(<vec>,<string>)` operator plots a text `<string>` at a specified position that is given by the position vector `<vec>`. This position vector can be given either in Euclidean *xy*-coordinates or in homogeneous coordinates.

*Example:* The line `drawtext((0,0),"Hello World")` prints the string "Hello World" with lower left corner at the position (0, 0).

*Modifiers:* The `drawtext` operator supports several operators for the modification of appearance and position.

| `size` | *<real>* | sets the text size |
|---|---|---|
| `color` | *[<real>,<real>,<real>]* | sets the text color |
| `alpha` | *<real>* | sets the text opacity |
| `xoffset` | *<real>* | set an *x* offset in pixels between text and base point |
| `yoffset` | *<real>* | set a *y* offset in pixels between text and base point |
| `offset` | *[<real>,<real>]* | set an *xy* offset in pixels between text and base point |
| `align` | *"left", "right", "mid"* | determines where the text alignment should be |
| `bold` | *<bool>* | flag for bold text |
| `italics` | *<bool>* | flag for italics text |
| `family` | *<string>* | specifying the font family |

Besides the available font families also the three strings `serif`, `sansserif` and `monospaced` are allowed, that always produce corresponding standard font types.

*Example:* The code

```
x=1..10;
forall(x,i,
  drawtext((i,0),"Text",
  size->2*i+15,
  bold->true,
  color->(1-i/10,0,i/10))
)
```

produces the picture below.

If the string expression contains a <newline> chacracter then this line-break is literally interpreted. By this it is easy to produce multilined text, as the following piece of code shows.

```
drawtext((0,0),size->18,
"In Cinderella 'newlines' in Text
are really used as line terminators.
So this text will appear as a
multilined text.")
```

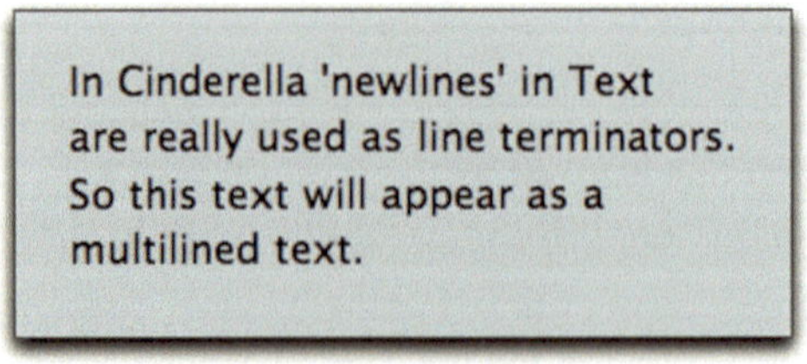

**Available font families: `fontfamilies()`**

*Description:* This operator produces a list of all font families that are available on your computer. Usually this will be quite a lot. The following piece of code displays the names of some of these font families in their own font style.

```
families=fontfamilies();
i=0;
while(length(families)>i,i=i+1;
    drawtext((mod(i,5)*7,round(i/5)),families_i,family->families_i);
)
```

**Unicode Characters: `unicode(<string>)`**

*Description:* Cinderella can display any Unicode Character in text strings. To access a unicode character the statement `unicode(<string>)` accepts a unicode description (a number) given as a string of digits. It returns a string with the corresponding unicode character. By default the base of the unicode description is 16. So hexadecimal descriptions are usually accepted as input. The base can be changed by a modifier.

The base of the unicode description can be changed with the `base` modifier.

*Example:* Both `unicode("0041")` and `unicode("65",base->10)` produce the character "A".

**Test whether a string can be displayed: `candisplay(<string>)`**

*Description:* This operator tests, whether a given string can be displayed in the currently chosen font. It returns a boolean value.

**Drawing tables: `drawtable(<vec>,<list>)`**

One- and two-dimensional lists can be easily drawn in a geometric view as tables. In the simplest form one has only to provide the list and a position where the table should be drawn. Modifiers can be used to fine tune the appearance of the table.

*Example:* The following code produces the picture below:

```
x=1..10;
table=apply(x,(#,#^2,#^3,#^4));
drawtable((0,0),table);
```

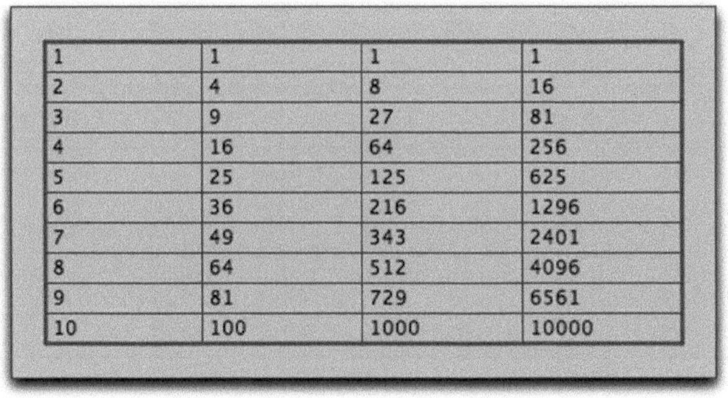

| 1 | 1 | 1 | 1 |
|---|---|---|---|
| 2 | 4 | 8 | 16 |
| 3 | 9 | 27 | 81 |
| 4 | 16 | 64 | 256 |
| 5 | 25 | 125 | 625 |
| 6 | 36 | 216 | 1296 |
| 7 | 49 | 343 | 2401 |
| 8 | 64 | 512 | 4096 |
| 9 | 81 | 729 | 6561 |
| 10 | 100 | 1000 | 10000 |

*Modifiers:* The `drawtable` operator supports several modifiers to control the graphical appearance of the table. A summary of the modifiers is given in the following table:

| `width` | *<int>* | the width of the cells in pixels |
|---|---|---|
| `height` | *<int>* | the width of the cells in pixels |
| `flip` | | exchanges the role of rows and columns |
| `border` | *<bool>* | turns on/off drawing of lines and borders |
| `size` | *<real>* | sets the text size |
| `color` | *[<real>,<real>,<real>]* | sets the text color |
| `alpha` | *<real>* | sets the text opacity |
| `offset` | *[<real>,<real>]* | sets an *xy* offset in pixels between text and base point |
| `align` | *"left", "right", "mid"* | determines the horizontal text alignment within a cell |
| `back` | *<bool>* | turns on/off the drawing of a table background |
| `back` | *[<real>,<real>,<real>]* | turns on the drawing of table background and sets it to an RGB color |
| `backalpha` | *<real>* | sets opacity of the table background |

*Example:* The following code is a more elaborate example using the `drawtable` operator. modifiers are used to create a nice appearance of the tables. A second table is used to create a heading for the table.

```
x=1..9;
tab=apply(x,(#,#^2,#^3,#^4));
tab1=("x","$x^2$","$x^3$","$x^4$");
linecolor((0,0,.8));
drawtable((0,0),tab,
  width->50,
  height->22,
  back->(1,0,0),
  backalpha->0.1,
  align->"right",
  size->12
);
linecolor((0,0,0));
```

```
drawtable((0,8.1),tab1,flip->true,
  width->50,
  height->33,
  back->(0,0,1),
  backalpha->0.4,
  align->"mid",
  size->16,
  color->(1,1,1)
);
```

| x | $x^2$ | $x^3$ | $x^4$ |
|---|---|---|---|
| 1 | 1 | 1 | 1 |
| 2 | 4 | 8 | 16 |
| 3 | 9 | 27 | 81 |
| 4 | 16 | 64 | 256 |
| 5 | 25 | 125 | 625 |
| 6 | 36 | 216 | 1296 |
| 7 | 49 | 343 | 2401 |
| 8 | 64 | 512 | 4096 |
| 9 | 81 | 729 | 6561 |

## 11.5 TeX Rendering

Starting with version 2.4 of Cinderella the text rendering became significantly more powerful. It is now also possible to render formulas using a variant of the TeX typesetting language [12]. This language is standard for mathematical documents and allows for the description of very complex formulas. Currently Cinderella covers about 95% of the formatting capabilities of the TeX formula language. We here only report on the most important formatting issues. We also report on the major differences to standard TeX.

TeX rendering can also be used in texts for element captions and in usual geometric text objects.

### 11.5.1 Activating TeX Rendering

Within a usual string the TeX formula rendering can be activated by enclosing the formula by dollar signs: $...here is the formula text...$. The following statement produces a formula that already covers some interesting features of TeX rendering:

```
drawtext(
  (0,0),
  "Sum formula: $\sum_{i=1}^n i^2 = { 2\cdot n^3+ 4\cdot n^2 +n\over 6 }
);
```

$$\text{Sum formula: } \sum_{i=1}^{n} i^2 = \frac{2 \cdot n^3 + 3 \cdot n^2 + n}{6}$$

The TeX renderer tries to do its best even if uncommon font families are chosen for the rendering of formulas.

```
drawtext(
  (0,0),
  "Sum formula: $\sum_{i=1}^n i^2 =
               { 2\cdot n^3+ 3\cdot n^2 +n\over 6 }$",size->20,
  family->"Lucida Calligraphy"
);
```

$$\textit{Sum formula: } \sum_{i=1}^{n} i^2 = \frac{2 \cdot n^3 + 3 \cdot n^2 + n}{6}$$

## 11.5.2 Subscripts and Superscripts

Perhaps the most simple and most common usage of TeX rendering is to equip labels with subscripts or superscripts. This can be done by the using the _ and the signs, respectively. If the sub- or superscripts are again more complicated formulas then they have to be enclosed in curly brackets. Unlike in usual TeX sub- or superscripts that only consist of numbers do not have to be enclosed in curly brackets. The following piece of code exemplifies various usages of sub- and superscripts.

```
textsize(20);
drawtext((0,0), "$A_1$");
drawtext((2,0), "$A_123$");
drawtext((4,0), "$A_1^12$");
drawtext((6,0), "$A_{1_2}^{1/2}$");
drawtext((8,0), "$A_{1_2}^{\sqrt{x^2+y^2}}$");
```

$$A_1 \qquad A_{123} \qquad A_1^{12} \qquad A_{1_2}^{1/2} \qquad A_{1_2}^{\sqrt{x^2+y^2}}$$

## 11.5.3 Special Formula Elements

The TeX renderer is capable of rendering special formula signs like sums, square roots and integrals. The following sample gives an impression of how these objects can be rendered.

```
textsize(20);
drawtext((0,0), "$\sum_{i=1}^n (i^2+1)");
drawtext((3,0), "$\sqrt{x^2+y^2}");
drawtext((6,0), "$\int_a^b f(x)dx");
```

$$\sum_{i=1}^{n}(i^2+1) \quad \sqrt{x^2+y^2} \quad \int_a^b f(x)dx$$

TeX control commands must be preceded by a \ sign. Observe that \sum and \int automatically generate a reasonable placement of upper and lower indices. Besides \sum and \int, there are also other symbols available that treat upper and lower indices in a special way. Here comes a complete list of them:

\prod, \coprod, \bigcup, \bigcap, \bigwedge, \bigvee, \bigoplus, \bigotimes, \bigodot, \biguplus, \int, \iint, \iiint, \oint

$$\sum \prod \coprod \bigcup \bigcap \bigwedge \bigvee \bigoplus \bigotimes \bigodot \biguplus \int \iint \iiint \oint$$

## 11.5.4 Brackets

The Cinderella TeX implementation allows for the use of four types of brackets in formulas:

- Round bracktes: (....)

- Square braclets: [....]

- Curly brackets: \{...\}

- Vertical lines: |...|

Since Curly brackets have a special semantic meaning in TeX formulas they have to be preceded by a backslash. In formulas it is often necessary to use brackets of different sizes. For this purpose each bracket can be preceded by one of the modifying commands \big, \Big, \bigg, \Bigg. They produce brackets of different sizes as the following piece of code exemplifies.

```
drawtext((0,0),size->16,
"$\Bigg( \bigg( \Big( \big( (\ldots)
         \big) \Big) \bigg) \Bigg)$");
drawtext((5,0),size->16,
"$\Bigg[ \bigg[ \Big[ \big[ [\ldots]
         \big] \Big] \bigg] \Bigg]$");
drawtext((10,0),size->16,
"$\Bigg\{ \bigg\{ \Big\{ \big\{ \{\ldots\}
         \big\} \Big\} \bigg\} \Bigg\}$");
```

```
drawtext((15,0),size->16,
"$\Bigg| \bigg| \Big| \big| |\ldots|
          \big| \Big| \bigg| \Bigg|$")
```

There is also a more versatile way of creating a huge bracket. Using the command \left and \right one can generate brackets that fit the size of the enclosed formula exactly, as shown in the next example:

```
drawtext((0,0),size->16,
"$\left[\sum_{i=1}^n \left({\sqrt sin(i)\right)\right]^2$")
```

The \left and \right statements must be properly nested. If one wants to suppress an opening or closing brackets one can use \left. or \right..

### 11.5.5 Formulas with Special Layout

Some formulas like for instance *fractions* force a special layout that places the different parts of the formulas at special locations. Cinderella TeX Statements that support this kind of rendering are

\frac, \over, \choose, \binom

Their usage is shown in the next example:

```
drawtext((0,0),size->16,"${1+n^2\over 1-n^2}$");
drawtext((3,0),size->16,"${2\choose 3}$");
drawtext((6,0),size->16,"$\frac{a+b}{x^2}$");
drawtext((9,0),size->16,"$\binom{a+b}{x^2}$");
```

### 11.5.6 Whitespace

Usual blanks and newlines are only used as syntactic separators in formulas and do not have any influence on the layout of the formula. To introduce whitespace the commands \, \;, \quad, \qquad, \! are used. They produce a whitespace measured in units of an "m" of the current font.

- \qquad: whitespace of 2.0 units

- \quad: whitespace of 1.0 unit

- \;: whitespace of 5/18 units

- \,: whitespace of 3/18 units

- \!: negative whitespace of -5/18 units

```
drawtext((0,0),size->16,"$A\!A A \,A\;A\quad A \qquad A$")
```

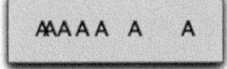

### 11.5.7 Over- and Underlining

Cinderella TeX supports several statements that allow to put decorations like arrows and lines on top of or below a formula. The supported commands are

```
\overline, \underline, \overleftarrow, \overrightarrow, \vec,
\hat, \tilde
```

The arguments of these statements have to be enclosed in curly brackets. Here are a few formulas that use these features:

```
drawtext((0,0),size->16,"$\overline{A}\;\cap\;\overline{B}\;=\;
                 \overline{A\;\cup\; B}$");
drawtext((6,0),size->16,"$|\overrightarrow{(x,y)}|\;= \;
                 \sqrt{x^2+y^2}$");
drawtext((13,0),size->16,"$\tilde{X}+\hat{Y}\;=\;
                 \underline{X\oplus Y}$");
```

$$\overline{A} \cap \overline{B} = \overline{A \cup B} \qquad |\overrightarrow{(x,y)}| = \sqrt{x^2+y^2} \qquad \tilde{X}+\hat{Y} = \underline{X \oplus Y}$$

### 11.5.8 Vectors and Matrices

Matrices and Vectors are essential for mathematical formulas. They can also be rendered by Cinderella's TeX implementation. The most basic way to introduce a matrix (or a vector) is by using the \begin{array}{...}.....\end{array} statement. An array consists of a sequence of rows seperated by \\. Each row consists of the entries of the row separated by &. The second pair of curly brackets in the statement above contains formatting information for each column. Here

- r means right align,

- l means left align,

- c means center align.

In the following example the `array` statement is combined with enclosing brakets to generate a matrix-like object.

```
drawtext((0,0),
"$M\;=\;\left(
\begin{array}{lcr}
 1+1&2&3\\
 1&2+2&3\\
 1&2&3+3\\
 \end{array}
\right)
$"
,size->20);
```

$$M = \begin{pmatrix} 1+1 & 2 & 3 \\ 1 & 2+2 & 3 \\ 1 & 2 & 3+3 \end{pmatrix}$$

Arrays can also be used to create formulas with several cases, as the following example shows:

```
drawtext((0,0),
"$sign(x)\;:=\;\left\{
\begin{array}{ll}
  1&if \quad x>0\\
 -1&if \quad x<0\\
  0&if \quad x=0\\
 \end{array}
\right.
$"
,size->20);
```

$$sign(x) := \begin{cases} 1 & if \quad x>0 \\ -1 & if \quad x<0 \\ 0 & if \quad x=0 \end{cases}$$

Cinderella also supports matrix operators that do not need the formatting informations and that generate the enclosing brackets automatically. The five types of admissible matrices are illustrated in the code below:

```
drawtext((0,0), "$\begin{matrix} a-\lambda & b\\
                        c & b-\lambda\\ \end{matrix}$",);
drawtext((4,0), "$\begin{pmatrix}a-\lambda & b\\
                        c & b-\lambda\\ \end{pmatrix}$");
drawtext((8,0), "$\begin{bmatrix}a-\lambda & b\\
                        c & b-\lambda\\ \end{bmatrix}$");
```

```
drawtext((12,0),"$\begin{Bmatrix}a-\lambda & b\\
                        c & b-\lambda\\ \end{Bmatrix}$");
drawtext((16,0),"$\begin{vmatrix}a-\lambda & b\\
                        c & b-\lambda\\ \end{vmatrix}$");
```

$$\begin{matrix} a-\lambda & b \\ c & b-\lambda \end{matrix} \quad \begin{pmatrix} a-\lambda & b \\ c & b-\lambda \end{pmatrix} \quad \begin{bmatrix} a-\lambda & b \\ c & b-\lambda \end{bmatrix} \quad \begin{Bmatrix} a-\lambda & b \\ c & b-\lambda \end{Bmatrix} \quad \begin{vmatrix} a-\lambda & b \\ c & b-\lambda \end{vmatrix}$$

### 11.5.9 Color

Cinderellas TeX supports colored text via a special command \color{...}. The currently predefined colors are:

white, black, red, green, blue, darkred, darkgreen, darkblue, magenta, yellow, cyan, orange

Using the color statement influences color of the subformula in which it it used.

```
drawtext((0,0),size->20,color->(0,0,0),
  "Sum formula: $
\sum_{\color{darkgreen}i=1}^{\color{darkgreen}n} {\color{darkred}i^2}
\quad = \quad
{\color{blue}{ 2\cdot n^3+ 4\cdot n^2 +n\over 6 }}$"
);
```

$$\text{Sum formula: } \sum_{i=1}^{n} i^2 \quad = \quad \frac{2\cdot n^3+4\cdot n^2+n}{6}$$

### 11.5.10 Plain Text

Sometimes it may be useful to use a passage of normal text within a formula. This can be done by the \mbox{....} statement. The following example illustrates its usage:

```
drawtext((0,0),size->20,color->(0,0,0),
  "$
\sum_{\mbox{All i not equal to j}}(i^2+j^2)
$"
);
```

$$\sum_{\text{All i not equal to j}} (i^2+j^2)$$

## 11.5.11  Special Characters

In mathematical formulas one needs many special characters. The following list gives an overview of all characters that are associated to special TeX statements.

### 11.5.11.1  Greek letters

| | | | | | | |
|---|---|---|---|---|---|---|
| α | \alpha | β | \beta | γ | \gamma | δ | \delta |
| ε | \epsilon | ε | \varepsilon | ζ | \zeta | η | \eta |
| θ | \theta | ϑ | \vartheta | ι | \iota | κ | \kappa |
| λ | \lambda | λ | \lamda | μ | \mu | μ | \my |
| ν | \nu | ν | \ny | ξ | \xi | o | \omicron |
| π | \pi | ϖ | \varpi | ρ | \rho | ϱ | \varrho |
| σ | \sigma | ς | \varsigma | τ | \tau | υ | \upsilon |
| υ | \ypsilon | φ | \phi | χ | \chi | ψ | \psi |
| ω | \omega | A | \Alpha | B | \Beta | Γ | \Gamma |
| Δ | \Delta | E | \Epsilon | Z | \Zeta | H | \Eta |
| Θ | \Theta | I | \Iota | K | \Kappa | Λ | \Lambda |
| Λ | \Lamda | M | \Mu | M | \My | N | \Nu |
| N | \Ny | Ξ | \Xi | O | \Omicron | Π | \Pi |
| P | \Rho | Σ | \Sigma | T | \Tau | Υ | \Upsilon |
| Υ | \Ypsilon | Φ | \Phi | X | \Chi | Ψ | \Psi |
| Ω | \Omega | | | | | | |

### 11.5.11.2  Arrows

| | | | | | | |
|---|---|---|---|---|---|---|
| ← | \leftarrow | → | \rightarrow | → | \to | ↔ | \leftrightarrow |
| ⇐ | \Leftarrow | ⇒ | \Rightarrow | ⇔ | \Leftrightarrow | ↦ | \mapsto |
| ↩ | \hookleftarrow | ↼ | \leftharpoonup | ↽ | \leftharpoondown | ↪ | \hookrightarrow |
| ⇀ | \rightharpoonup | ⇁ | \rightharpoondown | ⟵ | \longleftarrow | ⟶ | \longrightarrow |
| ⟷ | \longleftrightarrow | ⟸ | \Longleftarrow | ⟹ | \Longrightarrow | ⟺ | \Longleftrightarrow |
| ⤶ | \longmapsto | ↑ | \uparrow | ↓ | \downarrow | ↕ | \updownarrow |
| ⇑ | \Uparrow | ⇓ | \Downarrow | ⇕ | \Updownarrow | ↗ | \nearrow |
| ↘ | \searrow | ↙ | \swarrow | ↖ | \nwarrow | ⇝ | \leadsto |
| ⇠ | \dashleftarrow | ⇇ | \leftleftarrows | ⇆ | \leftrightarrows | ⇚ | \Lleftarrow |
| ↞ | \twoheadleftarrow | ↢ | \leftarrowtail | ⇋ | \leftrightharpoons | ↰ | \Lsh |
| ↫ | \looparrowleft | ↶ | \curvearrowleft | ↺ | \circlearrowleft | ⇢ | \dashrightarrow |
| ⇉ | \rightrightarrows | ⇄ | \rightleftarrows | ⇛ | \Rrightarrow | ↠ | \twoheadrightarrow |
| ↣ | \rightarrowtail | ⇌ | \rightleftharpoons | ↱ | \Rsh | ↬ | \looparrowright |
| ↷ | \curvearrowright | ↻ | \circlearrowright | ⊸ | \multimap | ↿ | \upuparrows |
| ⇊ | \downdownarrows | ↿ | \upharpoonleft | ↾ | \upharpoonright | ⇃ | \downharpoonleft |
| ⇂ | \downharpoonright | ⇝ | \rightsquigarrow | ↭ | \leftrightsquigarrow | ← | \leftarrow |
| ↑ | \uparrow | → | \rightarrow | ↓ | \downarrow | ↔ | \leftrightarrow |
| ↕ | \updownarrow | ↖ | \nwarrow | ↗ | \nearrow | ↘ | \searrow |
| ↙ | \snarrow | ⇐ | \Leftarrow | ⇑ | \Uparrow | ⇒ | \Rightarrow |
| ⇓ | \Downarrow | ⇔ | \Leftrightarrow | ⇕ | \Updownarrow | ⇖ | \Nwarrow |
| ⇗ | \Nearrow | ⇘ | \Searrow | ⇙ | \Snarrow | | |

### 11.5.11.3  Math characters

| Sym | Command | Sym | Command | Sym | Command | Sym | Command |
|---|---|---|---|---|---|---|---|
| ≤ | \leq | ≪ | \ll | < | \prec | ≺ | \preceq |
| ⊂ | \subset | ⊆ | \subseteq | ⊏ | \sqsubset | ⊑ | \sqsubseteq |
| ∈ | \in | ⊢ | \vdash | \| | \mid | ⌣ | \smile |
| ≥ | \geq | ≫ | \gg | > | \succ | ≻ | \succeq |
| ⊃ | \supset | ⊇ | \supseteq | ⊐ | \sqsupset | ⊒ | \sqsupseteq |
| ∋ | \ni | ⊣ | \dashv | ∥ | \parallel | ⌢ | \frown |
| ∉ | \notin | ≡ | \equiv | ≐ | \doteq | ~ | \sim |
| ≃ | \simeq | ≈ | \approx | ≅ | \cong | ⋈ | \Join |
| ⋈ | \bowtie | ∝ | \propto | ⊨ | \models | ⊥ | \perp |
| ≍ | \asymp | ≠ | \neq | ± | \pm | × | \times |
| ∪ | \cup | ⊔ | \sqcup | ∨ | \vee | ⊕ | \oplus |
| ⊙ | \odot | ⊗ | \otimes | △ | \bigtriangleup | ◁ | \lhd |
| ⊴ | \unlhd | ∓ | \mp | ÷ | \div | ∖ | \setminus |
| ∩ | \cap | ⊓ | \sqcap | ∧ | \wedge | ⊖ | \ominus |
| ⊘ | \oslash | ○ | \bigcirc | ▽ | \bigtriangledown | ▷ | \rhd |
| ⊵ | \unrhd | ◁ | \triangleleft | ▷ | \triangleright | ⋆ | \star |
| ∗ | \ast | ∘ | \circ | · | \bullet | ⋄ | \diamond |
| ⊎ | \uplus | † | \dagger | ‡ | \ddagger | ≀ | \wr |
| ∑ | \sum | ∏ | \prod | ∐ | \coprod | ∫ | \int |
| ⋃ | \bigcup | ⋂ | \bigcap | ⨆ | \bigsqcup | ∮ | \oint |
| ⋁ | \bigvee | ⋀ | \bigwedge | ⨁ | \bigoplus | ⨂ | \bigotimes |
| ⨀ | \bigodot | ⨄ | \biguplus | … | \dots | ⋯ | \cdots |
| ⋮ | \vdots | ⋱ | \ddots | ℏ | \hbar | ℓ | \ell |
| ℜ | \Re | ℑ | \Im | ℵ | \aleph | ℘ | \wp |
| ∀ | \forall | ∃ | \exists | ℧ | \mho | ∂ | \partial |
| ′ | \prime | ∅ | \emptyset | ∞ | \infty | ∇ | \nabla |
| △ | \triangle | □ | \Box | ◊ | \Diamond | ⊥ | \bot |
| ⊤ | \top | ∠ | \angle | √ | \surd | ◊ | \diamondsuit |
| ♡ | \heartsuit | ♣ | \clubsuit | ♠ | \spadesuit | ¬ | \neg |
| ♭ | \flat | ♮ | \natural | ♯ | \sharp | Ϝ | \digamma |
| ϰ | \varkappa | ℶ | \beth | ℸ | \daleth | ℷ | \gimel |
| ⋖ | \lessdot | ⩽ | \leqslant | ≦ | \leqq | ⋘ | \lll |
| ≲ | \lesssim | ≶ | \lessgtr | ⪋ | \lesseqgtr | ≼ | \preccurlyeq |
| ⋞ | \curlyeqprec | ≾ | \precsim | ⋐ | \Subset | ⊏ | \sqsubset |
| ∴ | \therefore | ⌣ | \smallsmile | ⊴ | \vartriangleleft | ⊴ | \trianglelefteq |
| ⋗ | \gtrdot | ≧ | \geqq | ⋙ | \ggg | ≿ | \gtrsim |
| ≷ | \gtrless | ⪌ | \gtreqless | ≽ | \succcurlyeq | ⋟ | \curlyeqsucc |
| ≿ | \succsim | ⋑ | \Supset | ⊐ | \sqsupset | ∵ | \because |
| ∥ | \shortparallel | ⌢ | \smallfrown | ⊵ | \vartriangleright | ⊵ | \trianglerighteq |
| ≑ | \doteqdot | ≓ | \risingdotseq | ≒ | \fallingdotseq | ≖ | \eqcirc |
| ≗ | \circeq | ▲ | \triangleq | ≏ | \bumpeq | ≎ | \Bumpeq |
| ∼ | \thicksim | ≈ | \thickapprox | ≊ | \approxeq | ∽ | \backsim |
| ⊨ | \vDash | ⊩ | \Vdash | ⊪ | \Vvdash | ϶ | \backepsilon |
| ∝ | \varpropto | ≬ | \between | ⋔ | \pitchfork | ◀ | \blacktriangleleft |
| ▷ | \blacktriangleright | ∔ | \dotplus | ⋉ | \ltimes | ⋓ | \Cup |
| ⊻ | \veebar | ⊞ | \boxplus | ⋈ | \boxtimes | ⋋ | \leftthreetimes |
| ⋎ | \curlyvee | · | \centerdot | ⋊ | \rtimes | ⋒ | \Cap |
| ⊼ | \barwedge | ⊟ | \boxminus | ⊡ | \boxdot | ⋌ | \rightthreetimes |
| ⋏ | \curlywedge | ⊺ | \intercal | ⨯ | \divideontimes | ∖ | \smallsetminus |

| ⊖ \circleddash | ⊙ \circledcirc | ⊛ \circledast | ℏ \hbar |
|---|---|---|---|
| ℏ \hslash | □ \square | ∎ \blacksquare | Ⓢ \circledS |
| △ \vartriangle | ▲ \blacktriangle | ∁ \complement | ▽ \triangledown |
| ▼ \blacktriangledown | ◇ \lozenge | ◆ \blacklozenge | ★ \bigstar |
| ∠ \angle | ∡ \measuredangle | ∢ \sphericalangle | ` \backprime |
| ∄ \nexists | ⅁ \Finv | ∅ \varnothing | ð \eth |
| ℧ \mho | ∣ \vert | ∥ \Vert | ∁ \C |
| ε \vareps | H \H | ℑ \Im | ℓ \ell |
| ℕ \N | ℙ \P | ℚ \Q | ℜ \Re |
| ℝ \R | ℤ \Z | ± \pm | ∓ \mp |
| ★ \star | ∗ \ast | • \bullet | · \centerdot |
| ℵ \aleph | ∈ \in | ∉ \not\in | ∋ \ni |
| ∌ \not\ni | \ \backslash | ∖ \setminus | / \slash |
| ∀ \forall | × \times | ∩ \cap | ∪ \cup |
| · \cdot | ∞ \infty | ⇒ \implies | ⋮ \vdots |
| ⋱ \ddots | ⋯ \cdots | … \ldots | |

### 11.5.11.4 Unicode

In the rare case that a certain character is not provided by the standard TeX commands there is also a way to include a unicode character in a TeX formula. This can be done by using one of the two TeX statements \unicode{...} or \unicodex{...}. The first of these statements expects a decimal number that specifies the unicode character. The second statement expects a hexadecimal number. The following (slightly advanced) example first composes a string of unicode statements using the sum(...) function. Then it renders the resulting string as a TeX formula

```
chess=sum(0..11,i,"\;\unicode{"+(9812+i)+"}");
drawtext((0,0),size->30,"$"+chess+"$");
```

The resulting picture is shown below:

## 11.6 Image Manipulation and Rendering

Cinderella.2 introduces various ways to deal with images in a geometric construction. On the one hand it is possible to load images in a media database and use them freely within CindyScript. Images can be arbitrarily moved, rotated, scaled, perspectively transformed, of even transformed by a function. On the other hand it is possible to use the drawing functions of CindyScript to create custom images which are in turn used as blueprint for other purposes.

### 11.6.1 The Media Browser

Prior to all image rendering there is the need to load a specific image into a Cinderella construction. This is done via the Media Browser from which you can access all images you need for a construction. The Media Browser is opened by choosing the menu item *File/Media Browser*. You can load images by pressing the + button and remove them again by pressing the - button. Once an image is loaded you see its internal name and a preview of the image. By default, the image name is the original file name, but you can change the internal name by double clicking and editing it. Under this name the image can be accessed from within CindyScript.

*The Media Browser*

The images in the media browser can also be used as images for points and lines as well as background of the view port. This functionality is available via the inspector (p. 153).

### 11.6.2 Drawing and Transforming Images

**Drawing an image: `drawimage(<pos>,<imagename>)`**                    `drawimage`

*Description:* This operator takes an image (from the Media Browser) and draws it at a position given by the first argument. The reference point is by default the center of the image. However, this can be altered by modifiers. Also the scaling and positioning of the image can be altered.

*Examples:* The following code draws the image called `myimage` at the position given by the point *A*.

```
drawimage(A,"myimage"))
```

The following code shows various usages of the `drawimage` command. It creates the graphics given below.

```
drawimage(A,"myimage");
drawimage(B,"myimage",scale->1.5,angle->30°);
drawimage(C,"myimage",ref->"lb");
```

*Reference points:* Reference points can be specified using the modifiers `ref`, `refx`, `refy`. While the `ref` modifier expects a two-dimensional information, the last two refer to the two coordinate directions separately. The reference information may be given in three different ways:

- Absolute to the pixel coordinates of the original image: If the image for example was originally 400 by 800 pixels then the modifier `ref->[100,200]` will specify the position in the middle of the left lower quarter of the image.

- As relative quotient with respect to the image's dimensions: In this case the position has to be specified by a percentage value (given as a string). So for instance, `ref->["25%", "25%"]` would again place the reference point in the middle of the left lower quarter of the image - but now independent of the original image's dimensions.

- Symbolically as a pair of letters: Here in the *x*-direction the letters `l`, `c`, `r` represent *left', center and right. In the* y *direction the letters* `b`, `c`, `t` *represent* bottom', *center* and *top*. So the lower left corner may be specified either by `ref->"lb"` or by `ref->["l","b"]`.

All the position information can be used separately for the horizontal and vertical direction, for example with `refx->"l"`, `refx->100` or `refx->"10%"`. The following picture exemplifies the positions of a few reference points.

*Modifiers:* The modifiers of this function are listed below.

| | | |
|---|---|---|
| `alpha` | *0.0 ... 1.0* | opacity of the image |
| `angle` | *real* | rotation angle around the reference point |
| `rotation` | *real* | same as `angle` |
| `scale` | *real* | scaling |
| `scale` | *vec* | separate scaling in both directions |
| `scalex` | *real* | scaling in *x*-direction |
| `scaley` | *real* | scaling in *y*-direction |
| `flipx` | *boolean* | vertical reflection |
| `flipy` | *boolean* | horicontal reflection |
| `ref` | *see above* | *xy*-position of reference point |
| `refx` | *see above* | *x*-position of reference point |
| `refy` | *see above* | *y*-position of reference point |
| `rendering` | *"fast" or "nice"* | specify the rendering quality |

**Drawing an image: `drawimage(<pos>,<pos>,<imagename>)`**                    `drawimage`

*Description:* Draws a copy of an image whose position and size is specified by two reference points. By default, these reference points are the two corners of the bottom edge. It is also possible to specify the position of the reference points within the image. The syntax for specifying a reference point is the same as in the previous command.

*Examples:* The simplest usage of the operator is given in the following example.

```
drawimage(A,B,"myimage"))
```

More advanced usages can be found in the following example. They generate the picture given below. Observe how in the third row the two reference points are specified individually.

```
drawimage(A,B,"MyImage");
drawimage(C,D,"MyImage",aspect->1);
drawimage(E,F,"MyImage",flipx->true);
drawimage(G,H,"MyImage",refx1->"20%",refy1->"50%",ref2->"rt");
```

*Modifiers:*

| | | |
|---|---|---|
| **alpha** | *0.0 … 1.0* | opacity of the image |
| **flipx** | *boolean* | vertical reflection |
| **flipy** | *boolean* | horicontal reflection |
| **aspect** | *real* | specify the aspect ratio |
| **ref1** | *see above* | *xy*-position of reference point one |
| **refx1** | *see above* | *x*-position of reference point one |
| **refy1** | *see above* | *y*-position of reference point one |
| **ref2** | *see above* | *xy*-position of reference point two |
| **refx2** | *see above* | *x*-position of reference point two |
| **refy2** | *see above* | *y*-position of reference point two |
| **rendering** | *"fast" or "nice"* | specify the rendering quality |

*Example:* A more advanced example is given below. It iteratively maps two points (stored the variables a and b) by a transformation matrix. In each step a corresponding image is drawn. This code creates a logarithmic spiral of images. One important issue may arise if many images are drawn in a construction. One may have to select between *nice* or *fast* drawing of the images. For this there is a modifier rendering that may be set either to "nice" or to fast. By default it is set to prefer the *nice* rendering.

```
a=(1,0);
b=(2,-1);
w=30°;
m=((cos(w),-sin(w)),
   (sin(w),cos(w)))*0.9;

repeat(100,
  drawimage(a,b,"MyImage");
  a=m*a;
```

```
  b=m*b;
);
```

**Drawing an image: `drawimage(<pos>,<pos>,<pos>,<imagename>)`**     `drawimage`

*Description:* This operator is similar to the last one. Now three reference points are specified to define an affine transformation of the image. By default, the reference points are set to the bottom/left, the bottom/right and the top/left corner of the picture. Again the position op the reference points can be altered by modifiers.

*Example:* The simplest usage of this command is given by the following piece of code.

```
drawimage(A,B,C,"MyImage"))
```

This operator is very well suited to draw images under geometric transformation like reflections or rotations. For this the picture has to be drawn with respect to the mapped reference points. An example of this technique is given in the code below

```
drawimage(A,B,C,"MyImage"))
drawimage(A',B',C',"MyImage"))
```

*Modifiers:* The command supports the same modifiers as the previous one and the following additional ones.

| ref3 | *see above* | *xy*-position of reference point three |
|------|-------------|----------------------------------------|
| refx3 | *see above* | *x*-position of reference point three |
| refy3 | *see above* | *y*-position of reference point three |

**drawimage**

**Drawing an image: `drawimage(<pos>,<pos>,<pos>,<pos>,<imagename>)`**

*Description:* Again this command is more general than the previous one. This time the four corners (or more generally reference points) are used to specify a projective transformation of the image. By default the position of the reference points are the corners of the image taken in counterclockwise order starting at *left/bottom.*

*Example:* The simplest usage is given by the following piece of code.

```
drawimage(A,B,D,C,"myimage")
```

The following picture shows the result of this statement. For clarity a projective grid has been added to the picture.

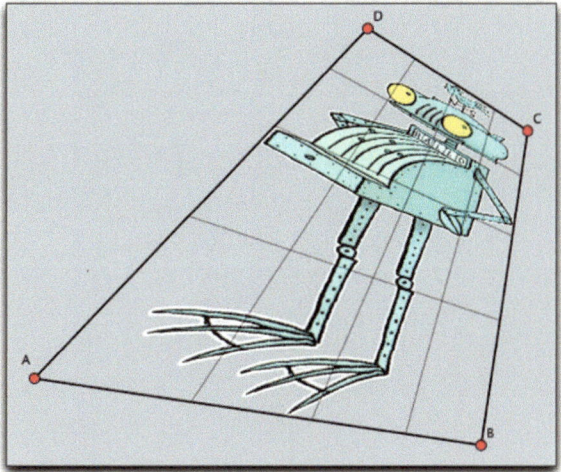

*Modifiers:* The command supports the same modifiers as the previous one and the following additional ones.

| | | |
|---|---|---|
| **ref4** | *see above* | *xy*-position of reference point four |
| **refx4** | *see above* | *x*-position of reference point four |
| **refy4** | *see above* | *y*-position of reference point four |

### Deforming an image: `mapimage(<imagename>,<function>)`               `mapimage`

*Description:* This command is extremely powerful for deforming images by functions. Usually the function is assumed to be a function that maps a two dimensional vector to a two dimensional vector. By using the `complex->true` modifier it is also possible to use functions that map the complex plane to the complex plane. The range of the function must be specified by the modifiers `xrange` and `yrange`. If they are not specified the ranges are assumed to be between 0.0 and 1.0. The rectangles defined by these ranges are identified with the rectangular region of the image. From there every image point is mapped to create the resulting image. The functionality is essentially similar to the `mapgrid` command.

*Example:* Here the *x*-axis of the image is deformed using a function that modifies its width by adding a *sin*-function. In the code the function is specified first. Then it is used in the `mapimage` command. For better reference a grid has been added to drawing that illustrates the deformation.

```
f(z):=(z_1*(sin(z_2)+1.3),z_2);
mapimage("MyImage",f(#),
xrange->(-1,1),
yrange->(0,pi),resolution->30
);
mapgrid(f(#),color->(0,0,0),alpha->0.3,
xrange->(-1,1),
yrange->(0,pi),resolution->30
);
```

*A diet for Ritter Rost*

This example is similar to the previous one but now a complex function is used to specify the mapping.

```
r=complex(A);
f(x):=exp(r*x);
mapimage("myimage",f(#),
  complex->true,
  xrange->(0,1),
  yrange->(0,pi),resolution->30
);

mapgrid(f(#),complex->true,
xrange->(0,1),yrange->(0,pi),color->(0,0,0),alpha->0.5
,resolutiony->30,resolutionx->10,step->10,size->1);
```

*Modifiers:* The function supports the following modifiers.

| | | |
|---|---|---|
| **alpha** | *0.0 ... 1.0* | opacity of the image |
| **xrange** | *vec* | start and end in *x*-direction |
| **yrange** | *vec* | start and end in *y*-direction |
| **complex** | *boolean* | use a complex function |
| **resolution** | *int* | quality of the resulting picture |

**Getting dimensions of an image: `imagesize(<imagename>)`**                    `imagesize`

*Description:* Returns the original size of an image. This is a pair of integer values that refers to the pixel width and height of the original image.

**Getting pixel data: `imagergb(<imagename>,<int>,<int>)`**                    `imagergb`

*Description:* The function `imagergb(<imagename>,x,y)` delivers the raw data of the color information of the pixel at original position *(x,y)*. The operator returns a four-dimensional vector with the raw data of the color. The first three entries represent the *rgb*-value with each entry ranging from 0 to 255. The last entry represents the alpha value.

*Example:* The following piece of (slightly elaborate) code first asks for the dimensions of an image and then samples the image in both directions. It plots a point with the corresponding color and opacity and by this creates a very rough copy of the image.

```
drawimage(A,"MyImage",scale->2);
dim=imagesize("MyImage");
forall((0..dim_1/10)*10,i,err(i);
  forall((0..dim_2/10)*10,j,
```

```
    col=imagergb("MyImage",i,j);
    draw((i,-j)*.03,color->(col_1,col_2,col_3)/255,
                    alpha->col_4,
                    border->false)
  )
)
```

### 11.6.3 Creating Custom Images

So far all image operation referred to images that were preloaded via the Media Browser. There is also the possibility to create images within Cinderella. Once these images are created one can pipe all drawing operations coming from CindyScript to such an image using the `canvas` commands. The result of the drawing operations are no longer directly visible. However, once the custom image is filled with content one can use it and draw it to draw it on screen with one of the `drawimage` operations. The `canvas` operations are built in analogy to the `drawimage` operations. You use them to place the canvas (the custom image) at an arbitrary place in the plane, even using transformations. This concept is extremely powerful and in the context of this manual we will only sketch its basic usage.

createimage **Creating a custom image: `createimage(<imagename>,<int>,<int>)`**

*Description:* The operator `createimage(<imagename>,width,height)` creates an image buffer of the specified dimensions. Initially such an image is fully transparent and contains no drawings. The image buffer is accessible under the specified name in the Media Browser and can furtheron be used by `drawimage(...)` operations.

clearimage **Erasing an image: `clearimage(<imagename>)`**

*Description:* This operator removes all content from an image. After using this operator the image still exists, but it does no longer contain any drawings. It is completely transparent.

**Removing an image: `removeimage(<imagename>)`**

*Description:* This operator removes the image from the Media Browser. After this the image can no longer be accessed.

## 11.6.4 Painting on a canvas

Images that are creating by a script may be used as a canvas to which drawing commands from CindyScript may be piped. This is a very powerful concept, but it needs a little explanation.

Imagine you have a piece of code that creates some drawing, say:

```
linesize(3);
repeat(10,i,
  drawcircle(K,i*.5);
);
drawtext(K+(-2,2),"These are",size->20,color->(1,1,0));
drawtext(K+(-2,-2),"some circles",size->20,color->(1,1,0));
```

This drawing will look like the picture below on the left. Now we want to use this drawing code to generate a custom bitmap. For this we first must create such a bitmap.

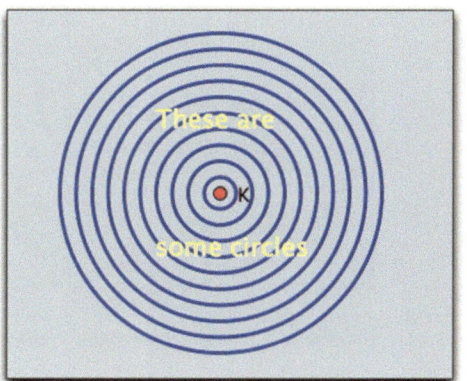
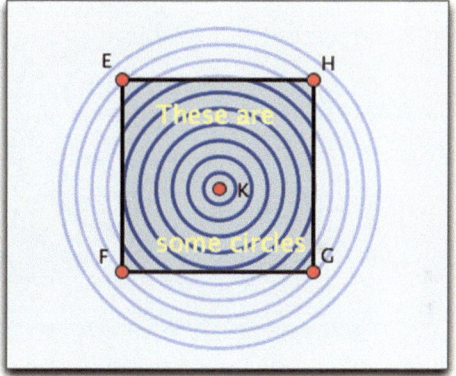

We can do this by creating an image in the *initialization* part of the script with the following code:

```
createimage("image",400,400)
```

Now we specify a region in which this bitmap should be used as a canvas. In our example we do so by drawing a rectangle with certain corner points.

The code for drawing on the canvas is included as an argument to a `canvas` function in the following way:

```
canvas(F,G,E,"image",
    //....here comes the drawing code....
)
```

The position of the canvas is specified as a parallelogram generated by the corner points *F*, *G* and *E*. The picture above on the right illustrates which part of the original drawing will be captured by the canvas. Actually, all the drawing operations that are inside the `canvas` will not have any directly visible effect, so the above picture on the right is just for illustration purposes.

Next, the image can be painted at any location using a `drawimage(...)` command. The complete resulting code may look like

```
clearimage("image");
canvas(F,G,E,"image",
 linesize(3);
 repeat(10,i,
   drawcircle(K,i*.5);
 );
 drawtext(K+(-2,2),"These are",size->20,color->(1,1,0));
 drawtext(K+(-2,-2),"some circles",size->20,color->(1,1,0));
);

drawimage(A,B,C,D,"image");
```

The `drawimage(...)` command used here produces a projective transformation of the generated canvas. It results in the picture shown below on the left

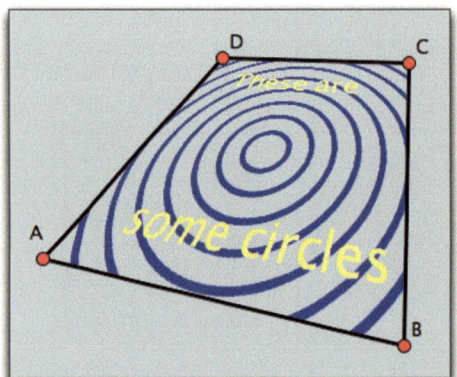 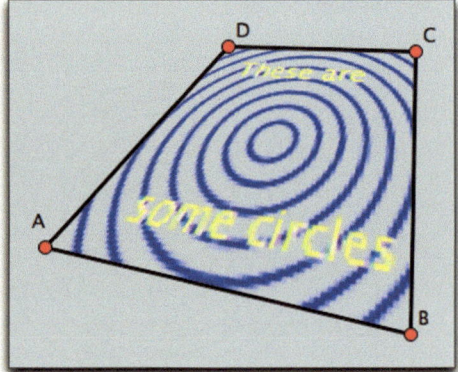

Notice that the image in the canvas is only a bitmap. Therefore it must be specified with suitable resolution. If in the initialization one would have used only an 80x80 bitmap, the picture on the right would result.

Once a canvas is created, it may be used without any further effort in different places and orientations many times. This is illustrated by the following piece of code and the corresponding image.

```
drawimage(A,P,M,O,"image");
drawimage(B,P,M,Q,"image");
drawimage(D,R,M,O,"image");
drawimage(C,R,M,Q,"image");
```

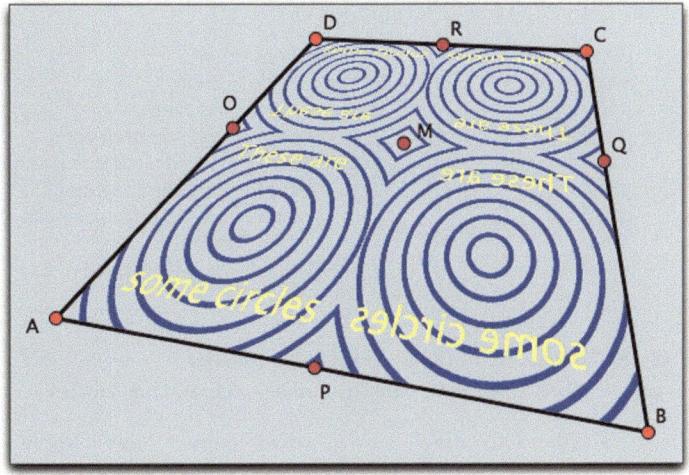

## Painting on a canvas with one reference point:                          `canvas`
`canvas(<pos>,<imagename>,<drawing code>)`

*Description:* This operator uses `<imagename>` as a canvas and positions it with respect to one reference point identically to the rules of the `drawimage(<pos>,<imagename>)` operator. All drawing statements in `<drawing code>` are piped to the canvas. The positioning modifiers are identical to those of `drawimage(<pos>,<imagename>)`.

*Modifiers:* The command has several modifiers.

| | | |
|---|---|---|
| **angle** | *real* | rotation angle around the reference point |
| **rotation** | *real* | same as `angle` |
| **scale** | *real* | scaling |
| **scale** | *vec* | separate scaling in both directions |
| **scalex** | *real* | scaling in *x*-direction |
| **scaley** | *real* | scaling in *y*-direction |
| **flipx** | *boolean* | vertical reflection |
| **flipy** | *boolean* | horicontal reflection |
| **ref** | *see above* | *xy*-position of reference point |
| **refx** | *see above* | *x*-position of reference point |
| **refy** | *see above* | *y*-position of reference point |

## Painting on a canvas with two reference points:                         `canvas`
`canvas(<pos>,<pos>,<imagename>,<drawing code>)`

*Description:* This operator uses `<imagename>` as a canvas and positions it with respect to two reference point identical to the rules of the `drawimage(<pos>,<pos>,<imagename>)` operator. All drawing statements in `<drawing code>` are piped to the canvas. The positioning modifiers are identical to those of `drawimage(<pos>,<pos>,<imagename>)`.

*Modifiers:* The command has several modifiers.

| | | |
|---|---|---|
| `flipx` | *boolean* | vertical reflection |
| `flipy` | *boolean* | horizontal reflection |
| `aspect` | *real* | specify the aspect ratio |
| `ref1` | *see above* | xy-position of first reference point |
| `refx1` | *see above* | x-position of first reference point |
| `refy1` | *see above* | y-position of first reference point |
| `ref2` | *see above* | xy-position of second reference point |
| `refx2` | *see above* | x-position of second reference point |
| `refy2` | *see above* | y-position of second reference point |

**canvas**

**Painting on a canvas with three reference points:**
`canvas(<pos>,<pos>,<pos>,<imagename>,<drawing code>)`

*Description:* Similar to the above statement. Now three reference points are used. The positioning modifiers are identical to those of `drawimage(<pos>,<pos>,<pos>,<imagename>)`.

*Modifiers:* The command supports the same modifiers as the previous command and in addition.

| | | |
|---|---|---|
| `ref3` | *see above* | xy-position of reference point three |
| `refx3` | *see above* | x-position of reference point three |
| `refy3` | *see above* | y-position of reference point three |

## 11.7 Shapes

Besides elementary drawing operations for lines, polygons, and circles, CindyScript also offers the possibility to combine these objects into more complicated objects called shapes. Shapes are not directly visible, but may be used for filling, outline drawing or clipping. They can be combined using logical operations like union, intersection or set difference. We first list the elementary operators and then explain their usage in a more elaborate example.

### 11.7.1 Shape primitives

**circle**

**A circular shape: `circle(<point>,<radius>)`**

*Description:* This operator creates a circular shape of a given `<radius>` around a given `<point>`.

**polygon**

**A polygonal shape: `polygon(<list>)`**

*Description:* This operator creates a polygonal shape defined by a list of positions given in `<list>`.

**A half-plane shape: `halfplane(<line>,<point>)`** `halfplane`

*Description:* This operator creates a half-plane shape. The half-plane is defined by its supporting line `<line>` and a point `<point>` contained in the half-plane, which determines on which side of the supporting line the half-plane lies. `<line>` may either be a three-dimensional vector of homogeneous coordinates, or a reference to a line object.

**The shape of the screen: `screen()`** `screen`

*Description:* This operator creates a rectangular shape that is large enough to cover all active drawing surfaces.

## 11.7.2 Combining shapes

Primitive shapes can be combined to form new shapes using three logical operators.

`<shape1>++<shape2>` creates the union of two shapes
`<shape1>~~<shape2>` creates the intersection of two shapes
`<shape1>-<shape2>` creates the difference of two shapes

## 11.7.3 Using shapes

Shapes can be used for filling, outline drawing and clipping.

**Fill a shape: `fill(<shape>)`** `fill`

*Description:* This operator fills a shape with a specified color.

*Modifiers:* This operator supports the following modifiers:

| | | |
|---|---|---|
| `color` | *[<real1>,<real2>,<real3>]* | sets the fill color to an RGB value |
| `alpha` | *<real>* | sets the opacity to the value `<alpha>` |

**Draw a shape: `draw(<shape>)`** `draw`

*Description:* This operator draws the outline of a shape.

*Modifiers:* This operator is a further polymorphic extension to the usual draw operator and supports the same modifiers.

**Set clippath: `clip(<shape>)`** `clip`

*Description:* This operator sets the clip path to the given shape. All subsequent drawing operations are clipped with respect to this clip path. The clip path is pushed on the appearance stack, thus it can be removed again by statements like `grestore()` or `greset()`.

### 11.7.4 Examples

The following examples illustrate the use of shapes. The following code defines three different shapes, a quadratic one and two circular ones. These are then combined into a more complex shape using logical operations. The complex shape is filled and outlined. The original shapes are outlined by a thinner line.

```
shape1=circle(E,4);
shape2=circle(F,4);
shape3=polygon([A,B,C,D]);
color((0,0,0));
shape=shape1++shape2++shape3;
fill(shape,color->(1,0.8,0));
draw(shape,size->3);
draw(shape1);
draw(shape2);
draw(shape3);
```

The resulting image is shown as the first picture below. The other three images are created using other combinations of the shapes.

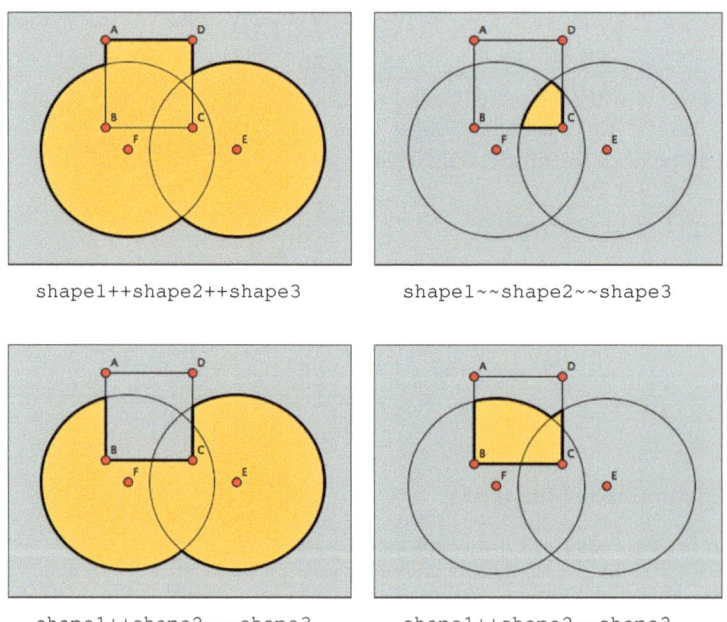

shape1++shape2++shape3          shape1~~shape2~~shape3

shape1++shape2——shape3          shape1++shape2~~shape3

The following code snippet demonstrates the usage of shapes as a clipping path. First of all, four circles and their union are defined as a shape. Then this shape is used as a clipping path for another drawing (concentric circles around point *E*).

```
r=3;
shape=circle(A,r)++circle(B,r)++circle(C,r)++circle(D,r);
clip(shape);
repeat(60,i,
  drawcircle(E,i/2,color->(0,.6,0),size->3);
);
greset();
```

```
draw(shape,color->(1,1,1),size->2);
```

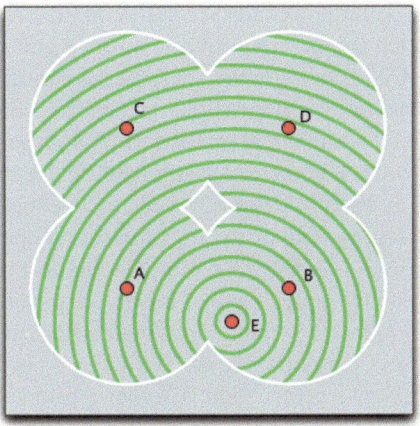

Shapes can become rather complicated objects, as the following iterative example shows. A word of caution may be appropriate here: The more complicated shape become, the more computational resources they consume. The use of overly complex shapes can slow down Cinderella considerably.

```
a=complex(A);
b=complex(B);
z=complex(C);
shape=circle((0,0),0);
repeat(50,
  shape=shape++circle(gauss(a),|a-b|);
  a=a*z;
  b=b*z;
);
fill(shape,color->(1,0.7,0));
draw(shape,color->(0,0,0),size->2);
```

## 11.8 The Coordinate System of CindyScript

There is an important feature of CindyScript that immediately changes the appearance of drawings treated by the script in a controlled and global way. Usually the coordinate system of CindyScript is the same as the coordinate system of the geometric construction. However it is possible to transform the coordinate system by special operators. Then all drawing is performed with respect to this modified reference frame. Care has to be taken when using the transformations: if many of them are applied it may be difficult to determine, where an actual drawing is performed. In order to still make an easy use of the transformation operators CindyScript provides two operators `gsave` and `grestore`. Similar to PostScript like languages these operators push/pop the actual state of the drawing engine to a stack. This state contains (besides the graphical default appearance information) the present coordinate transformation. So a temporal use of coordinate systems may be enclosed by a `gsave()`...`grestore()` construction. We first introduce the operators and than give a combined example.

*Caution:* In the current version of Cinderella 2.6 the application of the transformations is not yet implemented for circles. For circles only euclidean transformations (rotations, translations, reflections, scalings) are supported. Affine and projective transformations will be provided in a later release.

`translate`
### Translating the coordinate system: `translate(<list>)`

*Description:* This operator assumes that `<list>` is of the form `[<real>,<real>]` and translates the drawing coordinate system by this vector.

`rotate`
### Rotating the coordinate system: `rotate(<real>)`

*Description:* This operator takes a real number `<real>` and rotates the current drawing coordinate system by an angle determined by this number. The anlge is given in rad. If one wants to use angles in degree one can do this by the $^\circ$ operator (this operator multiplies a number by `pi/180`. So `rotate(30°)` rotates the coordinate system by $30°$.

`scale`
### Scaling the coordinate system: `scale(<real>)`

*Description:* This operator takes a real number `<real>` and scales the current drawing coordinate system by this factor.

*Examples:* The following table shows the effect of applying diverse transformations before invoking a code that draws of a square with vertex coordiantes `[0,0],[0,1],[1,1],[1,0]`.

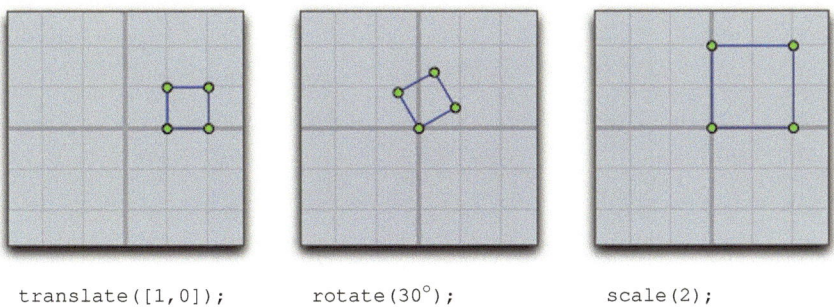

```
translate([1,0]);        rotate(30°);              scale(2);
```

The following table shows different combinations of transformations. The order of the operations may seem a little bit counterintuitive. It results from the fact that the operators transforms the coordinate system rather than the objects that are drawn.

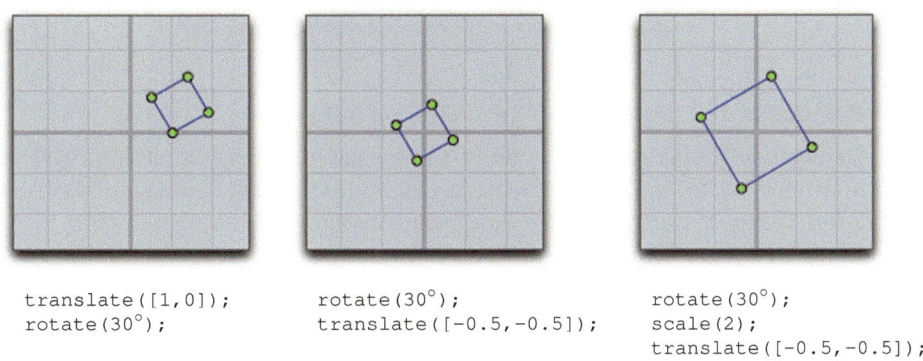

```
translate([1,0]);     rotate(30°);               rotate(30°);
rotate(30°);          translate([-0.5,-0.5]);    scale(2);
                                                 translate([-0.5,-0.5]);
```

Recursive or iterated application of transformations can lead to surprising effects. The picture below was generated by the following piece of code (assuming that square is a list of objects that draw the unit square).

```
repeat(90,
drawall(square);
translate((1,1));
scale(0.92);
rotate(30°);
)
```

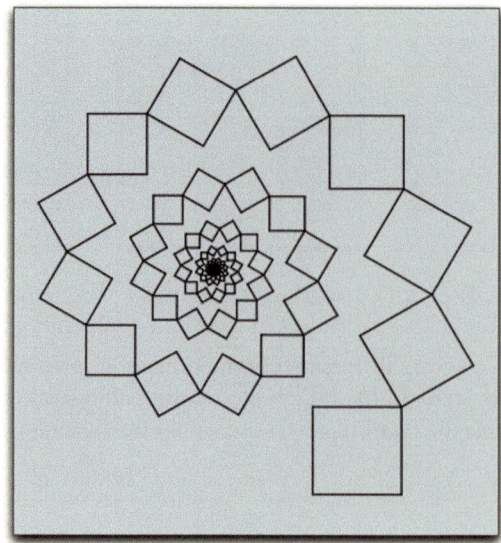

setbasis **Relating to a projective basis: `setbasis(<basis>)`**

*Description:* In the geometric part of Cinderella you can set a basis to which all drawing is related (p. 138). These bases can be translational bases, similarity bases, affine bases, or even projective bases. Such basis operations can also be used in CindyScript. The `setbasis(<basis>)` operator sets the drawing basis to a basis defined in Cinderella. The argument has to be the label of the basis in Cinderella. After applying the setbasis operator all prior coordinate transformations in CindyScript are obsoleted. However, they can be stored by `gsave` and restored by `grestore`.

*Example:* In the following example a basis `Bas0` was defined to be a projective basis related to the points A, B, C and D. This means that with respect to this basis the the corners `[0,0]`, `[0,1]`, `[1,1]`, `[1,0]` of the unit quadrangle are the points A, B, C and D. The first line of the code applies the basis transformation. The next three lines draw a grid within the unit square. The resulting image is shown below.

```
setbasis(Bas0);
x=(0..10)/10;
drawall(apply(x,([#,0],[#,1])));
drawall(apply(x,([0,#],[1,#])));
```

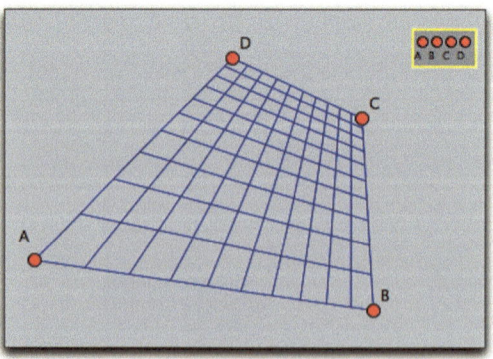

*See also:* Bases in Cinderella (p. 138)

**Relating to a translation basis: `setbasis(<vec1>)`**                    `setbasis`

**Relating to a similarity basis: `setbasis(<vec1>,<vec2>)`**              `setbasis`

**Relating to an affine basis: `setbasis(<vec1>,<vec2>,<vec3>)`**         `setbasis`

**Relating to a projective basis: `setbasis(<vec1>,<vec2>,<vec3>,<vec4>)`**  `setbasis`

*Description:* it is also possible to relate the internal drawing basis of CindyScript directly to a basis specified by points. This avoids the explicit creation of a basis in Cinderella. This can be easily done by providing points of the base frame of the basis.

*Example:* The following piece of code together with the picture below demonstrates this feature. in the code `sq` is first defined as a macro to draw a square grid. The grid is drawn with respect to several bases.

```
sq:=(draw([0,0],[1,0]);
    draw([0,0.25],[1,0.25]);
    draw([0,0.5],[1,0.5]);
    draw([0,0.75],[1,0.75]);
    draw([0,1],[1,1]);
    draw([0,0],[0,1]);
    draw([0.25,0],[0.25,1]);
    draw([0.5,0],[0.5,1]);
    draw([0.75,0],[0.75,1]);
    draw([1,0],[1,1]));

setbasis(A);
sq;
setbasis(B,C);
sq;
setbasis(D,E,F);
sq;
setbasis(G,H,L,K);
sq;
```

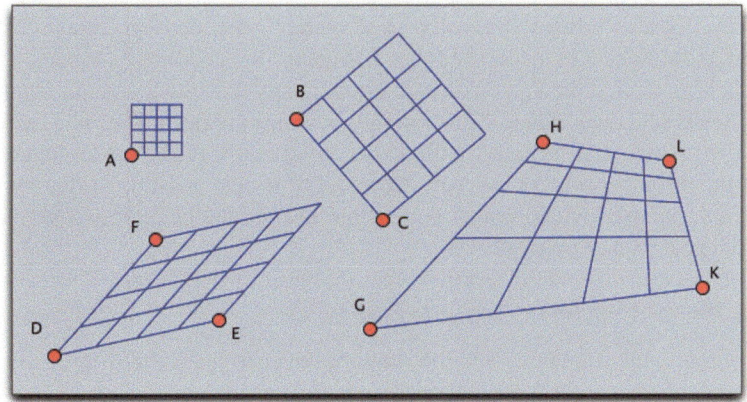

## 11.9 The Appearance and Basis Stack

Performing transformations of the coordinate system influences the global behavior of drawing operations that are performed. Very often such a change of the coordinate system is only intended for a certain drawing operation after which the coordinate system should be set to its old state again. CindyScript has a mechanism that serves for such services.

The `gsave()` operator stores all informations in the graphic state (coordinate transformations, sizes, colors, opacity) to a stack. While the `grestore()` operator reverses this effect by popping the information from the stack. Thus within a sequence `gsave()...grestore()` any coordinate transformations can be made without having effect to the remaining code.

**gsave**

### Storing the graphics state: `gsave()`

*Description:* This operator stores all informations that may influence the current graphics operations to a stack. The informations include the coordinate system, sizes for points, lines and text as well as colors and opacities.

**grestore**

### Restoring the graphics state: `grestore()`

*Description:* This operation pops the last stored graphics state from the stack and restores it to the drawing engine.

**greset**

### Resetting the graphics state: `greset()`

*Description:* This operator clears the graphics information to a stack and sets the coordinate system, all sizes, colors and the opacity back to the initial value.

*See also:* Appearance of Objects (p. 282)

## 11.10 Layers

In a sense there is a third dimension in drawings generated by CindyScript. Each drawing is associated to a specific layer. Layers are specified by numbers. When the drawing is generated the layers are painted in the order of the associated numbers. If no layer was specified then all Cindy script drawings are performed in a background layer, that is below all geometric elements. By default, all layers are cleared right before the *draw* part of CindyScript. However, it is also possible to mark a layer to be not *auto cleared* which makes it possible to produce background graphics that have to be calculated only once.

**layer**

### Setting the drawing layer: `layer(<int>)`

*Description:* This statement sets the drawing layer to a specified level. By default the layer is cleared right before the *draw* event takes place.

*Example:* The following piece of code draw four overlapping circles in the usual drawing order.

```
cir(x,y):=(
  fillcircle((x,y),3,color->(1,.8,0));
  drawcircle((x,y),3,color->(0,0,0),size->3);
);
cir(0,0);
cir(1,1);
cir(2,2);
cir(3,3)
```

If one calls a layer statement before each individual drawing, it is possible to reverse the order in which the circles are drawn onto the screen, as the next example demonstrates.

```
cir(x,y):=(
  fillcircle((x,y),3,color->(1,.8,0));
  drawcircle((x,y),3,color->(0,0,0),size->3);
);
layer(6);
cir(0,0);
layer(5);
cir(1,1);
layer(4);
cir(2,2);
layer(3);
cir(3,3)
```

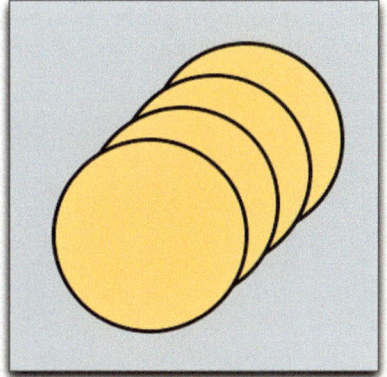

usual drawing                    drawing with reversed layers

### Clearing a layer: `emptylayer(<int>)`

emptylayer

*Description:* This function clears the given layer and sets it as the current active one.

### Remove all drawings in a layer: `clrscr()`

clrscr

*Description:* This statement immediately clears all drawings that were performed in a screen. It is sometimes useful to use this statement to make sure that a drawing is really in an empty state. In particular it may become necessary if the `repaint()` statement is called from inside other scriptslots, or if the autoclear flag of a layer is set to false.

**autoclearlayer**

**Automatic clearing of layers: `autoclearlayer(<int>,<boolean>)`**

*Description:* The autoclear flag of a layer can be set using this function. By default, every layer is cleared automatically during construction recalculations. You can turn this of by setting the autoclear flag to `false`.

If a layer is not cleared automatically, all drawing commands will be preserved. Cinderella does not store a bitmap, but actually redoes all drawings on every step, so you might run into performance problems if you add more and more drawing commands to a layer.

*Example:* You can use non-clearing layers to draw background or foreground graphics once in the *init* slot instead of drawing them every time in the *draw* slot of the construction (p. 395).

```
autoclearlayer(-4,false);
layer(-4);
repeat(1000,
    p = [random(20)-5,random(20)-5];
    color([random(),random(),random()]);
    fillpolygon(apply([[0,0],[1,0],[1,1],[0,1]],p+#),alpha->.2);
);
layer(0);
```

The code above in the *init* slot produces a random background once on layer -4. As the autoclear flag is set to false, this background will stay during all other operations, as demonstrated below. If you want to achieve the same effect in the *draw* slot you have to store all random polygons and their colors to recreate them on every cycle.

*A random background*

**screenbounds**

**Determining screen bounds: `screenbounds()`**

*Description:* If you work only with a finite part of the projective plane and do not use a spherical view (p. 116) it makes sense to ask for the screen bounds. This function returns a list of four points in homogeneous coordinates that define a rectangle covering all visible views.

This function is best used with only one Euclidean view for a construction.

**Determining screen resolution: `screenresolution()`**

*Description:* In contrast to hyperbolic and spherical views, Euclidean views have the same screen resolution everywhere. This function gives the number of pixels between the origin and point `[0,1]`. If several Euclidean views for the same construction are open, the function will return the maximum of all individual screen resolutions.

*Example:* You can create a checkerboard pattern by coloring pixels. The code below will color every other pixel independently of the current zoom. Please be aware that we do not recommend to work on the pixel level, but there may be situations that justify this. In most cases, the `colorplot` function (p. 298) is easier to use and a lot faster.

```
upperleft=(screenbounds()_1).xy;
lowerright=(screenbounds()_3).xy;
width in pixels=screenresolution()*(lowerright.x-upperleft.x);
height in pixels=screenresolution()*(upperleft.y-lowerright.y);

repeat(width in pixels/2, x,
 start->upperleft.x, stop->lowerright.x,
 repeat(height in pixels/2, y,
   start->upperleft.y, stop->lowerright.y,
   draw((x,y),border->false,size->.5,color->[0,0,0]);
 )
)
```

# Chapter 12

# Geometric Operators

The geometric operators provide high-level access to several elementary geometric operations. They can be applied either directly to objects of the geometric construction or to vectors (lists of numbers) that represent the coordinates for the geometric objects. If the operator returns a geometric object, it can also be directly drawn with the draw operator.

## 12.1 Lists and Coordinates

Coordinates for lines are always homogeneous coordinates (i.e., list of three numbers that are the parameters `[a,b,c]` of the line with equation `a*x+b*y+c=0`). Coordinates of points can be either Euclidean (list of two numbers `[x,y]`) or homogeneous (list of three numbers `[x,y,z]` that represent the point `[x/z,y/z]`). Return values will always be in homogeneous coordinates. In the text below we will indicate pointlike arguments as `<point>`, and linelike arguments as `<line>`, without further mentioning that the coordinates have to of the kind described above.

Since both object types `<point>` and `<line>` can be represented as lists of three numbers, one needs a way to distinguish these two cases. Internally, a list carries flags that indicate whether it has an intrinsic geometric meaning. One can interrogate the operator `geotype(<list>)` to obtain this information. This operator will return either `"Point"`, `"Line"`, or `"None"`. If such a vector has an intrinsic geometric meaning, then the `draw` operator will automatically render it as such an object.

## 12.2 Elementary geometric operators

**Intersection of two lines: `meet(<line1>,<line2>)`** `meet`

*Description:* This operator calculates the intersection of two lines. It returns a point in homogeneous coordinates.

`join`                        **Joining two points: `join(<point1>, <point2>)`**

*Description:* This operator calculates the line joining two points. The result is a line in homogeneous coordinates.

`parallel`                    **Calculating a parallel: `parallel(<point>, <line>)`**

`parallel`                    **Calculating a parallel: `parallel(<line>, <point>)`**

*Description:* This operator takes a point and a line as input (no matter in which order) and calculates the line parallel to the input line and through the input point. The line that is returned by the operator is again represented in homogeneous coordinates. This operator refers to Euclidean geometry. Hyperbolic and spherical computations are not supported. This operator can also be abbreviated by `para(...)`.

`perpendicular`               **Calculating an orthogonal line: `perpendicular(<point>, <line>)`**

`perpendicular`               **Calculating an orthogonal line: `perpendicular(<line>, <point>)`**

*Description:* This operator takes a point and a line as input (no matter in which order) and calculates the line orthogonal to the input line and through the input point. The line that is returned by the operator is again represented in homogeneous coordinates. This operator refers to Euclidean geometry. Hyperbolic and spherical computations are not supported. This operator can also be abbreviated by `perp(...)`.

*Combined example:* The following code creates the picture below. Observe that the operator generates an implicit typing of the return values, so that the objects are rendered correctly automatically.

```
A=[1,1];
B=[2,5];
C=[7,2];
a=join(B,C);
b=join(C,A);
c=join(A,B);
ha=perpendicular(A,a);
hb=perpendicular(B,b);
hc=perpendicular(C,c);
X=meet(ha,hb);
drawall([a,b,c,d,ha,hb,hc,X,A,B,C]);
```

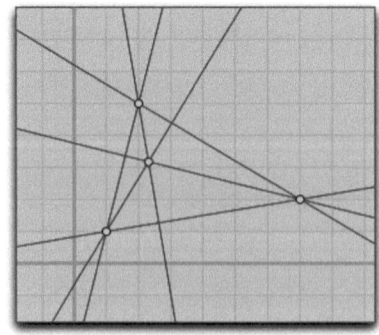

**Calculating an orthogonal vector: `perpendicular(<list>)`**                    `perpendicular`

*Description:* If the `perp` operator is invoked with one argument, it assumes that the input list consists of two numbers. Such a list `[a,b]` is converted by this operator to the list `[-b,a]`, which is the input vector rotated 90° about the origin.

**The area of a triangle: `area(<point1>,<point2>,<point3>)`**                    `area`

*Description:* This operator calculates the (oriented) area of the triangle formed by the three vertices `<point1>`, `<point2>`, and `<point3>`. If the orientation of the points is counterclockwise, then the area is positive, while if it is clockwise, the area is negative. If the three points are collinear, then the area is zero.

## 12.3 Useful Linear Algebra Operators

The following operators from linear algebra are very useful in geometric contexts. They apply especially to numeric vectors of length 3. For other useful operators in that context (such as `dist`, matrix operations, and scalar product) please refer to Vectors and Matrices (p. 271).

**The determinant of three points: `det(<vec1>,<vec2>,<vec3>)`**                    `det`

*Description:* This operator calculates the determinant of a $3 \times 3$ matrix formed by the three three-dimensional vectors `<vec1>`, `<vec2>`, `<vec3>`. Unlike the general determinant method described in the section Vectors and Matrices (p. 271), this method is optimized for performance.

**The cross product of two points: `cross(<vec1>,<vec2>)`**                    `cross`

*Description:* This operator takes two three-dimensional vectors and calculates their cross product. The cross product is a three-dimensional vector that is orthogonal to the other two vectors.

## 12.4 Conversion and Typing

**The type of an object: `geotype(<list>)`**                    `geotype`

*Description:* This operator determines whether an object has an explicit geometric meaning. It returns a string that is either `"POINT"`, `"LINE"`, or `"NONE"`.

Applying this operator to a list of two numbers will always return `"Point"`. Applying this operator to a list of three numbers will return `"Point"`, `"Line"`, or `"None"`, depending on whether the internal meaning of this list is set to `"Point"` or `"Line"`. Values that come from geometric objects of a Cinderella construction will always be assigned the corresponding geometric type. The output of the `meet` operator is always a "Point". The output of the operators `join`, `parallel`, and `perpendicular` is always a "Line". Furthermore, the geometric meaning can be explicitly set using the operators `line` and `point`.

point

**Declaring points: point(<vec>)**

*Description:* This operator explicitly sets the geometric type of a vector of three numbers to "Point". If the argument is not a list of three numbers, the operator has no effect.

line

**Declaring lines: line(<vec>)**

*Description:* This operator explicitly sets the geometric type of a vector of three numbers to "Line". If the argument is not a list of three numbers, the operator has no effect.

complex

**Points to complex numbers: complex(<point>)**

*Description:* This operator takes a point and converts it into a complex number. Here the usual coordinate system of the Euclidean plane is identified with the Gaussian complex plane. The point [x,b] is converted to the complex number a+i*b.

gauss

**Complex numbers to points: gauss(<point>)**

*Description:* This operator is the opposite of the previous one. It converts a complex number a+i*b to a list of two numbers [a,b].

crossratio

**Cross ratio of four points or lines: crossratio(<vec>,<vec>,<vec>,<vec>)**

*Description:* Calculates the geometric cross ratio of four points. If the four points are collinear, the usual cross ratio in the real projective plane is calculated. If the points are not collinear, then the cross ratio of the corresponding points in the complex projective line is calculated. For collinear finite points these two results coincide.

crossratio

**Cross ratio of four numbers: crossratio(<numb>,<numb>,<numb>,<numb>)**

*Description:* Calculates the cross ratio (A/B)/(C/D) of four real or complex numbers. An extremely useful geometric invariant.

## 12.5 Geometric Transformations and Bases

One can deal with geometric transformations on an explicit algebraic level. Transformations are best represented by $3 \times 3$ matrices. A real transformation corresponds to a matrix multiplication of the homogeneous coordinates of a point by the corresponding transformation matrix. There are several operators for the calculation of these transformation matrices.

linereflect

**Reflection in a line: linereflect(<line>)**

*Description:* Returns a matrix that represents a reflection in the line <line>.

*Example:* The following code takes the line *a*, creates the reflecting transformation, and maps point *C* by multiplying its homogeneous coordinates by the matrix. The result is the green point in the figure.

```
m=linereflect(a);
draw(m*C.homog);
```

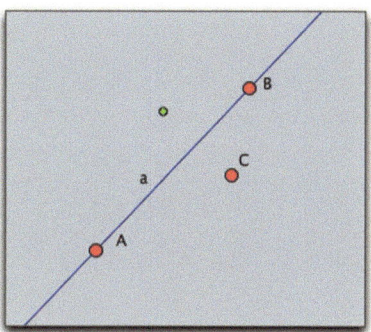

### Reflection in a point: `pointreflect(<point>)`                `pointreflect`

*Description:* Returns a matrix that represents a reflection in the point `<point>`.

### Translation: `map(<point1>,<point2>)`                                    `map`

*Description:* Returns a matrix that represents a translation that maps `<point1>` to `<point2>`.

### Similarity: `map(<point1>,<point2>,<point3>,<point4>)`                    `map`

*Description:* Returns a matrix that represents a similarity that maps `<point1>` to `<point2>` and `<point3>` to `<point4>`.

### Affine transformation: `map(<point1>,<point2>,<point3>,<point4>,`        `map`
### `<point5>,<point6>)`

*Description:* Returns a matrix that represents an affine transformation that maps `<point1>` to `<point2>`, `<point3>` to `<point4>`, and `<point5>` to `<point6>`.

### Projective transformation: `map(<point1>,<point2>,<point3>,<point4>,`    `map`
### `<point5>,<point6>,<point7>,<point8>)`

*Description:* Returns a matrix that represents a projective transformation that maps `<point1>` to `<point2>`, `<point3>` to `<point4>`, `<point5>` to `<point6>`, and `<point7>` to `<point8>`.

# Chapter 13

# Calculus

CindyScript is a purely numerical language. No symbolic computations are performed. Nonetheless, it is possible to do some operations that can usually be done only in symbolic systems. For instance, one can numerically calculate the derivative of a function or construct a tangent to a given function. This section illustrates the use of these two functions.

## 13.1 Derivatives and Tangents

### Calculating a derivative: d(<function>, <var>)                                          d

*Description:* This operator creates a function that is the derivative of another function, which is passed as a first argument. The variable of <function> with respect to which the derivative should be calculated has to be the standard run variable #. The free variable of the derivative is given as second argument.

*Example:* The following code demonstrates the use of the operator:

```
f(x):=(x-3)*(x-2)*(x-1)*x*.4;
g(x):=d(f(#),x);
h(x):=d(g(#),x);
plot(f(x),size->2);
plot(g(x),color->(0.8,0,0));
plot(h(x),color->(0,0,0));
```

The output generated by this program is the following drawing. The blue line is the original function f(x), the red line is the first derivative g(x), and the black line is the second derivative h(x).

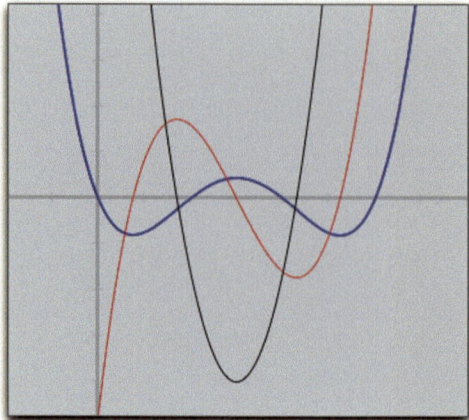

The second derivative was calculated as the derivative of the first derivative. However, a word of caution has to be said. The derivative operator is based entirely on numerical principles. If `f(x)` is the function that should be derived, then the corresponding derivative can be defined as `d(f(#),x)`. Here x denotes the position at which the derivative should be evaluated. The value of the derivative is then calculated by the formula

```
(f(x+eps)-f(x-eps))/2eps
```

for a sufficiently small number `eps`. This gives a reasonable approximation to the real derivative at this point. However, applying this operator several times in succession increases the error significantly. After about five iterations, the result is no longer usable. Thus one cannot expect to perform a reasonable calculation with a fifth derivative.

*Modifiers:* By default, the constant `eps` in the above formula is set to 0.0001. This value forms a reasonable compromise between reliability in higher derivatives and precision. This value can be altered using the modifier `eps-><number>`.

**tangent**

**Tangent: `tangent(<function>, <var>)`**

*Description:* This operator is very similar to the operator that calculates the derivative of a function. However, instead of calculating the value of the derivative, this operator calculates the homogeneous coordinates of a tangent to the function at a point. The point is entered as second argument.

*Example:* The following sample code calculates many tangents to a parabola:

```
f(x):=(x^2)/4;
repeat(250,start->-30,stop->30,x,
  t=tangent(f(#),x);
  draw(t,alpha->.3);
);
plot(f(x),size->3,color->(0,0,0));
```

The picture below shows the resulting image created by the program.

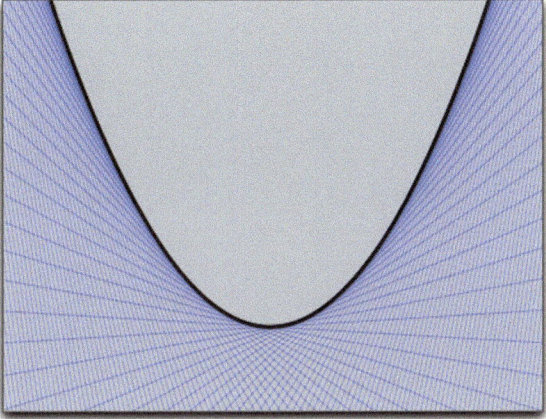

The result of the tangent function is a three-vector of homogeneous coordinates. In addition, the line attribute of this vector is set (see Geometric Operators (p. 349)) so that the coordinates are automatically drawn as a line.

## 13.2 Advanced calculations

**Guessing the value of a real number: `guess(<number>)`**                    **guess**

*Description:* The guess operator is perhaps one of the most sophisticated and powerful operators of CindyScript. It can be used to recover a symbolic meaning from a numerical floating-point value. The guess operator expects a number as input and returns a string. The string should be a symbolic expression describing the input number. The guess operator tries to generate a string of the form

```
a+b*sqrt(c) ,
```

where `a`, `b`, and `c` are rational numbers with numerator and denominator not larger than about 1000. If the input number is expressible in the described way with a reasonable numeric precision, then the guess operator will generate this expression. If not, the guess operator will return the original input number.

Thus it is possible to use the guess operator to discover whether a number is rational (then `b` is 0) or the solution of a quadratic equation.

*Examples:* This operator is sometimes extremely useful for finding hidden properties of geometric constructions. We illustrate this with a few examples. The first picture shows a construction in which the slope of the line and the coordinates of the intersection of two circles are "guessed":

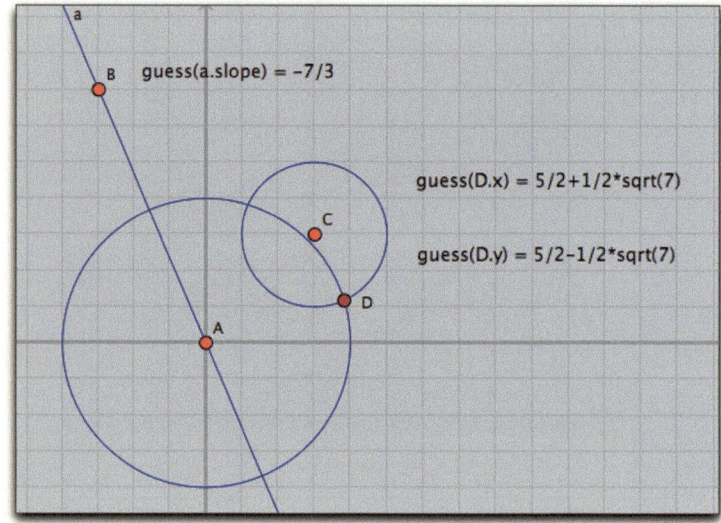

If the input coordinates and radii are integers, the resulting numbers will be rational or solutions of quadratic equations with relatively small coefficients. The next two examples show two serious applications of guessing. The first one compares two lengths in a regular pentagram. The length ratio will be exactly the golden ratio. The second example shows a nice connection between two areas of squares in a simple construction.

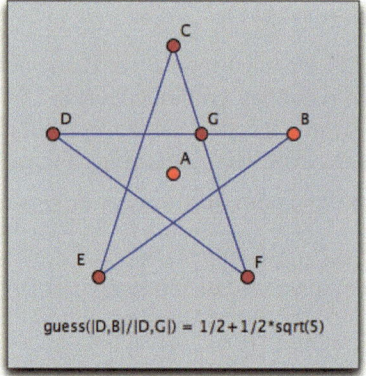
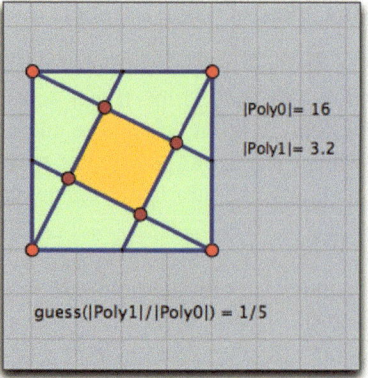

*Remark:* The workhorse behind the guessing operator is the so-called PSLQ algorithm, a truly ingenious algorithm that is able to discover (if real numbers $x$, $y$, and $z$ are given) integer dependencies $a{\cdot}x + b{\cdot}y + c{\cdot}z = 0$. If one wants to check whether the real number $x$ is the solution of a quadratic relation, one has to look for an integer relation of the form $a + b{\cdot}x + c{\cdot}x^2 = 0$. This is how the implementation of the guess operator works. The solution is then reconstructed from the calculated integral coefficients.

**Roots of a polynomial: `roots(<list>)`**

roots

*Description:* The function roots can calculate the roots of a univariate polynomial. As input it accepts the coefficients of the polynomial, ordered from lowest degree to highest degree. The results as well of the coefficients may be complex numbers.

*Examples:* For instance if one wants to calculate the roots of the polynomial $1+x^2=0$ one can simply call `roots([1,0,1])`. The resulting expression is `[-0-i*1,-0+i*1]`, since the polynomial has the two complex roots $+i$ and $-i$.

The code below calculates and draws the roots of a cubic polynomial given by its coefficients.

```
a=0.4;
b=-0.4;
c=-3;
d=-1;
f(x):=a*x^3+b*x^2+c*x+d;
plot(f(x),size->2);
r=roots([d,c,b,a]);
forall(r,draw((#,0)))
```

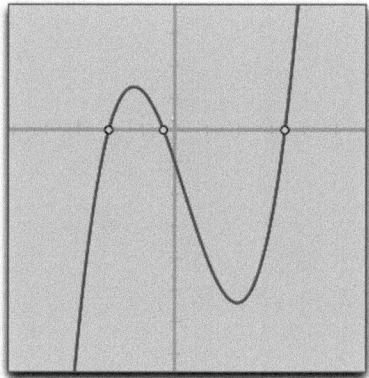

# Chapter 14

# Music Support: Syntherella

CindyScript offers a wide range of very versatile functions for the generation of audio output. These audio functions cover the generation of music (via a built-in synthesizer), as well as the generation of algorithmically generated audio samples (how, for instance, does a function sound?). They can be used to serve many different purposes, e.g. ...

- ...to enhance the appearance of usual Cinderella files

- ...as an integral part of certain Cinderella files (games, physical simulations, etc.)

- ...to demonstrate and study mathematical structures of music

- ...to demonstrate and study mathematical and physical structures of sound

The audio interface is subdivided into two essential parts:

**MIDI Audio**

The MIDI Functions (p. 362) can be used to create melody and instrument based audio signals. MIDI is a standard format for electronic music that is used for the communication between electronic instruments (like keyboards) and computers. The MIDI Functions (p. 362) of CindyScript represent an interface to a built-in synthesizer of your computer. It provides access to synthesized musical instruments (pianos, percussion instruments, strings, etc.) and to melodic scales and timing. Furthermore it gives access to sound characteristics like volume, reverb, balance, etc. The CindyScript MIDI interface is designed to provide a high-level and intuitive interface to this functionality.

**Sampled Audio**

The sampled audio interface tightly links mathematical functions and processes to audible experiences. It provides a streamlined set of features that allow to "play" mathematical functions, study harmonic spectra and create sound from algorithmic or physical simulations.

## 14.1 MIDI Functions

### 14.1.1 How Cinderella-MIDI Works

MIDI stands for *Musical Instrumental Digital Interface*. It is a standard protocol that is used for the communication between electronic instruments and computers. MIDI provides control over individual tones, sequences of tones, the sound of instruments, polyphony, volume and many other things which are essential for the perception of music. The CindyScript MIDI interface offers a streamlined set of statements for creating music and sound effects easily and flexibly. Before we dive into the details of these statements we will give a brief overview of the architecture and the capability of MIDI. This will make it easier to understand its interaction with Cinderella. Whenever easily possible we also mention how the MIDI statements are accessible in CindyScript.

### 14.1.2 Notes, Channels and Instruments

Roughly speaking the most fundamental functionality of MIDI is to send information to a music synthesizer and tell it to play a note. In its most simple form this piece of information just consists of a number between 0 and 127 that tells the synthesizer what note has to be played. In standard midi coding the number 60 represents the note C in the middle octave. Increasing or decreasing this number by one corresponds to a half-step. Thus 61 corresponds to C#, 60 to D and so forth. The following table associates notes and numbers for the range of two octaves:

C C# D D# E F F# G G# A B H c c# d e# e f f# g g# a b h c'
60 61 62 63 64 65 66 67 68 69 70 71 72 73 74 75 76 77 78 79 80 81 82 83 84

The corresponding CindyScript statement for playing a tone of a certain pitch is `playtone`. Try for example `playtone(60)`.

However, a note comprises far more than a certain pitch. It has an associated instrument by which it is played, a volume, a certain location in space and many other characteristics. Some of them are more of a static nature (like the instrument) and do not change from tone to tone. Others like the volume are more dynamic in nature and change frequently within the a piece of music. MIDI lets you control all these parameters in a way that is close to the characteristics of music. Imagine the playing of a piano. The main characteristics that changes the volume of a sound that is generated by hitting a key is the *velocity* with which the key is hit. With this velocity not only the volume changes but also the complete sound characteristics of the tone. MIDI does not only model the volume but also (and completely automatically) this change of sound characteristics. For this reason the parameter that controls the volume is called *velocity* rather than volume. In Cinderella this parameter is controlled by a modifier that ranges from 0.0 to 1.0 So with `playtone(60,velocity->0.5)` you get a C with a medium volume.

In the default setting of Cinderella such a note is played by a piano like instrument. There are two more concepts of MIDI that are essential for understanding the generation of sound: *instruments* and *channels*. A single MIDI device is capable of playing the instruments of an entire band with 16 players all at once. For this

MIDI offers 16 channels numbered from 0 to 15. Each of them can be associated to a specific instrument the instruments are numbered from 1 to 128 and cover a broad range of musical experiences (brass, mallets, strings, pianos, accordeons, guitars, percussion and many more). In CindyScript an instrument (say 25 which is an acoustic guitar) can be associated to a channel (say 3) either by `midichannel(3, intrument->25)` or by `instruments(25, channel->3)`. The standard channel in which a tone is played by `playtone(...)` is 0 but it can be changed by a modifier. So the code

```
midichannel(3,instrument->25); //guitar
midichannel(2,instrument->12); //vibraphone
playtone(60);
playtone(64);
playtone(67,channel->3);
playtone(72,channel->2);
```

plays a C-major chord whose two lowest tones are played by the piano and whose upper notes are played by a vibraphone and a guitar. Each channel can be used completely polyphonically. So it is no problem that the piano play two notes at the same time.

In CindyScript's MIDI implementation the 16 channels have preassigned instruments, to allow for convenient access to important sounds. The instruments associated to the channels are specified in the following table (the numbers behind the instruments indicate the instrument numbers):

0 : Acoustic Grand Piano (1)  8 : Synt. square wave (81)
1 : Glockenspiel (10)         9 : DRUMSET
2 : Vibraphone (12)           10 : FX 3, crystal (99)
3 : Marimba (13)              11 : Tinkle Bell (113)
4 : Xylophone (14)            12 : Flute (74)
5 : Woodblock (116)           13 : Acoustic Bass (33)
6 : Pizzicato Strings (46)    14 : Percussive Organ (18)
7 : Tremolo Strings (45)      15 : Alto Saxophone (66)

There are various other characteristics (reverb, balance, pressure, etc. ) of tones that can be accessed by modifiers. For a complete description we refer to the explanations for the various statements. We here only mention the most important one: *duration*. This parameter controls the time (in seconds) after which a note is released again. So `playtone(60,duration->0.1,velocity->1)` generates a short and loud C. The default for duration without explicit modifier is 1 second.

### 14.1.3 Timing and Sequencing

So far we have only dealt with tones that are played all at once. There are two ways for creating melodies in Cinderella. Either one creates the timing explicitly by using the `wait(...)` command of CindyScript or one uses the built-in MIDI sequencer. We first briefly describe how the method that does not rely on the sequencer works. For this we only have to separate individual notes by wait statements.

```
playtone(60,duration->3);      wait(500);
playtone(64,duration->2.5);    wait(500);
playtone(67,duration->2);      wait(500);
playtone(72,duration->1.5);
```

In the above example the timing is given explicitly so that all notes end at exactly the same time. One can also use this method to generate interesting melodies (or jingles) programmatically. Try out the following piece of code in CindyScript (it is worth the typing effort)

```
halftones=[61,63,66,68,70,73,75,78,80,82,85,87,90];
forall(60..90,i,
  playtone(i,velocity->if(contains(halftones,i),1,0.6),
           channel->if(contains(halftones,i),3,5));
  wait(100);
);
```

Generating notes this way has its disadvantages. First, the timing has to be described in detail by the CindyScript code. Second, during the `wait(...)` statements the execution of other CindyScript code is blocked, since CindyScript is *waiting*. For this reason (and several others) it is often by far more easier to use the built-in MIDI sequencer.

The MIDI sequencer is the second major feature of MIDI. It is a piece of program logic that allows to store MIDI events in a timed manner. Thus arbitrary pieces of music can be described and played. The MIDI sequencer associates each MIDI event with a timestamp that tells the synthesizer when a specific note has to be played. In CindyScript you can access the sequencer either via the `playmelody(...)` statement or via the `midiaddtrack(...);midistart()` commands. The first is simpler and more immediate, whereas the second is more powerful. We will briefly describe the first method here. The statement `plamelody(...)` simply expects a list of MIDI events. In its most basic form such a MIDI event is a list that just consists of a number that indicates the note followed by another number indicating the duration. So a broken C-major chord can for instance be coded as:

```
playmelody([[60,1],[64,1],[67,1],[72,4]]);
```

It is also possible to access the notes by their names given as strings instead of the numbers. By this the following piece of code

```
playmelody([
  ["c",1],["e",1],["f",1],["g",5],["c",1],["e",1],["f",1],["g",5],
  ["c",1],["e",1],["f",1],["g",2],["e",2],["c",2],["e",2],["d",5],
  ["e",1],["e",1],["d",1],["c",3],["c",1],["e",2],["g",2],["g",1],
  ["f",4],["f",1],["e",1],["f",1],["g",2],["e",2],["c",2],["d",2],
  ["c",5]
],speed->300,instrument->57)
```

plays a very simple version of "Oh, when the saints". Again the statement may be decorated by various modifiers. So here the instrument is chosen to be something that sound roughly like a trumpet (real trumpet players please forgive us). The speed is set to 300 *beats per minute*. A *beat* is the fundamental time unit for the sequencer. It is usually associated to a quarter note and there are usually four beats per measure or three beets per measure (if you have a valse). Also fractions of beats are allowed. They may be used to represent eight notes, triplets and so forth.

The Syntherella interface to the sequencer is capable of expressing lots of musical content on a rather high level. Details are explained later on.

### 14.1.4 Percussion

Cinderella uses the predefined setup called *General MIDI* which is a specific collection of synthesized instruments. There is one convention that is special (though very useful) in General Midi. Channel 10 plays a special role. It is reserved for percussion instruments and can be used to implement a drum set. So whenever you use channel 10 you must be aware that this cannel has a special meaning. The MIDI numbers that usually represent notes in this case are associated to specific percussion instruments (not all 128 keys are used here). The following table gives you an overview of the instruments associated to the keys.

| | | | |
|---|---|---|---|
| 35 Bass Drum 2 | 47 Mid Tom 1 | 59 Ride Cymbal 2 | 71 Short Whistle |
| 36 Bass Drum 1 | 48 High Tom 2 | 60 High Bongo | 72 Long Whistle |
| 37 Side Stick | 49 Crash Cymbal 1 | 61 Low Bongo | 73 Short Guiro |
| 38 Snare Drum 1 | 50 High Tom 1 | 62 Mute High Conga | 74 Long Guiro |
| 39 Hand Clap | 51 Ride Cymbal 1 | 63 Open High Conga | 75 Claves |
| 40 Snare Drum 2 | 52 Chinese Cymbal | 64 Low Conga | 76 High Wood Block |
| 41 Low Tom 2 | 53 Ride Bell | 65 High Timbale | 77 Low Wood Block |
| 42 Closed Hi-hat | 54 Tambourine | 66 Low Timbale | 78 Mute Cuica |
| 43 Low Tom 1 | 55 Splash Cymbal | 67 High Agogo | 79 Open Cuica |
| 44 Pedal Hi-hat | 56 Cowbell | 68 Low Agogo | 80 Mute Triangle |
| 45 Mid Tom 2 | 57 Crash Cymbal 2 | 69 Cabasa | 81 Open Triangle |
| 46 Open Hi-hat | 58 Vibra Slap | 70 Maracas | |

By this it is relatively easy to create a rhythm track that accompanies your music. The percussion instruments are also very useful to create sound effects in other Cinderella applications. For instance if you use CindyLab (p. 167) to create a Ping-Pong game, you can use the *Wood Blocks* (key 76 and 77) to create click effects when hitting the ball.

The following code shows a simple implementation of a Rock rhythm by using the sequencer. Observe that this program also uses the `"goto"` statement in `playmelody(...)` to rewind the track to the beginning.

```
playmelody(channel->9,speed->500,loop->8,
 [[35,3],[35,3],[35,3],[35,6],[35,1],["goto",0], //bass drum
  [-1,2],[42,4],[42,4],[42,4],[42,2],["goto",0], //high-hat
  [-1,4],[38,8],[38,4],["goto",0],               //snare
  [59,4],[59,4],[59,4],[59,4],["goto",0],        //ride cymbal
  [-1,13],[76,1],[62,1],[-1,2],["goto",0]        //percussion
]);
```

In the above code snippet several percussion instruments are overlaid in the same "melody" and form a complex rhythmic pattern. The note −1 is used as a pause.

### 14.1.5 Tracks and Pieces of Music

We will end our little introductory journey by creating a small piece of music consisting of a melody part and a drum pattern. MIDI can accept many different tracks for the sequencer. Each track may be associated with one player of a band. CindyScript offers a statement `midiaddtrack(...)` that silently adds a track to the sequencer without immediately playing it. All added track can be played by invoking the `midistart()` command. The following piece of code shows how our Rock rhythm can be combined with the "Oh, when the saints" melody to give a rocky version of this traditional piece of music. There are some specialties that will be explained in a moment.

```
midichannel(3,instrument->57);
midiaddtrack(channel->3,track->1,
  [["c",1],["e",1],["f",1],["g",5],["c",1],["e",1],["f",1],["g",5],
   ["c",1],["e",1],["f",1],["g",2],["e",2],["c",2],["e",2],["d",5],
   ["e",1],["e",1],["d",1],["c",3],["c",1],["e",2],["g",2],["g",1],
   ["f",4],["f",1],["e",1],["f",1],["g",2],["e",2],["c",2],["d",2],
   ["c",5]
]);
midiaddtrack(channel->9,track->2,stretch->1/2,offset->3,repeat->8,
  [[35,3],[35,3],[35,3],[35,6],[35,1],["goto",0], //bass drum
   [-1,2],[42,4],[42,4],[42,4],[42,2],["goto",0], //high-hat
   [-1,4],[38,8],[38,4],["goto",0],                //snare
   [59,4],[59,4],[59,4],[59,4],["goto",0],         //ride cymbal
   [-1,13],[76,1],[62,1],[-1,1]                    //percussion
]);
midistart(speed->250);
```

So in principle we add two tracks to the sequencer (one for the melody and one for the rhythm) and play it by using `midistart()`. There are some minor problems concerning timing and positioning that can be addressed using modifiers. First of all we take the "Oh, when the saints" track from the previous example as it is. We associate it with track 1 of the sequencer. Adding the Rock rhythm pattern from the other example to track 2 brings up several problems. First of all the timing does not fit. The beats in the melody are twice as long as the beats in the rhythm track. We can adjust this by using the modifier `stretch->1/2` equivalently we could have halved the position numbers of the rhythm track. Second "Oh, when the saints" has some pickup notes (the first three). So the rhythm track should not start immediately. We can fix this by using `offset->3`. Finally, we need all together eight repetitions of the rhythm pattern. We do so by saying `repeat->8`.

## 14.2 Single Tones

We first start with those statements that are used to generate single tones. Some of the tone characteristics can be addressed via modifiers.

**playtone**

**Playing a tone: `playtone(<int>)`**

*Description:* This is the simplest way to create a tone. It creates a tone that is played immediately when this statement is invoked. Without modifiers the current default channel with its associated instrument is used. The default duration is 1 second. The default velocity is `0.5`.

*Example:* The following code plays a broken C-major chord.

```
playtone(60);
wait(1000);
playtone(64);
wait(1000);
playtone(67);
```

```
wait(1000);
playtone(72, duration->4);
```

*Modifiers:* The command has several modifiers, most of them are self-explanatory.

| | | |
|---|---|---|
| **velocity** | *0.0 ... 1.0* | the volume of a tone (how fast a piano key is pressed) |
| **amp** | *0.0 ... 1.0* | identical to `velocity` |
| **duration** | *\<real>* | duration of the tone in seconds |
| **channel** | *0..15* | selecting the channel that is played |
| **reverb** | *0.0 ... 1.0* | reverb effect |
| **balance** | *-1.0 ... 1.0* | left/right panorama of tone |
| **bend** | *-2.0 ... 2.0* | bending a tone up to a whole note down or up |

By the `bend` modifier the pitch of the note can be altered by two half-steps up or down. One unit of bend corresponds to one half-step.

If duration is set to 0 (or smaller) then the tone is kept for an indefinite time and will be only stopped by the `stoptone(...)` statement.

**Stopping a tone: `stoptone(<int>)`**                                            **stoptone**

*Description:* Immediately stops the tone of the specified key. This might be useful if a tone has been started with indefinite length.

**Playing a tone of specified frequency: `playfrequency(<real>)`**                **playfrequency**

*Description:* This statement is very similar to `playtone`. However in this case the frequency (in Hz) is explicitly given by a real parameter. Sometimes it may be useful to use the more physical oriented function `playsin(...)` instead.

*Example:* `playfrequency(440)` plays a tone of 440 Hz with the currently chosen channel and instrument. The `playfrequency` statement is particularly interesting for simulating scales of non-european cultures, needed for instance in Indian Ragas or Javanesian Gamelan music.

*Modifiers:* The modifiers are identical to `playtone(...)`. However the `bend` modifier has no effect here.

## 14.3 Melodies

The Cinderella MIDI interface provides a rich set of instructions for handling the playing of sequenced music notes. In principle each note (or more generally each MIDI event) is associated to a certain timestamp that tells the sequencer when this note has to be played. Notes to be played are added as the elements of a CindyScript list. Each note is itself a small list consisting of the integer or a string representing the note to be played and its duration. Also control events like change of an instrument, or a change in volume are elements of this melody list. They are as well associated to a timestamp. Since all melody processing CindyScript functions refer to the same description of melodies by lists of events we explain this melody description language first.

## *14.3.1 Description of Melodies*

Melodies in CindyScript are described by lists of events. Each event either corresponds to a note that has to be played, a control command (like the choice of the chosen instrument), or a positioning command that describes where the next event will be added. The simplest for in which a melody may occur is a sequence of tones. The following piece of code describes invoking a simple C-major scale:

```
playmelody([["C",1],["D",1],["E",1],["F",1],
            ["G",1],["A",1],["H",1],["c",5]])
```

Alternatively the same scale could also be expressed by describing the notes by the corresponding MIDI integers rather than names:

```
playmelody([[60,1],[62,1],[64,1],[65,1],[67,1],[69,1],[71,1],[72,5]])
```

The second number in the short list describing a single note is its duration. The lengths of the durations are measured in *beats*. By default if the melody is played it play with a speed of 60 *beats per minute*. If necessary, this can be changed by a modifier of be the `midispeed(...)` command.

Internally while a melody is played roughly the following happens. When the melody starts the sequencer is set to its start position. The first note is played and the position of sequencer advances by the the duration associated to the note. Then the next note is played, the sequencer advances by the duration of the second note and so forth. Internally each beat is subdivided into many micro-steps that allow a precise positioning of each note. In fact the beat is subdivided into 240=5*4*2*2*2 microsteps. This allows for the *exact* positioning of eigth, sixteenth, thirty-seconds, triplets and quintuplets. For all practical applications these microsteps are so fine that one can simply consider them as a continuum. Hence a duration can simply be expressed by a real number. The approximation to the microposition is done automatically. In the following piece of code still for each tone of the scale a full beat is used. However the notes are repeated by subdividing each beat.

```
playmelody([["C",1],
            ["D",1/2],["D",1/2],
            ["E",1/3],["E",1/3],["E",1/3],
            ["F",1/4],["F",1/4],["F",1/4],["F",1/4],
            ["G",1/6],["G",1/6],["G",1/6],["G",1/6],["G",1/6],["G",1/6],
            ["A",1/8],["A",1/8],["A",1/8],["A",1/8],
            ["A",1/8],["A",1/8],["A",1/8],["A",1/8],
            ["H",1],["c",5]])
```

We will now describe the different elements of the melody description language.

*Tones and Chords:* We have already seen that tones can be encoded as well as integers or as string names. So playing a single note has the format `[<key>,<duration>]`, where `<key>` indicates the note (given as integer or string) and `<duration>` its duration by a positive real number. The following table describes how names are associated to the tones of the middle octave.

C  Cis Des D  Dis Es E  F  Fis Ges G  Gis As A  Ais B  H  c
48 49  49  50 51  51 52 53 54  54  55 56  56 57 58  58 59 60

The next higher octave corresponds to:

c   cis des d   dis es  e   f   fis ges g   gis as  a   ais b   h   c'
60  61  61  62  63  63  64  65  66  66  67  68  68  69  70  70  71  72

Tones of the lower octave can be further lowered by an octave by appending a prime. So for instance c″ corresponds to the MIDI note 24. Similarly tones of the higher octave can be raised by an octave by appending a prime. So c″′ corresponds to 96.

It is also possible to play chords instead of notes. For this the keys of the notes in a chord have to be collected have to be collected in a list. SO the format of a chord is [[<key1>,<key2>,<key3>,...],<duration>]. The following example shows a melody list that play chords of increasing complexity.

```
playmelody([["C",1],
           [["C","E"],1],
           [["C","E","G"],1],
           [["C","E","G","c"],5] ] )
```

Finally, one can use the number -1 or the strings "P" or "p" as a pause.

*Dynamics and phrasing:* There are also a number of melody events that can be used to influence the dynamics or the phrasing of the following notes. The either correspond to a single string or of a list with a key/value pair. The following statements are currently implemented:

- *"ppp":* piano-pianissimo

- *"pp":* pianissimo

- *"p":* piano

- *"mp":* mezzo-piano

- *"mf":* mezzo-forte

- *"f":* forte

- *"ff":* fortissimo

- *"fff":* forte-fortissimo

- *">":* accent on the next note

- *"staccato"* or *"st":* play the following notes very shortly (staccato style)

- *"legato"* or *"le":* play the following notes for a very long time (legato style)

- *["velocity",<real>]:* set the volume. <real> is between 0.0 and 1.0.

The following piece of code plays a scale beginning in a pianissimo staccato and ending with a loud legato:

```
playmelody([["st"],["pp"],["C",1],["D",1],["E",1],["F",1],["le"],
            ["f"],["G",1],["A",1],["H",1],["c",5]])
```

*Positioning:* The sequencer internally has a pointer that indicates the position at which a note is added. Usually the pointer advances by the duration of a note when a note is added. However, there are also statements to that allow to position the pointer directly and by this influence the timestamp at which the next note is added. When a track is added the pointer is set to the position *0*. By two statements "goto" and "gorel" the pointer can be moved to an absolute position or moved relatively. The exact syntax is as follows:

- *["goto", <real>]:* set the pointer to the position <real> measured from the beginning of the melody. Negative absolute positions are forbidden.

- *["gt ", <real>]:* same as "goto".

- *["gorel", <real>]:* set the pointer to the position calculated relatively by an offset of <real> to the actual position. The <real> may be negative. However moves resulting in negative absolute positions are forbidden.

- *["gr", <real>]:* same as "gorel".

The following piece of code adds a (more quiet) second voice by using the goto(...) statement.

```
playmelody([
    ["c",1],["e",1],["g",1],["a",1],["c'",4],["goto",0],["vel",0.3],
    ["g",1],["a",1],["c'",1],["e'",1],["a'",4],
],speed->100)
```

There are also more advanced ways of controlling the timing of notes that are added. They are closer to the usual musical notation. There are four statements:

- *["||:"]:* Beginning of a *da capo*.

- *[":||"]:* End of a *da capo*.

- *["1. "]:* First bracket.

- *["2. "]:* Second bracket.

Using these commands it is easily possible to transfer sheet notes with repetitions directly. The following line from "Oh, Susanna"

can be coded in the following way as a melody in CindyScript:

```
playmelody([
    ["c",.5],["d",.5],
    ["||:"],["e",1],["g",1],["g",1.5],["a",.5],
            ["g",1],["e",1],["c",1.5],["d",.5],
    ["1."],["e",1],["e",1],["d",1],["c",1],["d",3],
            ["c",.5],["d",.5],[":||"],
    ["2."],["e",1],["e",1],["d",1],["d",1],["c",4]
],speed->200)
```

*Intrument control:*

Finally, there are also a few statements to influence the sound and choice of an instrument inside a melody. By this it is possible to change the instrument while the melody is playing by control commands in the melody list. The commands are as follows:

- *["channel",<int>]:* changes the channel (0...15) that is currently used for the melody.

- *["ch",<int>]:* same as `channel`.

- *["instrument",<int>]:* changes the instrument (1...128) that is associated to the channel.

- *["inst",<int>]:*same as `instrument`.

- *[<int>,<int>,<int>,<int>]:* This statement (consisting simply of four integer numbers, gives access to other midi controls of instruments (consult a MIDI manual for details). The first number codes the controller, the second the channel to which it applies, the last two are data bytes.

When a track is restarted the instruments are reset to their original values.

The MIDI control language we described above now forms the basis of all other MIDI functions.

### Playing a melody: `playmelody(<list>)`                                     `playmelody`

*Description:* We have already used `playmelody(...)` in all previous examples on the melody language. It is the most direct was to play a melody that starts immediately. The list is assumed to be a melody described in the melody language. Invoking `playmelody(...)` adds the list to the sequencer and immediately plays it. It is important to know that when the melody is called all other tracks are erased from the sequencer.

*Modifiers:* In addition, the statement has modifiers for globally setting the channel, instrument, speed, etc.

| | | |
|---|---|---|
| **channel** | *0..15* | selecting the channel that is played |
| **instrument** | *1 ... 123* | selecting a specific instrument |
| **speed** | *<real>* | speed in beats-per-minute |
| **loop** | *0,1,2,3,4 ...* | repetition of the melody |
| **start** | *<real>* | start position (in beats) |

The modifier loop tells the sequencer to rewind and restart after the melody is ended. The number in this modifier specifies how often a rewind is done. So `loop->3` has the effect that the melody is played *four* times. Setting `loop->-1` is interpreted as loop indefinitely. If this was called then the melody has to be stopped by adding a new (perhaps empty) melody: `playmelody([])`. Alternatively the `midistop()` command can be used.

### Adding a track to the sequencer: `midiaddtrack(<list>)`                    `midiaddtrack`

*Description:* In contrast to `playmelody(...)` this statement adds a melody to the sequencer, but does not immediately play it. By this it is possible to build a more complex composition first by adding several voices and start playing it later when the composition is completed. Starting the sequencer is done via the `midistart()` command. If a track is added while the sequencer is already running the sequencer is *not* restarted. Silently the track is replaced by the new track. By this it is possible to build scenarios in which the melodies that are played change dependent on certain algorithmic processes. Furthermore the command has several modifiers that allow for altering the timing of the added track relative to the timing of the sequencer. Thus melodies can be stretched, shrinked or shifted. Details are explained below.

*Modifiers:* The function supports the following modifiers:

| | | |
|---|---|---|
| `channel` | *0..15* | selecting the channel that is played |
| `instrument` | *1 ... 123* | selecting a specific instrument |
| `speed` | *\<real\>* | speed in beats-per-minute |
| `track` | *0..10* | selecting the track of the sequencer |
| `start` | *\<real\>* | start position (in beats) |
| `mode` | *"add", "replace", "append"* | the mode in which the track is added |
| `stretch` | *\<real\>* | a factor that expands the beatlength in the added piece of melody (relative to the sequencer) |
| `offset` | *\<real\>* | an offset that shifts the added piece of melody relative to the sequencer. |
| `repeat` | *\<real\>* | how often the melody is added |

A few modifiers need a bit of further explanation. Cinderella allows for a total of 16 tracks named 0 to 15. Each track resembles an individual independent voice. The track to which a melody is added can be selected by the `track` modifier.

When you add a track the `mode` modifier specifies whether the current track is replaced by the new track, whether it is appended (to the end) of the already existing track or added (with overlay of timestamps) to the existing track.

The timing of the new melody can be adjusted with respect to the already existing track in the sequencer. By the `stretch` modifier a number can be given that serves as a factor of all beat specification in the melody. Setting `stretch->0.5` play the track with twice its original speed. Similarly `offset` can specify the position at which the new track is added. Setting `offset->8` indicates that the track starts after the first eight beats of the sequencer. Finally, repeat specifies how often the track is added consecutively. Using `stretch`, `offset` and `repeat` can be very helpful when adding a drum pattern to an existing melody.

**midistart**

### Start the sequencer: `midistart()`

*Description:* This command starts the MIDI sequencer. Tracks must have been added in advance by the `midiaddtrack(...)` command.

*Modifiers:* Modifiers are similar to those of the `playmelody(...)` command. The following modifiers are allowed.

| | | |
|---|---|---|
| `speed` | *\<real\>* | speed in beats-per-minute |
| `loop` | *0,1,2,3,4 ...* | repetition of the melody |
| `start` | *\<real\>* | start position (in beats) |

**midistop**

### Stop the sequencer: `midistop()`

*Description:* This command immediately stops the MIDI sequencer.

**midispeed**

### Setting the speed of the sequencer: `midispeed(<real>)`

*Description:* By this command the playing speed of the sequencer can be changed. The speed is given by a real number that resembles the beats-per-minute. By default the speed is set to 60, which corresponds to one beat per second (this is by far to slow for most music pieces). The speed value can also be altered, if the sequencer is already running. By this it is possible to (for instance) associated the speed with the position of a point that controls the speed.

### Requesting the speed of the sequencer: `midispeed()`

*Description:* This statement returns the current speed of the sequencer.

### Setting the position of the sequencer: `midiposition(<real>)`

*Description:* By this command the position of the sequencer can be set to a specific position. By this a CindyScript program can explicitly control for instance the entry point of a composition that is played.

### Requesting the position of the sequencer: `midiposition()`

*Description:* This command returns the actual position of the sequencer. When the sequencer is running this value changes continuously.

## 14.4  Sound characteristics

With the following CindyScript statements you can select the default instruments, associate them to channels, and influence their sound characteristics.

### Choose an instrument: `instrument(<int>)`

*Description:* With this command you can associate an instrument to a channel. If no channel is specified the default channel is used. Each instrument is identified by an integer in the range 1...128. The correlation between instruments and the integers is explained in the next section on `instrumentnames()`. If an explicit channel is specified by a modifier the instrument of the corresponding channel is altered.

*Modifiers:* This function supports various modifiers that influence the tone characteristics of the instrument.

| | | |
|---|---|---|
| `velocity` | *0.0 ... 1.0* | the volume of a tone (how fast a piano key is pressed) |
| `duration` | *<real>* | duration of the tone in seconds |
| `bend` | *-2.0 ... 2.0* | bending a tone up to a whole note down or up |
| `channel` | *0..15* | selecting the channel that is played |
| `reverb` | *0.0 ... 1.0* | reverb effect |
| `balance` | *-1.0 ... 1.0* | left/right panorama of tone |

Here the modifiers `velocity` and `duration` influence the default velocity (volume) and duration with which a tone is played (for instance by `playtone(...)`).

*Example:* The following piece of code plays a gentle c on a *Glockenspiel* then a loud and short one on a trumpet and finally a long one (of medium volume) on a piano:

```
instrument(10,velocity->0.2);
playtone(96);
wait(1000);
instrument(57,duration->0.1,velocity->1);
playtone(60);
wait(1000);
instrument(1,duration->2,velocity->.5);
playtone(60);
```

## instrumentnames

**Getting available instruments: `instrumentnames()`**

*Description:* This statement return a list of all instrument names available on your computer. In the General MIDI database (that is probably preinstalled on your machine) you have the following instruments at hand.

Piano :
1 Acoustic Grand Piano
2 Bright Acoustic Piano
3 Electric Grand Piano
4 Honky-tonk Piano
5 Electric Piano 1
6 Electric Piano 2
7 Harpsichord
8 Clavi

Chromatic Percussion :
9 Celesta
10 Glockenspiel
11 Music Box
12 Vibraphone
13 Marimba
14 Xylophone
15 Tubular Bells
16 Dulcimer

Organ :
17 Organ
18 Percussive Organ
19 Rock Organ
20 Church Organ
21 Reed Organ
22 Accordion
23 Harmonica
24 Tango Accordion

Guitar :
25 Acoustic Guitar (nylon)
26 Acoustic Guitar (steel)
27 Electric Guitar (jazz)
28 Electric Guitar (clean)
29 Electric Guitar (muted)
30 Overdriven Guitar
31 Distortion Guitar
32 Guitar harmonics

Bass :
33 Acoustic Bass
34 Electric Bass (finger)
35 Electric Bass (pick)
36 Fretless Bass
37 Slap Bass 1
38 Slap Bass 2
39 Synth Bass 1
40 Synth Bass 2

Strings :
41 Violin
42 Viola
43 Cello
44 Contrabass
45 Tremolo Strings
46 Pizzicato Strings
47 Orchestral Harp
48 Timpani

Ensemble :
49 String Ensemble 1
50 String Ensemble 2
51 Synth Strings 1
52 Synth Strings 2
53 Voice Aahs
54 Voice Oohs
55 Synth Voice
56 Orchestra Hit

Brass :
57 Trumpet
58 Trombone
59 Tuba
60 Muted Trumpet
61 French horn
62 Brass Section
63 Synth Brass 1
64 Synth Brass 2

Reed :
65 Soprano Sax
66 Alto Sax
67 Tenor Sax
68 Baritone Sax
69 Oboe
70 English Horn
71 Bassoon
72 Clarinet

Pipe :
73 Piccolo
74 Flute
75 Recorder
76 Pan Flute
77 Blown Bottle
78 Shakuhachi
79 Whistle
80 Ocarina

Synth Lead :
81 Lead 1 (square)
82 Lead 2 (sawtooth)
83 Lead 3 (calliope)
84 Lead 4 (chiff)
85 Lead 5 (charang)
86 Lead 6 (voice)
87 Lead 7 (fifths)
88 Lead 8 (bass + lead)

Synth Pad :
89 Pad 1 (new age)
90 Pad 2 (warm)
91 Pad 3 (polysynth)
92 Pad 4 (choir)
93 Pad 5 (bowed)
94 Pad 6 (metallic)
95 Pad 7 (halo)
96 Pad 8 (sweep)

Synth Effects :
97 FX 1 (rain)
98 FX 2 (soundtrack)
99 FX 3 (crystal)
100 FX 4 (atmosphere)
101 FX 5 (brightness)
102 FX 6 (goblins)
103 FX 7 (echoes)
104 FX 8 (sci-fi)

Ethnic :
105 Sitar
106 Banjo
107 Shamisen
108 Koto
109 Kalimba
110 Bagpipe
111 Fiddle
112 Shanai

Percussive :
113 Tinkle Bell
114 Agogo Bells
115 Steel Drums
116 Woodblock
117 Taiko Drum
118 Melodic Tom
119 Synth Drum
120 Reverse Cymbal

Sound effects :
121 Guitar Fret Noise
122 Breath Noise
123 Seashore
124 Bird Tweet
125 Telephone Ring
126 Helicopter
127 Applause
128 Gunshot

**Choose a channel: `midichannel(<int>)`**

*Description:* This command selects the current default channel that is used for playing tones with `playmelody(...)` or `playtone(...)`. Via modifers it is possible to change the instruments and tone characteristics of the selected channel.

*Modifiers:* The modifiers here are very similar to the `instrument` modifiers.

| | | |
|---|---|---|
| `velocity` | *0.0 ... 1.0* | the volume of a tone (how fast a piano key is pressed) |
| `duration` | *<real>* | duration of the tone in seconds |
| `bend` | *-2.0 ... 2.0* | bending a tone up to a whole note down or up |
| `instrument` | *0..15* | selecting the instrument that is played |
| `reverb` | *0.0 ... 1.0* | reverb effect |
| `balance` | *-1.0 ... 1.0* | left/right panorama of tone |

**Setting the volume of a channel: `midivolume(<real>)`**

*Description:* By this statement the overall volume of the MIDI sound of a channel is controlled. The parameter is a real number in the range 0.0...1.0. Usually this statement affect the volume of the default channel. However, by modifiers one can select the affected channel.

*Modifiers:*

| | | |
|---|---|---|
| `channel` | *0 ..15 or "all"* | the selected channel. |

The channel number is either explicitly specified or by using `"all"` all channels can be affected.

**Setting a controller of a channel: `midicontrol(<int>,<int>)`**

*Description:* By this a specific MIDI controller of a channel can be set. The first parameter specifies the number of the controller (0..127), the second parameter (0..127) specifies the data set to the controller. Controllers may affect characteristics like *reverb*, *balance*, and other characteristics specific to the instruments. The data values are taken in the range 0..127. For more details see a manual on General MIDI.

*Modifiers:* Modifier usage is similar to `midivolume`.


## 14.5 Three Little Pieces


In this final section on MIDI we want to demonstrate the use of the CindyScript MIDI in three small, but more advanced examples. The user is invited to take these examples as a starting point for further own experiments.

### 14.5.1 A Keyboard Piano

In this example we want to program a piano that can be played by hitting the keys of the computer keyboard. To do so one has simply to put the following code into the *timer tick* event of the CindyScript code window, press the play button that appears in the Cinderella view and play.

```
pairs=[[65,60],[83,62],[68,64],[70,65],[71,67],[72,69],[74,71],
       [75,72],[76,74],[59,76],[222,77],[92,79],[87,61],[69,63],
       [84,66],[90,68],[85,70],[79,73],[80,75],[93,78]];

l=keydownlist();
forall(pairs,p,
  if(contains(l,p_1)&!contains(ol,p_1),
     playtone(p_2,duration->-1,velocity->0.5));
  if(contains(ol,p_1)&!contains(l,p_1),stoptone(p_2));
);
ol=l;
```

The code makes use of the CindyScript operator `keydownlist()`. This operator gives a list of all (computer-internal) keycodes of keys that are pressed at this moment. The list `pairs` associates the key codes to notes that have to be played. The `forall` loop checks for every possibly hit or released key and specifically turns on or of off the note that is associated to this key by using `playtone` and `stoptone`. On a contemporary computer this code is quite free of latency and can be used as a real instrument.

### 14.5.2 Ping Pong With Sound

Our next example is really very tiny and uses just one `playtone(...)` statement in the right place. We want to create a Ping-Pong game with sound effects. By using the physics simulation facilities of CindyLab (p. 167) it is very easy to construct a physically reasonable interactive Ping-Pong table. The boundary of the table is created by physical *bouncers*. In the Inspector each of these bouncers can be associated with a script that is executed at the moment when a mass hits the bouncer. There we have to simply place a `playtone` statement that produces a "click" sound. That's it.

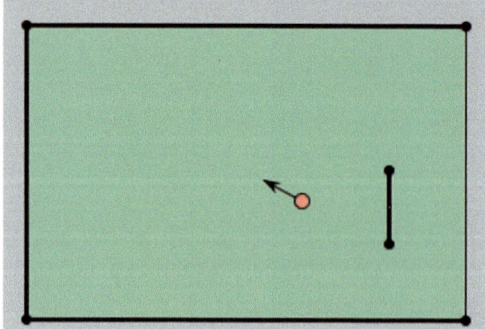
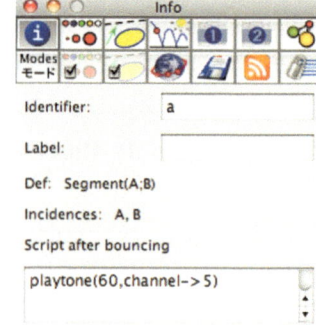

### 14.5.3 Let it Groove!

Finally, our last example exemplifies the precision of the sequencer timing. it simply plays the beginning of the famous Charlie Parker Jazz tune *Ornithology* (Charlies nickname was *Bird* and this title refers to his nickname). A typical thing for Jazz is its groovy timing, the Swing feeling. This comes from a certain sensible shift of of the notes with respect to the ground beat. In the following example code the variable g is used as a "Swing parameter". It shifts the beginning of each second note with resect to the ground beat. It is instructive to associate g with a movable point in a range from 0.0 to 1.0. By this one can very easily adjust the amount of Swing feeling used by the tune (in a range between 0.5 and 0.7 it sounds quite reasonable).

```
g=.6;
mel=[[62,2-g],[67,2+g],[69,2-g],[71,2+g],[72,2-g],[74,2+g],
     [71,2-g],[72,2+g],[74,2-g],[71,2+g],[67,3],[-1,9],
     [62,2-g],[67,2+g],[69,2-g],[70,2+g],[72,2-g],[74,2+g],
     [76,4],[77,4],[67,2-g],[69,2+g],[70,8],[74,2-g],
     [72,2+g],[69,4],[65,2-g],[70,2+g],[68,2-g],[69,2+g],
     [65,13],[-1,3]
];

drum=44;
beat=[[-1,2-g]];
apply(1..11,beat=beat++[[-1,4],[drum,4]]);
beat=beat++[[-1,4],[drum,4-2+g]];
midiaddtrack(mel ,channel->2,track->2);
midiaddtrack(beat,channel->9,track->1,velocity->.5);
midistart(speed->700);
```

## 14.6 Sampled-Audio Functions

The Sampled-Audio functions part of Syntherella provides a possibility to create specific wave forms that are played by the built-in speaker of a computer. By this it is possible to explore the interrelations of mathematics and the very fundamental structures of *sound*.

Syntherella provides essentially three specialized functions that cover different scenarios.

- *playfunction(...)* allows to form a sound wave by specifying its underlying function.

- *playlist(...)* plays a sound specified by a list of samples that specify the wave.

- *playsin(...)* plays a *sin(x)* wave with specified frequency. Harmonics can also be specified.

Before we will describe these statements in detail we will give a brief explanation (in a nutshell) of these statements and the subtleties that are relevant for the generation of an audio signal.

In all three functions the amplitude of the overall signal is restricted to the range (-1.0...1.0). Moreover the sum of all accumulated audio signals must stay in this range. We start our explanations with the `playsin(<real>)` function. In its simplest form it starts the audio-rendering of a sine wave of a given frequency. So `playsin(440)` generates a sine tone with 440Hz that is played for exactly one second. One might try to use the following piece of code to produce a simple chord:

```
playsin(440);
playsin(440*5/4);
playsin(440*3/2);
```

However the overall amplitude would exceed the possible range. Scaling down each signal by a factor of `1/3` will solve the problem. This can be achieved by the `amp-><real>` modifier. This modifier scales the amplitude of each signal. The resulting code will look as follows:

```
playsin(440,amp->1/3);
playsin(440*5/4,amp->1/3);
playsin(440*3/2,amp->1/3);
```

Everytime the function `playsin` is called a new tone is added to the already existing sounds. If the function `playsin` is called a second time while a tone is already playing the old tone still continues while a new tone is started. This may very soon lead to a situation where the dynamics range -1.0...1.0 is exceeded. There is a modifier `line-><int>` that helps to prevent such situations. Specifying `line->1` associates the tone to an output line that can carry at most one tone. So for instance the code

```
playsin(440,line->1);
wait(200);
playsin(550,line->1);
wait(200);
playsin(660,line->1);
```

starts to play a first tone with 440Hz, after 0.2 seconds replaces it with another tone of 550Hz and after another 0.2 starts another tone that is played for a full second. There is one more subtlety that is also resolved by the `line` modifier. When one tone is replaced by another you may a priori not take it for granted that the new tones starts in reasonable phase position relative to the first one. The `line` modifier takes care of this issue automatically. Rather than simply replacing one wave by another, it changes the timing of the underlying waves. It even goes one step further and makes a smooth transition between the two frequencies. The differences are shown in the images below.

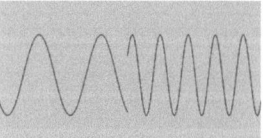 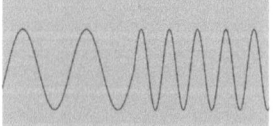 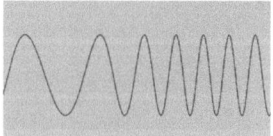

*Without phase correction*          *Phase corrected*                    *Smooth transition*

**Playing a periodic signal defined by overtones: `playsin(<real>)`**     `playsin`

*Description:* In its simplest form this operator plays the sound of a sine wave with constant amplitude and duration of one second. The duration, amplitude and sound characteristics can be altered by suitable modifiers.

*Modifiers:* The command has several modifiers.

| | | |
|---|---|---|
| `amp` | *0.0 ... 1.0* | global amplitude (volume) of the sample |
| `damp` | *<real>* | factor for exponential damping |
| `harmonics` | *<list>* | spectrum of the tone |
| `duration` | *<real>* | duration for playing |
| `stop` | *<real>* | same as duration |
| `line` | *<number> or <string>* | a sound-line associated to the tone |

*Examples:* The simplest usage of `playsin` is for instance given by

```
playsin(440)
```

This plays a sine wave tone of 440Hz. Changing this to

```
playsin(440,damp->3,stop->5)
```

creates a tone that lasts 5 seconds (`stop->5`) and has some exponential damping (`damp->3`). One can add also harmonics to make the sound more complex. Harmonics are specified by a list of amplitudes that specify the amplitude of the different overtones. The code

```
playsin(440,damp->3,stop->5,harmonics->[0.5,0.3,0.2,0.1])
```

produces a wave form of the type

$$0.5\sin(440\cdot 2\pi x)+0.3\sin(2\cdot 440\cdot 2\pi x)+0.2\sin(3\cdot 440\cdot 2\pi x)+0.1\sin(4\cdot 440\cdot 2\pi x)$$

Again this will be exponentially damped.

**Playing sample defined by a function: `playfunction(<funct>)`**     `playfunction`

*Description:* Takes a function as input and interprets it as a wave form. The time unit is such that one unit in the function correspond to one second.

*Modifiers:* The command has several modifiers.

| | | |
|---|---|---|
| `amp` | *0.0 ... 1.0* | global amplitude (volume) of the sample |
| `damp` | *<real>* | factor for exponential damping |
| `start` | *<real>* | start position of sample |
| `stop` | *<real>* | end position of sample |
| `duration` | *<real>* | duration for playing |
| `line` | *<number> or <string>* | a sound-line associated to the tone |
| `silent` | *<boolean>* | suppress playing |
| `export` | *<boolean>* | export sample data |

*Examples:* The following piece of code produces a sine wave of 440Hz.

```
playfunction(sin(440*x*pi*2))
```

The following piece of code produces an exponentially damped noise signal.

```
playfunction(random(),damp->8)
```

By specifying `start` and `stop` one can exactly determine the region of a function that is used for a sound sample. This region may even be very short. Using the `duration` modifier one can force that the sample is played over and over. Thi following piece of code samples exactly one sine wave and plays it over and over for one second.

```
playfunction(sin(1000*x*2*pi),stop->1/1000,duration->1)
```

The data generated by the function sampling can also be exported to an array. The sample rate is 44100 samples per second. The exporting can even be done silently. For instance, the following code creates a list of 44 sample points of a sine wave.

```
sample=playfunction(sin(1000*x*2*pi),stop->1/1000,
                    silent->true,export->true)
```

The list created by this statement can in turn be used by the `playwave` operator that is able to play sampled audio data. This could be done for instance by

```
playwave(sample,duration->1)
```

**playwave**

### Playing sample defined by a list: `playwave(<list>)`

*Description:* This statement creates a sound that is driven by a list that contains sample data. The vales of the samples are assumed to lie in the range *-1.0...1.0*. The sample rate is 44100 samples per second. Usually the sample is played over and over for one second. The play time may be modified by the `duration` modifier.

*Modifiers:* The command has several modifiers.

| | | |
|---|---|---|
| **amp** | *0.0 ... 1.0* | global amplitude (volume) of the sample |
| **damp** | *<real>* | factor for exponential damping |
| **duration** | *<real>* | duration for playing |
| **line** | *<number> or <string>* | a sound-line associated to the tone |

*Example:* The following piece of code produces three samples that are played by the `playwave` operator. Although an explicit duration is given after the time specified in the `wait` each sample is replace by the next one since all samples use the same line. Before the `playwave` function proceeds to the next sample it is made sure that the actual sample was completed.

```
sample0=apply(1..200,sin(#*2*pi/200));
sample1=apply(1..100,sin(#*2*pi/100));
sample2=apply(1..50,sin(#*2*pi/50));
playwave(sample0,duration->1,line->1);
wait(400);
playwave(sample1,duration->1,line->1);
wait(400);
playwave(sample2,duration->1,line->1);
```

**stopsound**

### Stopping sampled audio: `stopsound()`

*Description:* This statement immediately terminates all sampled sound audio.

# Chapter 15

# Special Operators

This section presents a collection of operators that do not fit into any of the other categories. Nevertheless, most of them are extremely useful. Some of them are even necessary for the interaction of Cinderella and CindyScript.

## 15.1 Interaction with Geometry

CindyScript can interact in different ways with a geometric drawing that was created with Cinderella. We already saw that it can read the numerical data and appearance of geometric elements. However it can also change the position of the free elements of a construction. CindyScript may even inquire and change the construction sequence by creating and deleting new geometric elements.

### 15.1.1 Moving Elements

The calculations within CindyScript can be used to control the positions of free elements in a Cinderella construction. One way of doing this is to explicitly set the position information of a free element. For instance, if `A` is a free point, the line `A.xy=[1,1]` sets this point to the coordinates `[1,1]`. Another way of moving an element is with the `moveto` operator.

**Moving a Free Element: `moveto(<geo>,<pos>)`**                                       `moveto`

*Description:* In this operator, `<geo>` is a free geometric object and `<pos>` (usually a vector) describes a position to which this object should be moved. Calling this operator simulates a move for this geometric object.

If `<geo>` is a free point, then `<vec>` can be a list `[x,y]` of two numbers or a list `[x,y,z]` of three numbers. The first case is interpreted as Euclidean coordinates, while the second case is interpreted as homogeneous coordinates and sets the point to `[x/z,y/z]`.

If `<geo>` is a free line, then `<vec>` has to be a list of three numbers `[a,b,c]`, and the line is set to the line described by the equation `a*x + b*y + c = 0`.

*Examples:* The following code lines summarize possible ways to move geometric elements (we also include the possibilities of moving elements by accessing their data fields):

```
//A is a free point
moveto(A,[1,4]);        //moves A to Euclidean coordinates [1,4]
A.xy=[1,4];             //moves A to Euclidean coordinates [1,4]
A.x=5;                  //sets the x coordinate of A to 5, lets the y
    coordinate unchanged
A.y=3;                  //sets the y coordinate of A to 3, lets the x
    coordinate unchanged
moveto(A,[2,3,2]);      //moves A to homogeneous coordinates [2,3,2]
A.homog=[2,3,2];        //moves A to homogeneous coordinates [2,3,2]

//a is a free line
moveto(a,[2,3,4]);      //moves a to homogeneous coordinates [2,3,4]
a.moveto=[2,3,4];       //moves a to homogeneous coordinates [2,3,4]

//b is a line through a point
a.slope=1;              //sets the slope of the line to 1

//C is a circle with free radius
C.radius=1;             //sets the radius of the circle to 1
```

## 15.1.2 Handles to Objects

**mover**

**Who has moved: `mover()`**

*Description:* This operator gives a handle to the element that is currently moved by the mouse.

**elementsatmouse**

**Elements close to the mouse: `elementsatmouse()`**

*Description:* This operator gives a list with handles to all the elements that are close to the current mouse position.

*Example:* The following script is a little mean. Putting it into the mouse move slot will make exactly those elements disappear that are close to the mouse. They reappear if the mouse moves away again.

```
apply(allelements(),#.alpha=1);
apply(elementsatmouse(),#.alpha=0);
repaint();
```

**incidences**

**Incidences of an object: `incidences(<geo>)`**

*Description:* This operator returns a list all the elements that are generically incident to a geometric element `<geo>`.

**element**

**Getting an element by name: `element(<string>)`**

*Description:* This operator returns the geometric object identified by the name given in `<string>`.

### 15.1.3 Creating and Removing Elements

**Creating a free point: `createpoint(<string>,<pos>)`**                 `createpoint`

*Description:* This operator creates a new point with label `<string>`. The point will beset to position `<pos>`. If an element with this name is already exists then no new element is created. However, if there already exists a free point with this name, then this point is moved to the specified position.

**Creating a geometric element: `create(<list1>,<string>,>list2>)`**                 `create`

*Description:* With this operator it is possible to generate arbitrary geometric elements that are functional in a geometric construction. Due to the fact that algorithms may create multiple outputs several subtleties arise. This function is meant for expert use only.

The first list contains a list `<list1>` of element names for the generated output objects of the algorithm. `<string>` is the internal name of the geometric algorithm. The second list `<list2>` is a list of the parameters that are needed for the definition. The following table shows a few possible creation statements.

```
create(["A"],"FreePoint",[[1,1,1]]);
create(["B"],"FreePoint",[[4,3,1]]);
create(["a"],"Join",[A,B]);
create(["X"],"CircleMP",[A,B]);
create(["Y"],"CircleMP",[B,A]);
create(["P","Q"],"IntersectionCircleCircle",[X,Y]);
create(["b"],"Join",[P,Q]);
create(["M"],"Meet",[a,b]);
```

This sequence of statements creates the fully functional construction shown below. Observe that in the sixth statement when two circles are intersected there must be a list of two output elements specified.

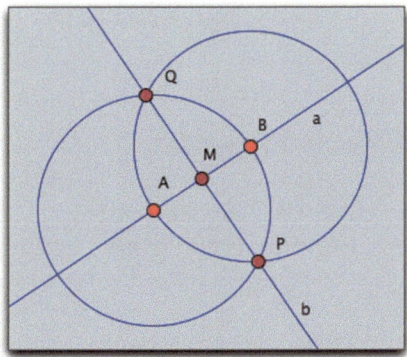

You can find the valid parameters for elements by constructing them manually and using the `algorithm` and `inputs` functions described below.

**Removing a geometric element: `removeelement(<geo>)`**                 `removeelement`

*Description:* Removes a geometric element together with all its dependent elements from a construction.

**inputs**

### Input elements of an element: `inputs(<geo>)`

*Description:* This operator returns a list all the elements that are needed to define the object `<geo>`. These may be other geometric, elements, numbers or vectors.

**algorithm**

### Algorithm of an element: `algorithm(<geo>)`

*Description:* This operator returns a string that resembles the algorithm of the definition the object `<geo>`.

*Example:* The following piece of code generates all information contained in a construction sequence.

```
els=allelements();
data=apply(els,([[#.name],algorithm(#),inputs(#)]));
```

Applied to the construction of a perpendicular bisector in the picture below it generates the following output:

```
[[["A"],"FreePoint",[[4,4,-4]]],
 [["B"],"FreePoint",[[4,-3,1]]],
 [["a"],"Join",[A,B]],
 [["C"],"Mid",[A,B]],
 [["b"],"Orthogonal",[a,C]]
]
```

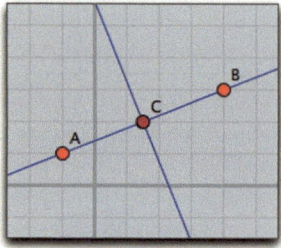

## 15.1.4 Accessing Element Properties

Element properties like color, size, etc. are conveniently accessible via operators like `.color`, `.size` etc. However, elements have by far more properties. All of them can be generically accessed by the following operators.

**inspect**

### List all inspectable properties: `inspect(<geo>)`

*Description:* Returns a list of names of all private properties of a geometric element.

*Example:* The operator `inspect(A)` applied to a the free point *A* returns the following list of property names.

```
[name,definition,color,color.red,color.blue,color.green,alpha,visibility
          ,
drawtrace,tracelength,traceskip,tracedim,render,isvisible,
```

```
text.fontfamily,plane,pinning,incidences,labeled,textsize,textbold,
    textitalics,
ptsize,pointborder,printname,point.image,
point.image.media,point.image.rotation,freept.pos]
```

**Accessing an inspectable property: `inspect(<geo>,<string>)`**          `inspect`

*Description:* Accesses an arbitrary inspectable property.

*Example:* One can access the color of a point *A* by `inspect(A,"color")`

**Setting an inspectable property: `inspect(<geo>,<string>,<data>)`**          `inspect`

*Description:* Setting the value of inspectable property.

*Example:* One can set the color of a point *A* to white by `inspect(A,"color",(1,1,1))`

**Forcing a repaint operation: `repaint()`**          `repaint`

*Description:* This operator causes an immediate repaint of the drawing surface. This operator is meant to be used whenever a script has updated a construction and wants to display the changes. It is not allowd to use this operator in the `draw` or in the `move` slot.

**Forcing a delayed repaint operation: `repaint(<real>)`**          `repaint`

*Description:* As `repaint` but with a time delay of as many milliseconds as given by he parameter

**Points on a locus: `locusdata(<locus>)`**          `locusdata`

*Description:* This operator returns a list of points in *xy*-coordinates that are all on a locus given by the name `<locus>` of a geometric element.

## 15.2 File Management

There is a number of operations that allow for the interaction of CindyScript with files that are stored elsewhere on the computer. Please note that these commands will not work with applets in HTML pages.

### 15.2.1 Reading Files

**Loading data: `load(<string>)`**          `load`

*Description:* This operator takes the argument `<string>`, which is considered to be a file name (possibly preceded by directory information). If the file name is legitimate, then the entire information contained in the file will be returned as a string. This operator is particularly useful together with the `tokenize` operator, which helps to analyze structured data. The data are read from the currently active directory, which can be set with the `setdirectory` operator.

*Example:* Assume that in the file `LoadDemo.txt` contains the data

`abc,gfdg;1,3,5.6,3.141;56,abc,xxx,yyy`

The following code reads the data and creates a list by tokenizing it with respect to ";" and ",".

```
x=load("LoadTest.txt");
y=tokenize(x,(";",","));
apply(y,println(#));
```

The resulting output is

```
[abc,gfdg]
[1,3,5.6,3.141]
[56,abc,xxx,yyy]
```

**import**

### Importing program code: `import(<string>)`

*Description:* This operator takes the argument `<string>`, which is considered to be a file name (including directory information). If the file name is legitimate, then the whole content of the file is assumed to be able to be parsed by CindyScript code, and it is immediately executed. In this way, one can load libraries with predefined functionality. It is advisable to use the `import` operator only in the "Init" section of CindyScript, since otherwise, the file will be read for each move.

**setdirectory**

### Setting the directory: `setdirectory(<string>)`

*Description:* This operator sets the directory for all subsequent file operations.

## 15.2.2 Writing Files

It is also possible to write files by a sequence of Cindy script commands. The usual cycle for writing is: Open a file – write to it – close the file. This can be done using the following commands.

**openfile**

### Opening a file: `openfile(<string>)`

*Description:* Opens a file with the specified name. The function returns a handle to the file that is needed for subsequent print operations.

**println**

### Println to a file: `println(<file>,<string>)`

*Description:* Identical to the `println(...)` command. However this command prints to the file specified by `<file>`.

**print**

### Print to a file: `print(<file>,<string>)`

*Description:* Identical to the `print(...)` command. However this command prints to the file specified by `<file>`.

**Print to a file: `closefile(<file>)`**                                    `closefile`

*Description:* This command finally closes the file.

*Example:* The following example illustrates a file write cycle:

```
f=openfile("myFile");
println(f,"Here are some numbers");
forall(1..15,print(f,#+" ");
println(f,"");
closefile(f);
```

This code generates a file with the following content:

```
Here are some numbers
1 2 3 4 5 6 7 8 9 10 11 12 13 14 15
```

### 15.2.3 Connection to HTML

**Opening a web page: `openurl(<string>)`**                                 `openurl`

*Description:* Opens a browser with the webpage given in <string>.

**Calling javascript: `javascript(<string>)`**                             `javascript`

*Description:* In exported applets this statement calls a statement in the Javascript environment of the browser. The statement is given by the content of the `<string>`. In the standalone application this statement does nothing.

*Example:* The following piece of script will cause a message window to pop up in the browser:

```
javascript("alert('Hi from Cinderella!!')");
```

### 15.2.4 Network Connections

The TCP commands of Cinderella are rudimentary at best, but they provide the basic functionality necessary for simple networking. You should be able to send and retrieve data over the internet.

**Open a TCP port: `openconnection(<string>,<int>)`**                       `openconnection`

*Description:* Opens a bidirectional tcp connection to the server specified by the first argument and the port specified by the second argument. The return value is a handle to this network connection.

**Write to a TCP connection: `print(<handle>,<string>)`**                   `print`

**Write to a TCP connection: `println(<handle>,<string>)`**                 `println`

*Description:* The `print` and `println` functions not only support writing to a file, but also to a network connection created by `openconnection`.

flush                   **Flush output to a TCP port: `flush(<handle>)`**

                        *Description:* Flushes the output buffer of the given connection.

readln                  **Read from a TCP connection: `readln(<handle>)`**

                        *Description:* Reads a line from the given connection. If no data can be read, this
                        command times out after 5 seconds.

closeconnection         **Close a TCP connection: `closeconnection(<handle>)`**

                        *Description:* Closes the connection given by the handle.

                        *Example:* In the following example we open a connection to a web server and read
                        the HTML code from there.

```
x=openconnection("cermat.org",80);
println(x,"GET /");
y="";
while(!isundefined(y),y=readln(x);println(y));
closeconnection(x);
```

## 15.3 Console Output

In the CindyScript input window you find a console for text output. For most practical purposes this will not be used for the final construction. However, it is extremely useful for debugging.

print                   **Printing text: `print(<expr>)`**

                        *Description:* This operator prints the result of evaluating `<expr>` to the console.

err                     **Printing text: `err(<expr>)`**

                        *Description:* Prints the result of evaluating `<expr>` to the console. If `<expr>` is a
                        variable, the variable name is printed as well. Very useful for debugging.

println                 **Printing text: `println(<expr>)`**

                        *Description:* This operator prints the result of evaluating `<expr>` to the console and
                        adds a newline character to the end of the text.

println                 **Printing a newline: `println()`**

                        *Description:* This operator prints a newline character to the console.

clearconsole            **Clearing the console: `clearconsole()`**

                        *Description:* Removes all text from the console.

**Conditional print: `assert(<bool>,<expr>)`**                                    `assert`

*Description:* This operator is mainly meant for convenience purpose when generating own error mesages. it is equivalent to `if(!<bool>,println(<expr>))`. It can be used to test wheter a condition is met and otherwise generate an error message.

*Example:* A typical usage of this operator is the following:

```
assert(isinteger(k),"k is not an integer");
```

**Output a status message: `message(<expr>)`**                                    `message`

*Description:* This operator shows the result of evaluating `<expr>` in the status line of the application, or in the status line of the browser for Cinderella applets.

## 15.4 Timing and Animations

### 15.4.1 Time and Date

CindyScript has an internal clock that provides access to the current date and time. The clock can be also used to synchronize some automated animations. Furthermore, an operator is provided that is synchronized with the current timestamp of a running animation or physics simulation.

**Accessing time: `time()`**                                                       `time`

*Description:* This operator returns a list `[h,m,s,ms]` of four integers. The four values correspond to "hour," "minute," "second," "millisecond" on the computer's clock.

*Example:* The following code produces a simple clock on a Cinderella view. The variable `t` contains the time information. The subsequent code is used to produce a clocklike drawing on the view. An auxiliary function `p(w)` is defined that produces points on the unit circle. The code must be placed in the "Tick" section of CindyScript in order for it to run continuously.

```
t=time();

p(x):=[sin(2*pi*x),cos(2*pi*x)];
O=[0,0];
S=p(t_3/60)*4;
M=p(t_2/60)*5;
H=p((t_1*60+t_2)/(12*60))*3.5;
draw(O,S);
draw(O,M,size->2);
draw(O,H,size->3);
apply(1..12,draw(p(#/12)*5));
apply(1..60,draw(p(#/60)*5,size->1));

drawtext((3,5),t);
```

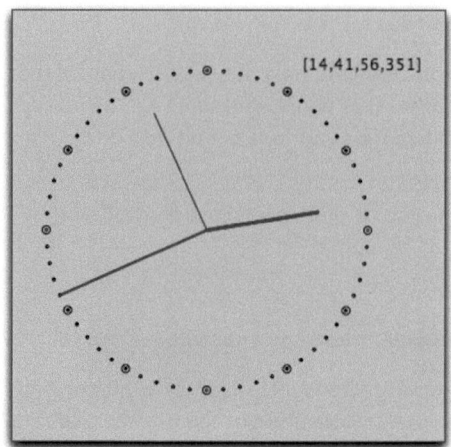

**date**          **Accessing date: date()**

*Description:* This operator returns a list [y,m,d] of three integers. The three values correspond to "year," "month," and "day" on the computer's calendar.

**seconds**       **Timestamp: seconds()**

*Description:* Returns the time elapsed since the last evaluation of resetclock(). The time is scaled in a way such that one unit corresponds to one second. The time's resolution is on the millisecond scale.

**resetclock**    **Resetting the internal seconds: resetclock()**

*Description:* Resets the value of the seconds() operator.

**simulationtime**  **Accessing the timestamp of a simulation: simulationtime()**

*Description:* This operator gives a handle to the running time clock synchronized with the progression of an animation or simulation.

*Caution:* This operator is still experimental.

**wait**          **Pause the script for a specified time: wait(<real>)**

*Description:* Will completely halt every script execution for a number of milliseconds as given by the parameter.

*Example:* The following code produces an acoustic jingle.

```
repeat(25,i,
  playtone(72+i);
  wait(100);
)
```

## 15.4.2 Animation Control

**Starting the animation: `playanimation()`**

*Description:* This statement starts the animation. Also physics simulation in Cindy-Lab (p. 167) depend on running animations.

**Pausing the animation: `pauseanimation()`**

*Description:* This statement pauses the animation.

**Stopping the animation: `stopanimation()`**

*Description:* This statement stops the animation. Stopping an animation also causes the geometric elements to be restored to there original position.

## 15.5 User Input

Sometimes it is necessary to handle user input by mouse or the keyboard explicitly. There are special evaluation times "Mouse Down," "Mouse Up," "Mouse Click," "Mouse Drag," and "Key Typed" for this (see Entering Program Code (p. 225)). These evaluation times are captured exactly when the corresponding events occur. If one wants to react to the corresponding event data, there are several operators that read the input data.

### 15.5.1 Mouse and Key

**Mouse position: `mouse()`**

*Description:* Returns a vector that represents the current position of the mouse if the mouse is pressed. The vector is given in homogeneous coordinates (this allows also for access of infinite objects). If one needs the two-dimensional euclidean coordinates of the mouse position one can access them via `mouse().xy`.

**Key input: `key()`**

*Description:* Returns a string that contains the last typed character.

**Is a certain key pressed: `iskeydown(<int>)`**

*Description:* This operator returns a boolean value that is true if a certain key is pressed. The key in question is specified by the integer in the argument. This operator can be used to determine for instance the *shift* key is pressed. The codes for keys are usually 65, 66, 66, .... for 'A', 'B', 'C',... Codes for 'shift', 'crtl' and 'alt' are usually 16, 17, 18.

**keydownlist**                **List of all pressed keys: keydownlist ()**

*Description:* This operator returns a list of the codes of all pressed keys. An interesting application of the keydown list id given in the chapter on MIDI functions, where you find an example keyboard piano (p. 375).

## 15.5.2 AMS Data on Gravity

On Apple hardware, CindyScript can access the gravity sensor of a laptop and determine its relative orientation in space. The gravity sensor returns a three dimensional vector.

**amsdata**                    **Getting raw AMS data: amsdata ()**

*Description:* This operator returns the raw data of the AMS sensor.

**calibratedamsdata**        **Getting calibrated AMS data: calibratedamsdata ()**

*Description:* This operator returns a calibrated version of the AMS sensor data. The calibrated data is a vector of unit length that represents the orientation of the computer in space.

## 15.5.3 Creating Custom Toolbars in a View

Cinderella can be used to export interactive worksheets to an html page. Very often it is desirable not only to export an interactive construction but also a set of construction tools along with it (like buttons for constructing points, lines or circles). By using the following set of CindyScript operations it is easily possible to create (and remove) custom toolbars that reside within an applet window.

Toolbars are in particular important for creating interactive student exercises. An example for this is given in Interactive Exercises (p. 433).

**createtool**                 **Creating a custom toolbar: createtool (<string>,<int>,<int>)**

**createtool**                 **Creating a custom toolbar: createtool (<list>,<int>,<int>)**

*Description:* Creates one or many toolbuttons in a Cinderella view. The first argument is either a string that describes a single construction tool or a list or matrix of strings that describe an entire toolbar. The other two arguments describe the position relative to a corner of the screen in pixel distances. Normally a createtoolbar statement is located in the *init* slot of the script editor.

The following string identifiers that correspond to the construction tools are available:

- *Moving:* "Move"
- *Points:* "Point", "Intersection", "Mid", "Center"
- *Lines:* "Line", "Segment", "Line Through", "Parallel", "Orthogonal", "Angle Bisector"
- *Circles:* "Circle", "Circle by Radius", "Compass", "Circle by 3", "Arc"
- *Conics:* "Conic by 5", "Ellipse", "Hyperbola", "Parabola"

- *Special:* "Polar Point", "Polar Line", "Polygon", "Reflection", "Locus"
- *Measure:* "Distance", "Angle", "Area"

It is also possible to add construction tools from CindyLab (p. 167):

- *Local:* "Mass", "Velocity", "Rubberband", "Spring", "Coulomb"
- *Environmentsl:* "Gravity", "Sun", "Floor", "Bouncer", "Magnet"

The position of the tools is fixed relative to the construction view. By default the upper left corner is chosen. By using the modifyer `reference` one can also choose the other corners. Allowed values for this modifier are "UL", "UR", "LL", "LR". Here the first letter stands for *upper/lower* and the second letter stands for *left/right*.

*Examples:* The simplest usage is for instance given by the following piece of code. The tool created by

```
createtool("Move",2,2);
```

More complicated examples that create toolbars with several tools are given below

```
createtool(["Move","Point","Line","Circle"],2,2);
```

```
createtool(["Move","Point","Line","Circle"],2,2,flipped->true);
```

```
createtool(
  [
    ["Move","Point","Line","Circle"],
    ["Parallel","Orthogonal","Circle by Radius","Compass"],
    ["Distance","Angle","Area","Polygon"],
  ]
,2,2,flipped->false);
```

```
createtool(

...same as example above...

,reference->"LR");
```

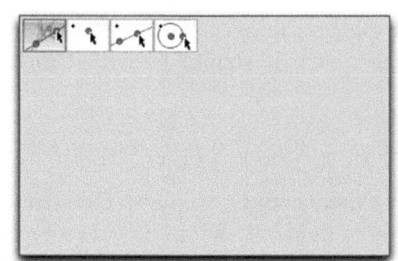

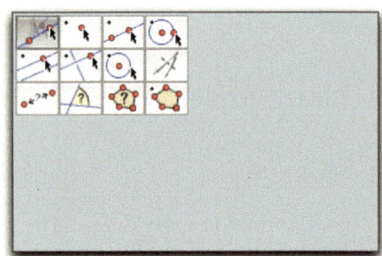 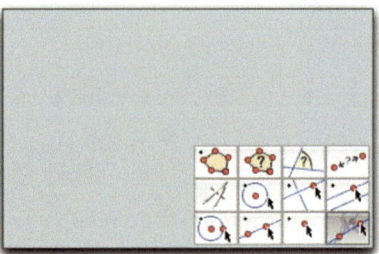

*Modifiers:* The createtool operator can handle the modifiers summarized in the following table:

| | | |
|---|---|---|
| **reference** | *<string>* | reference position |
| **flipped** | *<bool>* | `flipped->true` exchanges rows and columns |
| **space** | *<int>* | spacing (in pixels) between tools |

**removetool**

### Removing a tool from a custom toolbar: `removetool(<string>)`

*Description:* Removes a tool from the custom toolbar.

**removetools**

### Removing all custom toolbars: `removetools()`

*Description:* Removes all tools from the custom toolbar.

## 15.6 Interaction with CindyLab

**simulation**

### The simulation environment: `simulation()`

*Description:* This operator provides a handle to the simulation environment. The simulation environment offers several fields than can be used to access its global properties.

| Property | Writeable | Type | Purpose |
|---|---|---|---|
| friction | yes | real | total friction of the simulation |
| gravity | yes | real | total gravity of the simulation |
| kinetic | no | real | total kinetic energy of the simulation |
| ke | no | real | total kinetic energy of the simulation |
| potential | no | real | total potential energy of the simulation |
| pe | no | real | total potential energy of the simulation |

**addforce**

### Applying a force: `addforce(<mass>,<vec>)`

*Description:* Applying a force <vec> to an existing mass <mass>. This operator is useful to implement user defined force fields. It should be called in the `Integration Tick` slot.

**setforce**

### Applying a force: `setforce(<mass>,<vec>)`

*Description:* Setting the force <vec> for an existing mass <mass>. This operator is very useful to implement user defined force fields. It should be called in the `Integration Tick` slot.

**Probing particle forces: `force(<vector>)`** `force`

*Description:* The operator `force` is closely related to physics simulations in Cindy-Lab (p. 167). It can be used for testing the force that would affect a mass particle at a specific position. The vector `<vector>` represents the position. The operator returns a two-dimensional vector that is the force at this position. If no modifiers are used, the operator assumes that the probe particle has `mass=1`, `charge=1` and `radius=1` (see Free Mass (p. 175)).

*Example:* The following picture was generated using the `drawforces` operator and a color plot of the `force` operator. It shows the force field and force strength of the electrostatic field of two charges.

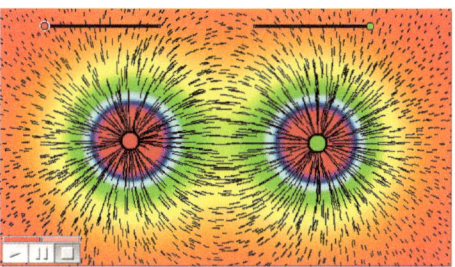

```
A.charge=(|C,G|-3)*3;
B.charge=(|E,H|-3)*3;
f(x):=max((0,min([x,1])));
colorplot([0.1,0.1,0.1]+hue(f(abs(force(#)/3))),(-10,-10),(20,10));
drawforces(stream->true,move->0.2,color->[0,0,0],resolution->10);
```

*Modifiers:* It is also possible to set the values of mass, charge and radius explicitly. Each of these values can be set by a modifier of the same name. If at least one of these values is set explicitly, then all unset values are set to zero. Thus `force([0,0],charge->2)` tests the force that would be present for a particle of `charge=2`, `mass=0`, and `radius=0` at point `[0,0]`.

## 15.7 Entering CindyScript Code

### 15.7.1 The CindyScript Editor

To enter CindyScript (p. 219) one can use the editor that is available from the menu *Scripting/Edit Scripts*. Here we explain briefly how to use the editor.

### 15.7.1.1  The Input Window

*The Script Editor*

The script editor shows a three pane view. On the left you see a an overview over all occasions (see below) and the associated scripts. On the right you see, below a panel that features a start, stop and help button as well as a field to enter script names, a large text area which is used to edit (i.e. type) scripts, and a smaller text area that shows any output from the scripts.

### 15.7.1.2  Occasions

Cinderella is highly interactive, and that is the reason for many "occasions" that are suited for triggering the execution of CindyScript (p. 219) commands. On the left side of the script editor you see the available occasions.

Usually, you write scripts for the "Draw" occasion. These are executed whenever the view (p. 116) is rendered. To edit a script, first click on "Draw".

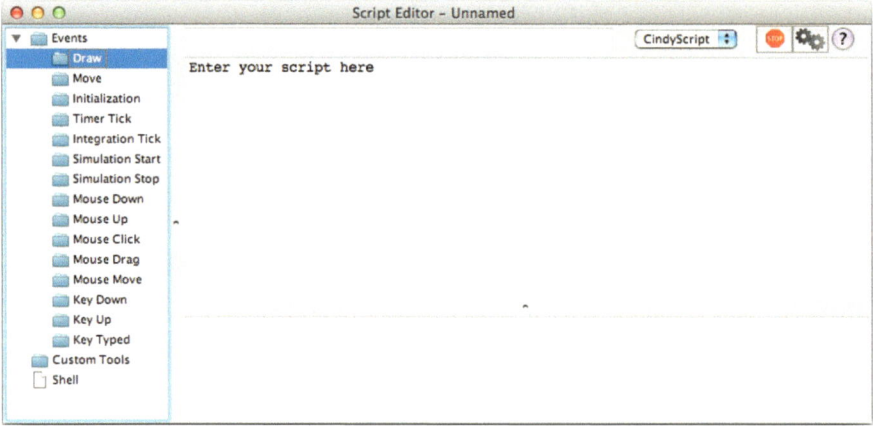

*Entering scripts for the draw occasion.*

The edit area will display the message "Enter your script here". Click there, and start entering text.

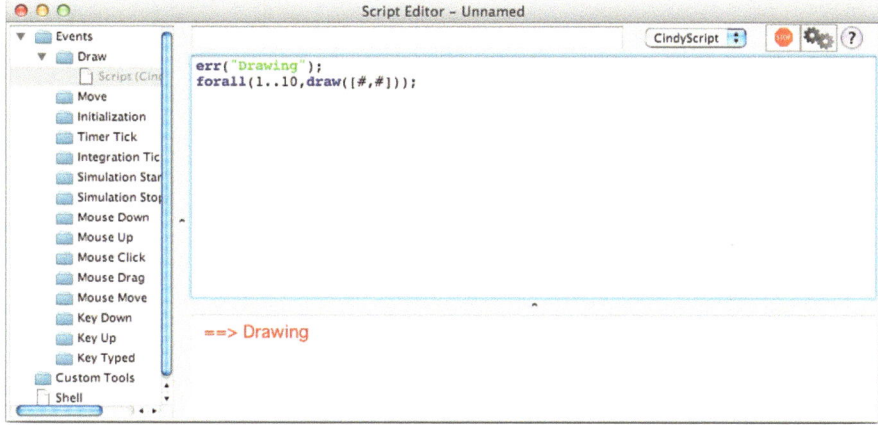

In the view you will notice a diagonal of green points that were created by the script.

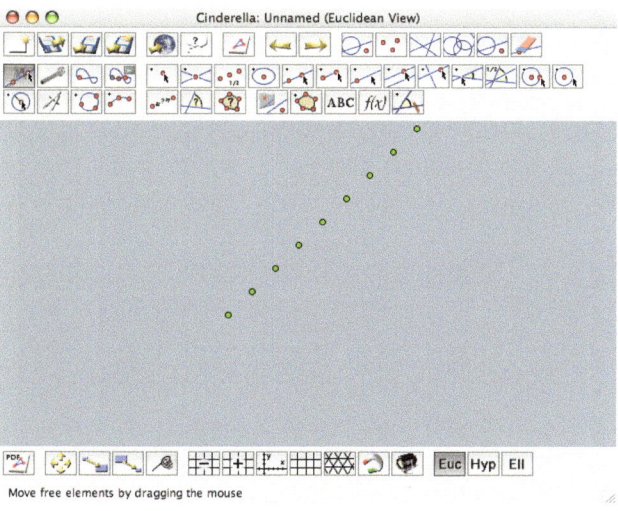

You find an overview over all occasions in the introduction to CindyScript (p. 219).

## 15.7.2  The Shell

You can also enter CindyScript (p. 219) commands and have them executed imme-diately. Just choose the "Shell" item from the left panel, and type the commands into the text area on the right. Pressing shift+enter will execute the command you typed, and you will see the in- and output in the lower text area. You can use shift-up and shift-down to scroll through a history of commands entered.

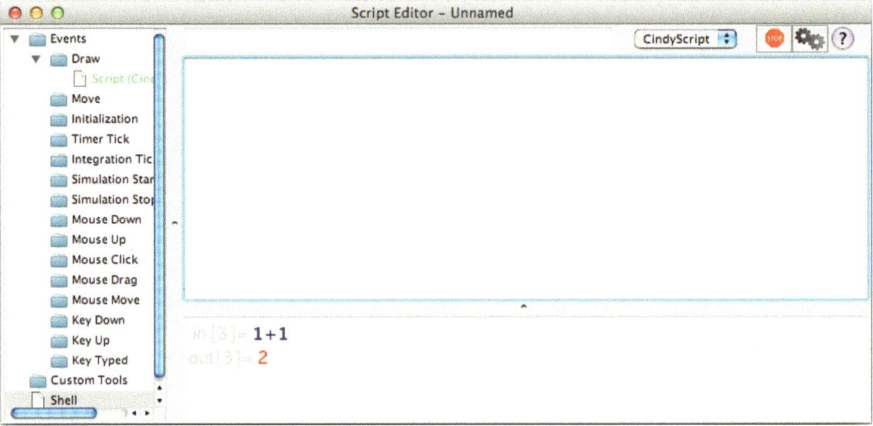

### 15.7.3  The Command Line

Sometimes it is very convenient to manipulate constructions using CindyScript commands. You can either do this using the shell window as described above, or, for short commands, you can use the command line integrated into the construction view. To enable it, choose "Scripting/Command Line" from the menu, or press the corresponding keyboard shortcut Ctrl-Enter (Windows and Unix) or CMD-Enter (Mac OS X).

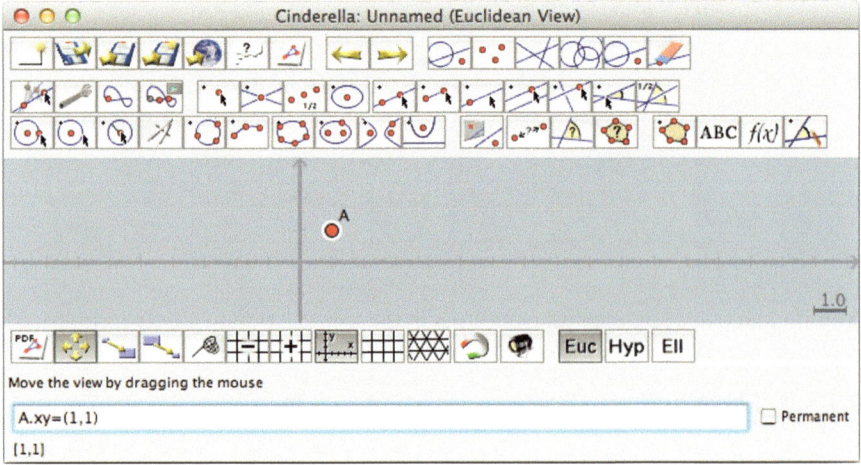

*Using the command line to move a point*

The command line is located below the status line. You can enter arbitrary CindyScript functions and evaluate them by pressing the enter key. If you press shift-enter, the code will be evaluated in the same way, but the command line text field will not be cleared. You can use this in case you want to issue several similar commands or if you look for the correct syntax by trial-and-error.

If you check the *permanent*-checkbox to the right of the command line, then the code will be stored in the draw occasion and thus will be evaluated whenever the construction is redrawn. You can access the code by opening the script editor as explained above.

As a quick example we show how to move a point *C* permanently to the midpoint of two points *A* and *B*: Draw the three points, activate the command line, enter `C.xy=(A+B)/2`, and check the *permanent* checkbox.

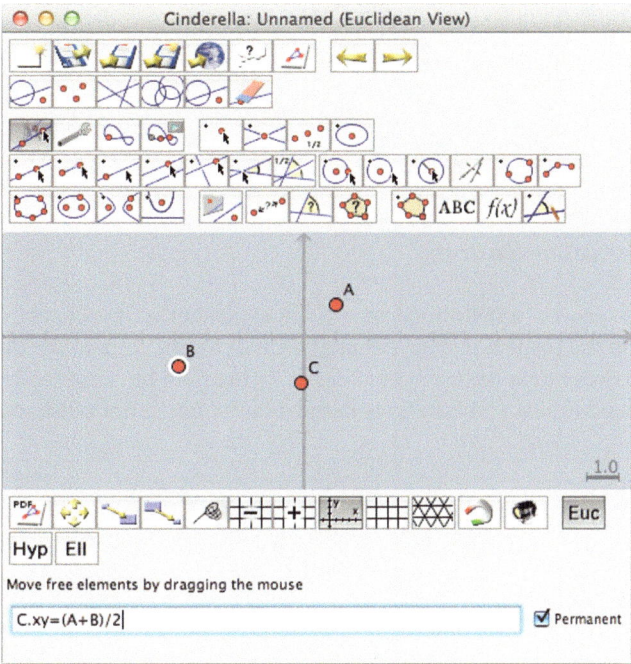

*Using the permanent checkbox*

After pressing enter, *C* will move to the midpoint of *A* and *B* and stays there, even if *A* or *B* moves. Checking the Script Editor reveals the automatically generated draw occasion script named CommandLine.

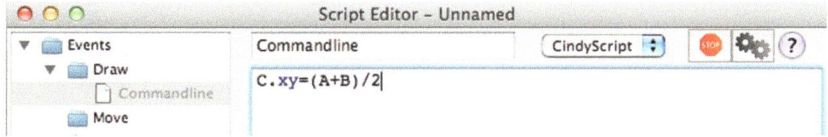

*The automatically generated command line script*

### 15.7.4 CindyScript and the Inspector

Many text input fields of the Inspector (p. 153) window accept CindyScript code as input. The script will only be evaluated once – if you want to make permanent changes you have to use either the command line or the draw occasion in the Script Editor. After pressing enter, you can still see your CindyScript (p. 219) code, but if the input field looses the input focus its value will be replaced with the evaluation result.

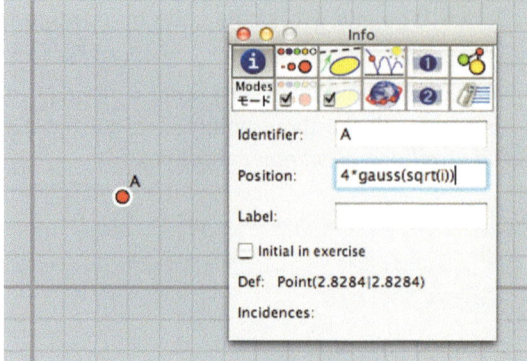

## 15.7.5  Clickable Buttons

Text objects can be transformed into a clickable button by checking the *Use as button* box in the inspector. You can attach CindyScript code to the button that will be executed every time the user clicks on it. As the field for script code is very small we recommend to just call functions defined in the *init* script of the construction.

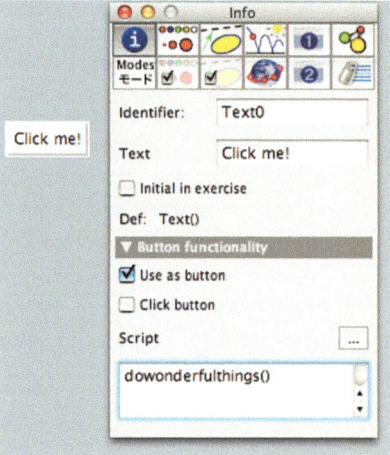

## 15.7.6  Other Languages

You can choose the programming language used to interpret a script using the choice box in the top panel. Available languages are CindyScript (p. 219), Python, JRuby and CDY, the internal language that is used to store constructions. However, currently we only support CindyScript, and this is also the only language you can use in Cinderella applets (p. 419).

# Chapter 16

# Tiny Code Examples

In this section we give a very short demonstration of different CindyScript programming techniques. Each example presents a small task and a CindyScript implementation that handles it.

### Center of Gravity

Setting a point $D$ to the center of gravity of three other points. Changing its color on a condition.

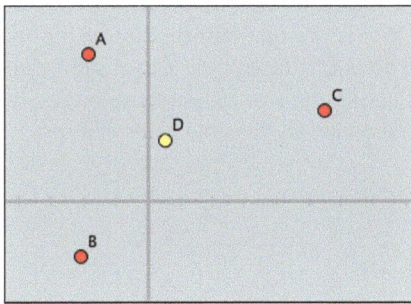

```
//in draw
D.xy=(A+B+C)/3;
D.color=if(D.x>0,
    (1,1,0),
    (0,0,1)
)
```

The code is straight forward. Observe that for setting the coordinate of $D$ the assignment is made to `D.xy`. The conditional at the end returns color values depending on the position of $D$.

**Selecting and Clustering**

Separating points above and below the *x*-axis, and drawing the two clusters.

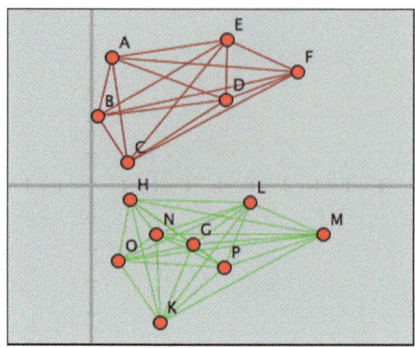

```
//in draw
pts=allpoints();
above=select(pts,p,p.y>0);
below=pts--above;
segs=pairs(above);
drawall(segs,color->(0.6,0,0));
segs=pairs(below);
drawall(segs,color->(0,0.6,0))
```

The points above the *x*-axis are distinguished by using the `select` operator with a suitable condition. The points below the *x* axes are calculated by using a set difference. The lines collecting the two clusters can be easily calculated by using the `pairs` operator that calculates all possible pairs in a set.

## Computing and Drawing a Convex Hull

Calculate the convex hull of all points in a drawing. Observe the use of the ~> operator to deal with numeric instabilities.

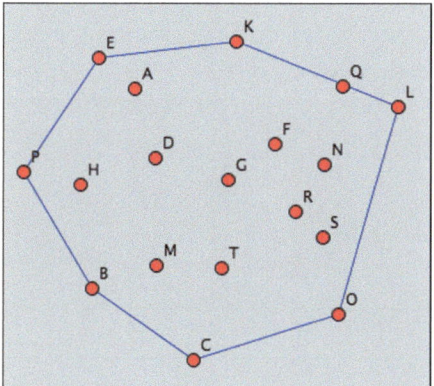

```
//in draw
pts=allpoints();
leftof(A,B):=select(pts,p,area(A,B,p)~>0);
rightof(A,B):=select(pts,p,area(A,B,p)~<0);

isedge(A,B):=(leftof(A,B)==[]%rightof(A,B)==[]);
segments=pairs(pts);
hull=select(segments,seg,isedge(seg_1,seg_2));
drawall(hull);
```

Calculating the convex hull here is performed by a very nice geometric principle: a segment in the convex hull has all the other points on one of its sides. Here two functions `leftof` and `rightof` are defined that separate the two sides of a segment. Based on this a predicate `isedge` is defined, that test whether a pair of points forms an edge of the convex hull. (Notice the usage of the fuzzy comparison `~<0` to avoid numerical instability). Finally, all edges of the convex hull are selected and drawn.

**Creating an Analogue Clock**

Create a clock that shows hours, minutes and seconds. For this, draw a few arrows whose endpoints have appropriate names and enter the following script.

```
//in timerstep
t=time();

p(x):=(sin(2*pi*x),cos(2*pi*x));
B.xy=p(t_3/60)*4;
C.xy=p(t_2/60)*5;
D.xy=p((t_1*60+t_2)/(12*60))*3.5;

apply(1..12,i,draw(p(i/12)*5));
apply(1..60,i,draw(p(i/60)*5,size->1));

drawtext((3,5),t);
```

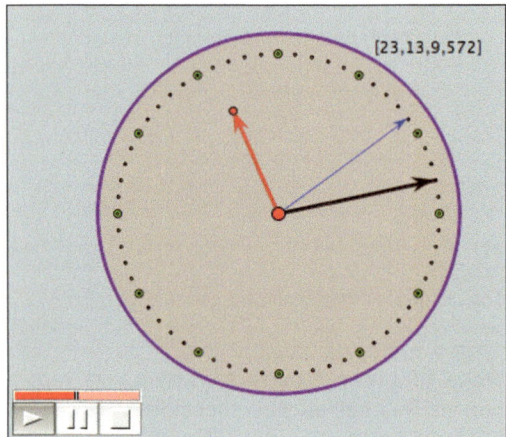

The dots are drawn by the script. The circle was drawn using a geometric object.

This clock takes advantage of the fact that the `time` operator gives access to the system time of the computer. The calculations are done in a way that the second hand jumps, while the other two hand seem to move contineously.

**Closest Point**

Mark the point closest to *A* by a large green dot.

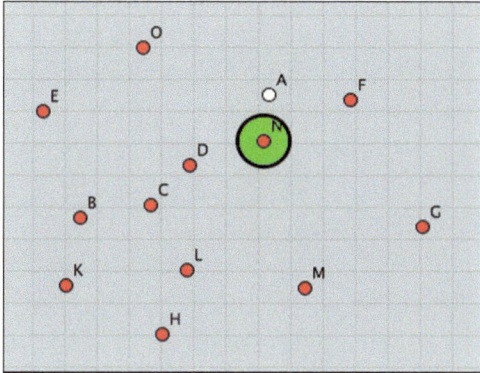

```
//in draw
pts=allpoints();
s=sort(pts,|#-A|);
p=s_2;
draw(p,size->20);
```

Observe that we have to use the second closest point in the script, since *A* is closest to itself. Another solution would be to remove *A* from the list before sorting it:

```
//in draw
pts=allpoints()--[A];
s=sort(pts,|#-A|);
p=s_1;
draw(p,size->20);
```

**Simple Ornament**

Draw a simple ornament by moving points. The color of the strokes should resemble the color of the points.

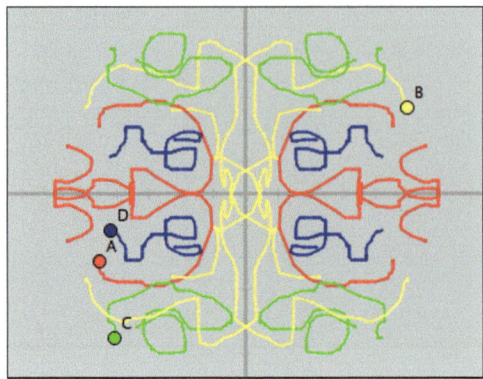

```
//in draw
forall(pts,p,
   p:"trace"=p:"trace" ++ [p.xy]
);

tr0=[[1,0],[0,1]];
tr1=[[-1,0],[0,1]];
tr2=[[-1,0],[0,-1]];
tr3=[[1,0],[0,-1]];

trs=[tr0,tr1,tr2,tr3];

forall(trs,t,
 forall(pts,p,
   connect((p:"trace")*t,color->p.color,size->2);
 );
);

//in init
pts=allpoints();
forall(pts,p,p:"trace"=[]);
```

Observe that each point stores its own trace in the key variable `"trace"`. The transformations are performed by two-dimensional matrix multiplications. By changing the values (and number) of the matrices one can easily crete much more complicated ornaments.

**Linear Regression**

Calculate and draw the line of linear regression to all points. Mark the squares corresponding to the underlying least square approximation.

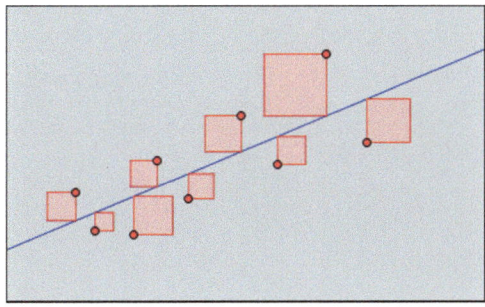

```
//in draw

//Least-square line
pts=allpoints();
m=apply(pts,(1,#.x));
y=apply(pts,#.y);
ma=transpose(m)*m;
mb=transpose(m)*y;
mainv=inverse(ma);
v=mainv*mb;
f(x):=v_2*x+v_1;
plot(f(x));

//Draw the squares

sq(x,y1,y2):=(
    d=y2-y1;
    p=((x,y1),(x,y2),(x+d,y2),(x+d,y1),(x,y1));
    drawpoly(p,color->(1,0.5,0.5),alpha->0.4);
    connect(p,color->(.8,0,0));
);

forall(pts,sq(#.x,#.y,(f(#.x))));
```

The code makes use of high level matrix calculations to find the linear regression line. The calculation follows the standard procedure for this calculation given in a basic course on linear algebra. By almost literally the same code it is even possible to calculate approximations (least square) by a whole set of base functions.

**Sunflower**

Create a nice version of the sunflower applet from the tutorial.

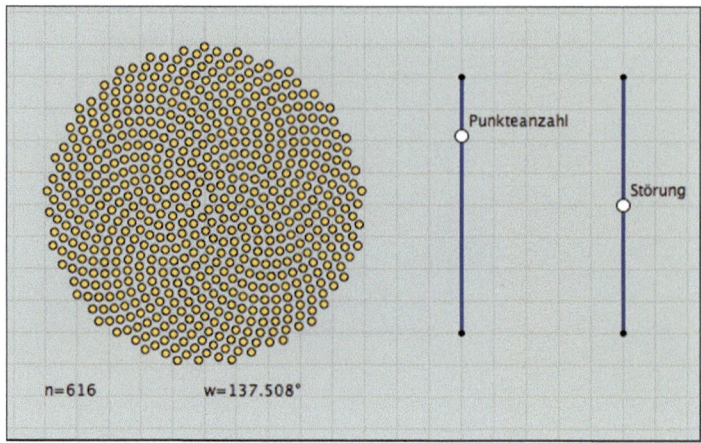

```
//in draw
n=round(100*|B,E|);
d=0.01*(|D,F|/|D,C|-.5);
ang=137.508°+d;
repeat(n,i,
  w=ang*i;
  r=sqrt(i)*.2;
  p=(sin(w),cos(w))*r;
  draw(p,color->(1,0.8,0));
);
drawtext((-5,-6),"n="+n);
drawtext((0,-6),"w="+180*ang/pi¡+"");
```

This code relies on the effect that a very certain angle $ang=137.508°$ produces a dense packing under the growth behavior of a sunflower. The example also illustrates that disturbing this angle a bit immediately leads to very prominent spirals, where the packing is not very isotropic.

### Function Plotter

Implement a function plotter with a freely enterable function and adjustable parameters.

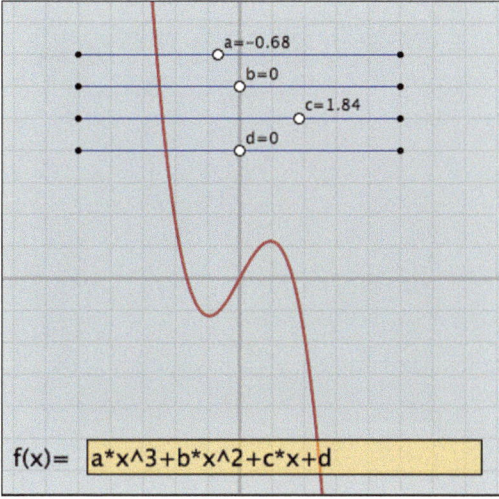

```
//in draw

//Parameter sliders
a=K.x;
b=L.x;
c=M.x;
d=N.x;
drawtext(K+(0.2,0.2),"a="+a);
drawtext(L+(0.2,0.2),"b="+b);
drawtext(M+(0.2,0.2),"c="+c);
drawtext(N+(0.2,0.2),"d="+d);

//Parse and plot
f(x):=parse(Text0.val);
plot(f(x),color->red(0.6),size->2);
```

The function is entered in a Cinderella input field called *Text0*. The text in this input field is directly parsed. The parameters are taken from the geometrically drawn sliders.

**Color Mixing**

Simulate additive color mixing of three light bulbs (red, green and blue). The lamps are be represented by geometric points *A*, *B* and *C*.

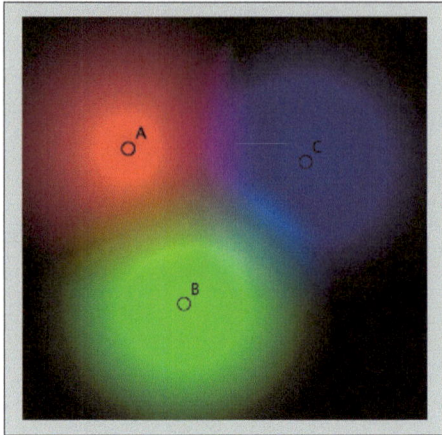

```
//in draw

colorplot(
  (1-|#,A|/4,
   1-|#,B|/4,
   1-|#,C|/4),
  (-5,-5),(5,5),pxlres->3
)
```

This color mixer is indeed extremely simple. It makes use of the fact that one can directly calculate the slots of an RGB color vector. The `colorplot` command does essentially all the rest of the work.

## Julia Sets

Calculate and draw the Julia set, a well known fractal, determined by the complex number represented by the point $C$.

```
//in draw

g(z,c):=z^2+c;

julia(z):=(
  iter=0;
  while(iter<100 & |z|<2,
    z=g(z,complex(C.xy));
    iter=iter+1;
  );
  1-iter/100;
);

colorplot(julia(complex(#))
   ,A,B,startres->16,pxlres->1);
```

Again using the `colorplot` command is the most powerful approach to this task. The function `julia` above iteratively calculates the color of each pixel. Using `startres` and `pxlres` ensures a fluent user experience.

**Swarm Simulation**

Simulate a swarm of fish. Each fish should try to swim in the same direction as its neighbors and try to swim to the approximate place where the neighbors are. All fish should avoid an obstacle given by a point $U$. To try this create some masses and make sure that one is labelled $U$ or change the according line in the script. Also build a fence with bouncers around the "aquarium."

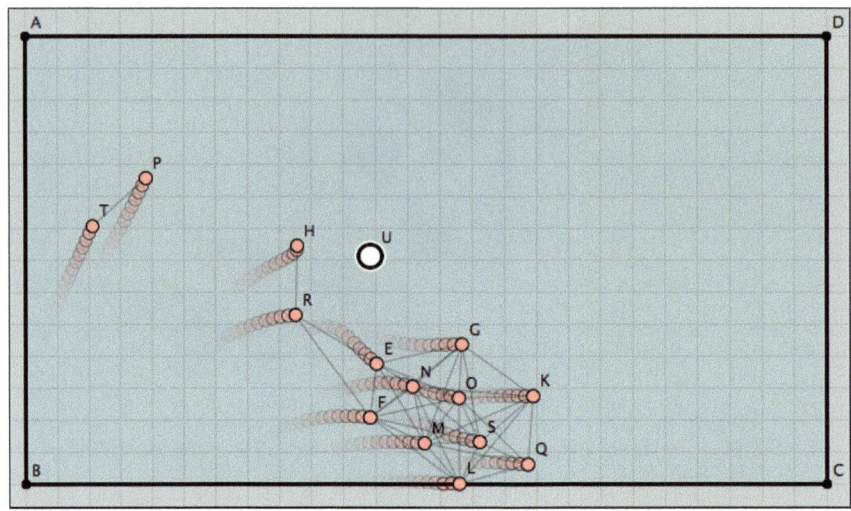

```
//in draw

ms=allmasses()--[U];
lim=0.5;
apply(ms,m,
  near=select(ms,p,|p-m|<4);
  avg=sum(near,m,m.v)/length(near);
  m.v=m.v+.2*(avg);
  if(|m.v|>lim,m.v=lim*m.v/|m.v|);
  //Draw connections
  apply(near,draw(#,m,color->(0,0,0),alpha->0.2));
);
```

Here all fish (as well as the obstacle) are modeled by masses with a positive charge. By this they push each other away. The script makes corrections to the velocity of each point. The behavior generated by this very simple algorithm is amazingly rich and looks quite convincing.

# Part V
# Advanced Topics

# Chapter 17

# Copy, Paste and Macros

The macro concept of Cinderella differs a little from macros available in other geometry software. We describe ways of using the macro capabilities below.

## 17.1 Copy/Paste

The easiest way to create a macro on the fly is to use the copy/paste actions. If you select several elements using the Move Mode (p. 71) and choose Edit/Copy from the menu (or the equivalent keyboard shortcut), Cinderella will copy the selected elements to the clip board. Additionally, intermediate construction steps that define the relations between the selected elements will be copied as hidden elements. When you paste the elements, these hidden elements will be pasted as well, and as a consequence the pasted elements will behave as they did before.

Here is an example. Start by using Add a Circle (p. 81) twice to create two points A and B and two circles that use A and B as center and B and A as point on the circle.

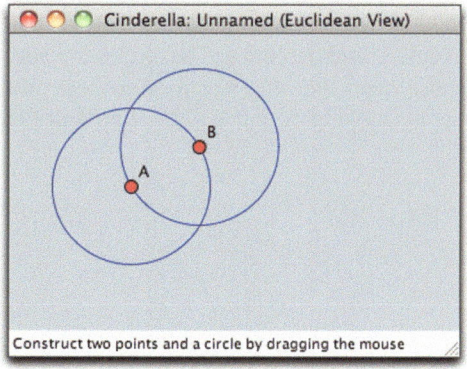

Next, use the Add a Line (p. 76) tool to add the perpendicular bisector of A and B.

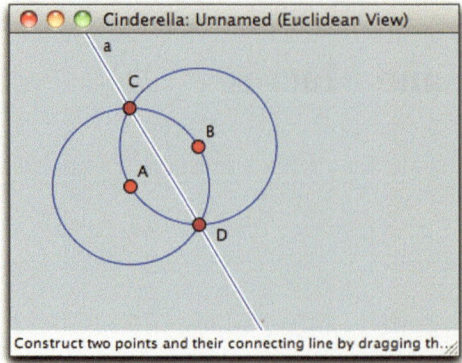

Now, switch to Move Mode (p. 71), press shift to add to the selection, and click A and B.

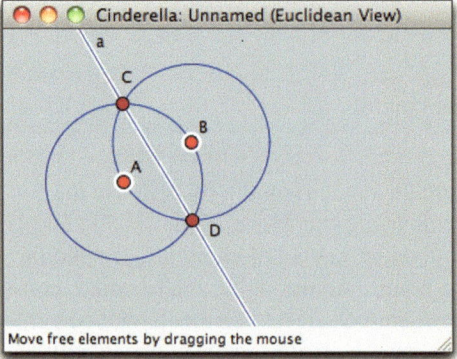

The three selected elements, A, B and the line a, can now be copied to the clipboard using the menu or the keyboard shortcut (CMD-C on the Mac or Ctrl-C on Windows).

Open a new Cinderella window (CMD-N or Ctrl-N) and paste the clipboard data. You will see points A and B and the line a. The circles are added as hidden elements – you cannot see or use them, but they are used to determine the position of line a. You can verify this by moving A and B – the line will behave as perpendicular bisector of A and B.

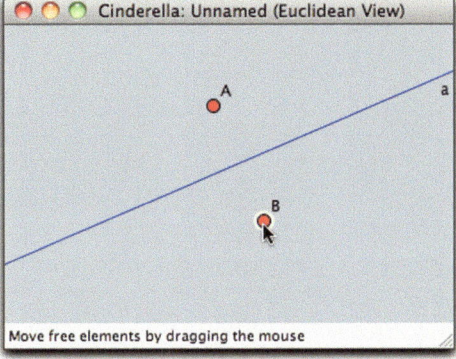

## 17.2  Redefine

What if you want to create the three perpendicular bisectors of a triangle? The solution above shows you how to speed up the creation of three perpendicular bisectors, but pasting the bisector construction three times is not satisfactory:

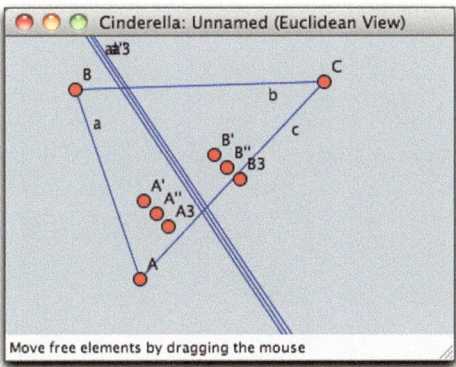

Note that you end up with three bisectors of three points of pairs (it's nice to know, though, that the pasted points have been renamed and moved a little). This is the time when the Redefine Point  mode is useful. Switch to this mode and move $A'$ to $A$, $B'$ to $B$, $B''$ to $C$, $A_3$ to $B$ and $B_3$ to $C$. The pasted points will be identified with the existing triangle points and you will end up with a proper construction containing only the triangle and its edge bisectors.

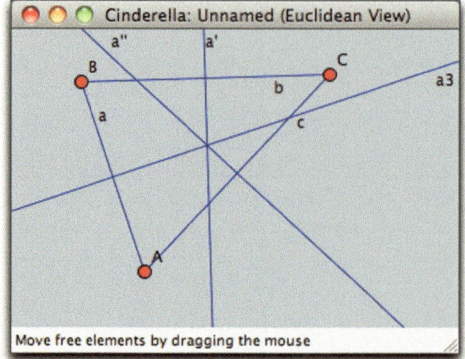

## 17.3  Creating own Tools

Working with copy/paste addresses most cases of repeating steps in a construction. If you want to use several partial constructions, you can also use a convenient way to create a new tool: Just use "Edit/Create Tool from Selection" after you selected the elements that define the new tool. Using the bisector example from above, we get the following dialog, where we can add a name and description for this tool. The icon is created automatically!

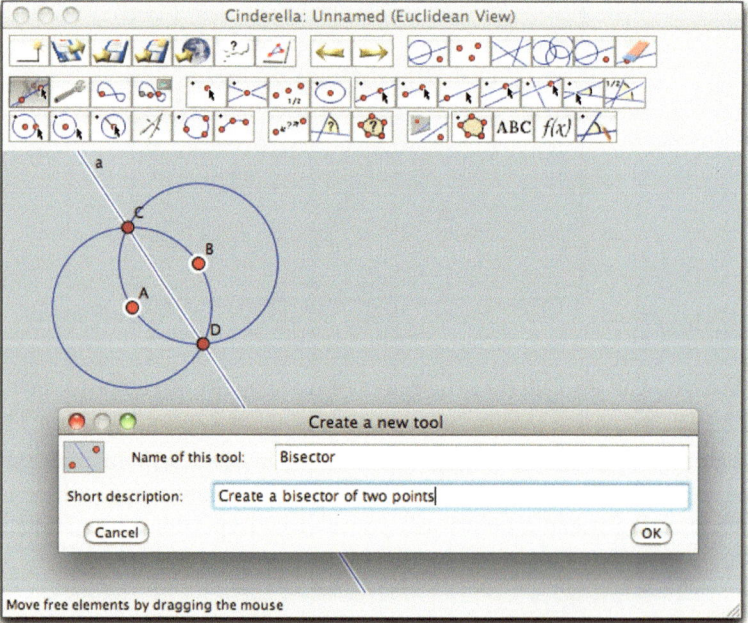

After you confirm the creation of the tool, it will be available in the toolbar. The tools you created can be saved to a file and loaded in other constructions using the corresponding menu entries in the Scripting menu.

# Chapter 18

# Creating Interactive Webpages

An exciting feature of Cinderella is its ability to create interactive webpages. You can publish any construction, even those using several views, within seconds, and without further knowledge about HTML.

This chapter explains the three scenarios for export: Plain examples, animations and construction exercises. You also find detailed information on the exported HTML code and instructions for post-processing of the web pages, e.g. adding explanatory text.

## 18.1 Glossary

If you are familiar with the World Wide Web and its technical background or if you do not want to bother with technicalities right now you can safely skip this section. As an aid for the further description we want to explain a few of the terms used below.

*HTML* is the page description language (or format) for web pages. It can be created and edited with any plain text editor, but it is much more convenient to use a special HTML editor. You can view the HTML code of a web page with the "view source" option in your web browser.

The HTML code mainly consist of the text which will be shown, enhanced by *tags* which describe the appearance and structure of this text. Here is an example:

```
<p>This is a paragraph containing some <b>bold text</b>
and some <i>italics text</i>.</p>
<p>This picture <img src="pappos.gif"> was produced
 with <a href="http://www.cinderella.de">Cinderella</a>!</p>
```

This fragment describes two paragraphs marked by <p>...</p>. The first paragraph contains two regions which should be typeset with special fonts, the first one, marked by ..., in bold, the second one in italics. You might recognize the easy structure of HTML, most of the elements are marked with opening (<something>) and closing tags (</something>).

The second paragraph in the example shows how to *include an image* using the tag. This tag does not have a closing companion, but it uses an option (src=...) to reference the image file. You also find a *hyperlink*, which is the most powerful element in HTML. You name a location which can be reached by clicking on the phrase included by ....

Hyperlinks are usually given by an URL, an uniform resource locator. These describe resources by the protocols which give access to them. For the WWW most communication is done with the Hypertext Transfer Protocol, short for http. This explains the http:-part of the WWW addresses, which can be left out since most browsers assume it as a default. Other examples are ftp, the File Transfer Protocol, or the prefix file: which describes files that reside locally on your harddisk.

A special tag is reserved for *Java integration* into web pages. Whenever a Java compatible browser encounters an applet tag it tries to load the java program referenced by the code option and runs it inside a rectangle on the web page. The size of the rectangle is given by the width and height options. These programs are called *Applets*, as opposed to standalone applications. The diminutive is a little misleading, since applets can be as powerful as full applications.

The applet tag can also contain an archive option which describes the location of the java code. For Cinderella we provide an archive called cindyrun.jar which contains all the code needed for showing and manipulating constructions. This is an example of how the applet tag produced by Cinderella's web export functions could look like:

```
<applet code    = "de.cinderella.CindyApplet"
        archive = "cindyrun.jar"
        width   = 435
        height  = 231>
<param...
</applet>
```

You find many <param> tags which pass additional parameters like the filename of the construction to the applet. Never change or delete these parameters without exactly knowing what they mean.

## 18.2 Exporting Plain Examples

This is the easiest way to create a webpage with Cinderella. Whenever you have created a configuration you can use the export button to create an interactive webpage showing this construction in exactly the views that you are currently using. Each view will be a separate applet, and these applets communicate using a kernel ID which you find as a parameter of the applet.

The construction itself is not saved in the HTML code, but in a separate file with the extension ".cdy". Whenever you create a web page you are prompted to save your file, if you have not done so before. The applet expects the file in the same directory as the html file of the web page. All this is automized by the Export dialog that pops up when you press the export button.

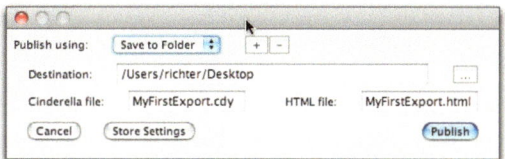

*Export dialog*

This dialog window asks for file names for the Cinderella file and the corresponding HTML file. The HTML file should end in ".html" or ".htm", depending on your local standards. If you do not supply one of these extensions, Cinderella will assume ".html" as a default. The dialog also makes sure that the Cinderella runtime library *cindyrun2.jar* is available in your export folder. This step finishes the web export, and you should be able to view the result in a Java-1.6 compatible web browser.

The runtime library "cindyrun2.jar" contains the necessary code to show and manipulate constructions. It is essential to have this file at the same place were your construction and HTML files reside. You also find this file in the installation directory so that you can copy it into the directory containing the interactive web page, if necessary for some reasons.

If you experience any problems, be sure to check whether:

- The file ending in ".htm" or ".html" exists and is readable.

- The file mentioned as value in <param name=filename> section of the html file exists and is readable (you should be able to load it with Cinderella)

- The file "cindyrun2.jar" is present in the same directory as the two files above and is referenced in the archive option of the applet tag.

- You are using a Java-1.6 compatible browser. We recommend using the most recent version of Internet Explorer, Firefox, Safari or similar browsers.

The exported construction is always in move mode. That means that movable elements can be dragged around within the applet rectangle. If you want to prevent elements from being movable, please use the pinning option in the Inspector (p. 153).

## 18.3 Exporting Animations

Exporting automatic animations is as easy as exporting interactive examples. Unlike earlier versions of Cinderella, Cinderella.2 does not distinguish between animations and ordinary constructions. Using the "Autostart Animations" option available in the Inspector (p. 153) you can make sure that an animation is automatically running when a user visits your page. It is also possible to hide the animation control panel in the applet by checking the corresponding option in the Inspector (p. 153)

While exporting animations please keep in mind that the potential visitors of your web page might have slower computers than you. You should adjust the animation speed accordingly.

## 18.4 Creating Interactive Exercises

The creation of interactive Exercises in the form it was available in Cinderella 1.4 is no longer present in the current release. However there are possibilities to create similar (and even more flexible) exercises also with Cinderella.2.6. For an instruction of how to do this please consult the section Interactive Exercises (p. 433).

## 18.5 Post-Processing

The web site containing your construction or exercise is very basic. Cinderella does not try to be a full-featured web editor. You can use any other web editor for post-processing the HTML files.

The width and height parameters of the applets can be changed, if you want to. We recommend that the view sizes should be set to the correct size before you export the construction.

Never change the kernelID parameter of the applet, it is important for inter-applet communication. The order and placement of the applets can be changed arbitrarily. You can also merge two different HTML pages, which gives you the possibility of showing different constructions on one web page.

If you have several Cinderella-enabled pages on your web site you can use a single *cindyrun2.jar* for all these. Then you will have to change the archive parameter of all your applets to reflect the location of your central *cindyrun2.jar*. It should be sufficient to include the complete URL of its location in the archive tag.

## 18.6 Legal Issues

Whether you are using the free version of Cinderella or you have purchased a pro license, we allow the non-commercial redistribution of the cindyrun2.jar runtime necessary for embedding Cinderella applets on web pages. The exact terms are available in the software license available on our website.

Basically, it all boils down to the fact that you should not try to make money out of your examples, that is, sell them or put them onto a commercial online service. You can certainly use them for teaching (even if you get paid for this). If you want to publish a book or CD-ROM which uses Cinderella, you should contact the Cinderella authors to get a written permission.

Whenever you are unsure about your license, please contact us in writing or via eMail. You can find contact details on the Cinderella website.

# Chapter 19

# Scribbling

Cinderella contains a specialized mode for using dynamic geometry on pen-driven devices, in particular interactive whiteboards. We call this mode *Scribbling*. In Scribble Mode, you just draw using the pen and Cinderella attempts to recognize what you intended to do with the stroke. Cinderella interprets every *scribble* that you do – a scribble means your pen movement from the moment you put it down until you pull it up again. While you are scribbling, Cinderella gives you a visual feedback in the upper center of the view on what would happen if you end the stroke right then.

## 19.1 Creating Elements

One class of scribbles are those that create new elements into the construction. The meaning of a scribble may be influenced by the elements that are selected when you start the scribble. The following elements can be created:

### 19.1.1 Points

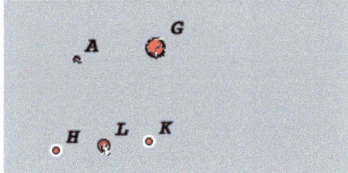

Draw a small scribble to create a Free Point. The size of the new point will be similar to the size of the scribble you drew. You can draw a point-scribble over an existing point to change its size. Preselect two points and draw a small scribble in the middle of those to create a Midpoint.

### *19.1.2  Lines*

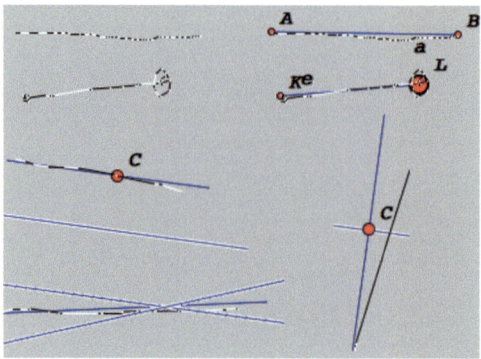

Draw a line to create a *line through two points*. The endpoints are created if they do not exist. Short scribbles create lines that are cut off at the endpoints; scribbles over 80% of the window are not. Preselect an existing line and draw a scribble parallel or orthogonal to it to create a *Parallel* or *Orthogonal*. Preselect two lines and draw an *Angular Bisector* to create one.

### *19.1.3  Polylines*

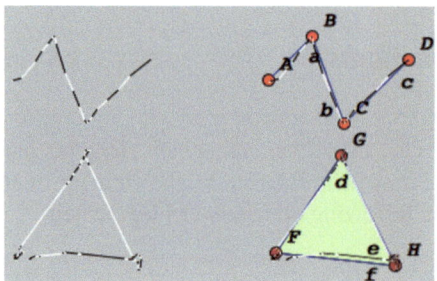

Draw multiple lines to create a sequence of segments. If the last point is near the first point, a *Polygon* is created.

### *19.1.4  Circles*

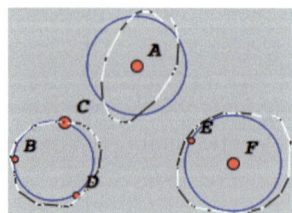

Draw a circle to create a *Circle By Radius*. A midpoint is created if it does not exist. Preselect a midpoint for greater tolerance. If the scribble passes exactly 3 existing points to create a *Circle By 3 Points*. Pass exactly 1 point to create a *Circle By Midpoint and Border Point*.

## 19.2  Gestures

Gestures are those scribbles that do not create a new element but influence those that are already there.

### 19.2.1  Moving and Selecting

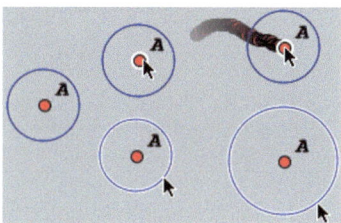

Tap an object once to select or deselect it. Tap in an empty space to deselect all. Move a selected object by dragging it. This applies to *Free Points* as well as other free elements, such as *Circle by Radius*.

### 19.2.2  Changing the Name

Tap an element twice within a short time period to open a dialog that allows you to change the element name.

### 19.2.3  Undo and Redo

Draw a horizontal, straight, quick scribble over 30% of the window width to the left to undo the last action; to the right to redo an action.

### 19.2.4  Delete All

Draw a quick, large gesture, crossing out all of the construction, to delete all elements.

### 19.2.5  Inspect

A straight line down, then back up to the startpoint, opens the Inspector.

### *19.2.6  Right Click*

A straight line up, then back down simulates a right mouse button click. Usually, this brings up the context menu; this can be adjusted in the *Modes / Language*-Tab of the Inspector.

## 19.3  Customizing the Scribble Mode

You can customize the scribble mode. In the *Info* tab of the Inspector, a *Settings...* button will appear when you have selected Scribble mode:

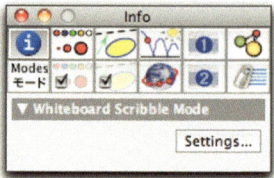

Press this button to bring up the Scribble Mode configuration interface. It has three tabs: *Recognize, Forms* and *Gestures*. In the *Recognize* tab, you can enable and disable the various recognition modes to facilitate the recognizing of the remaining modes:

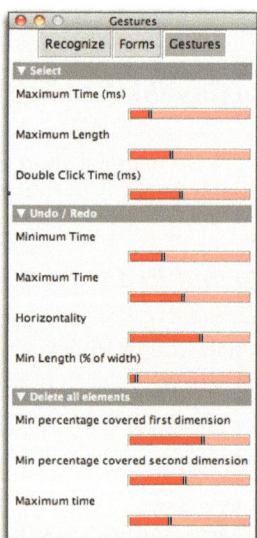

The other two tabs let you set various parameters of the recognition modes. You can use this to fine-tune Scribble Mode to your particular hardware device; the settings you do here are remembered across different Cinderella sessions on the same computer.

# Chapter 20

# Extensions

There are several ways to extend the functionality of Cinderella. Other scripting languages can be used, plugins can be written, applets can communicate with Javascript, etc.

This section gives a brief overview of some of these possibilities. Some of these features are still subject to change at the time of this writing. Thus we recommend to refer to the online manual as a reliable reference for these features.

## 20.1 Plugins

With CindyScript it is possible to enhance and customize the behavior of Cinderella in an almost unlimited variety of ways. Still it may be desirable from time to time to be able to use self-written Java code in connection with Cinderella. The Cinderella Plugin Structure gives a way to handle such situations. There may be several reasons that make it desirable to use a plugin. Here are some of them

- *Performance:* For some tasks CindyScript may simply be too slow, so that one wants to use some custom optimized code.

- *Native libraries:* Sometimes one wants to use native libraries (like for instance the JOGL library for dealing with 3D graphics). In these cases a plugin may be used to link Cinderella to this library.

- *Licensing issues:* Plugins make it possible to write and to offer extensions for Cinderella that run under a specifically chosen license (for instance GNU GPL). We will make use of this ourselves in the future and publish open source extensions to Cinderella. These plugins will be available for download soon under....

- *Third-party code:* Plugins also form a good basis to include Java code written by others into a Cinderella project.

At the time of this writing, Plugins are only available in the standalone version of Cinderella and are not applicable in applets. This will be fixed in the near future.

### 20.1.1 Plugin Architecture

For detailed descriptions of the plugin architecture see our website at
http://cinderella.de/plugins. There you may also find example plugins as well as
suitable make files to create them. Here we only describe the workflow very roughly.

A plugin itself is a `.jar` file that contains executable Java classes. The core link
of the plugin is a Java file (compiled to a class file) that exports all the functionality of the plugin to Cinderella. Within Cinderella, the plugin can be accessed using CindyScript. The core file of the plugin must extend the Java class
`CindyScriptPlugin`. This parent class is available through the `cindy2.jar` file
in the Cinderella application.

The code for a plugin may typically look as follows

```java
import de.cinderella.api.cs.CindyScript;
import de.cinderella.api.cs.CindyScriptPlugin;
import java.awt.*;
import java.util.ArrayList;
import java.util.Arrays;


public class ExamplePlugin extends CindyScriptPlugin {

    public String getName() {
        return "Example Plugin";
    }

    public String getAuthor() {
        return "Ulrich Kortenkamp and Juergen Richter-Gebert";
    }

    @CindyScript("sayHello")
    public String testFunction() {
        return "Hello from Plugin";
    }

    @CindyScript("square")
    public double quadrieren(double x) {
        return x * x;
    }

    @CindyScript("grayvalue")
    public double getGray(Color c) {
        return (c.getBlue() + c.getRed() + c.getGreen()) / 3.;
    }

    @CindyScript("testarray")
    public String writeArray(ArrayList<Double> al) {
        return Arrays.toString(al.toArray());
    }

}
```

The fragments given by code like `@CindyScript("square")` declare the name
under which the function is accessible in CindyScript. The plugin can be used in
CindyScript after it has been loaded with the `use` function:

**Loading a plugin: `use(<string>)`**

*Description:* This function loads a plugin with the name given as argument. It is good practice to call this function in the initialization slot of CindyScript.

So the above plugin may be used within CindyScript as follows:

```
use("ExamplePlugin");
println(sayHello());
println(square(4));
println(grayvalue((0.7,0.4,0.1)));
println(testarray([1,2,3,4,5]));
```

This will create the output

```
Hello from plugin
16
102.3333
[1.0, 2.0, 3.0, 4.0, 5.0]
```

Observe that all class casts are performed automatically. For a detailed description of the casting rules see the online documentation.

## 20.1.2 Available Plugins

At the time of this writing there are only few plugins available. We publish all user contributed plugins on our website at http://cinderella.de/plugins, so please check there for current versions. Here we briefly present two of the them.

### 20.1.2.1 Cindy to LegoNXT Communication

This plugin written by Michael Schmid provides a high-level interface to Lego Mindstorms NXT block. With this plugin all inputs and outputs of the Lego computer can be accessed and controlled. Thus it is possible to use Cinderella as a controller for Lego robots.

Among others this plugin implements statements like `nxtforward(...)`, `nxtturnright(...)`, or `nxtgetlight()`. Using these statements it is for istance directly possible to use a CindyScript program to influence the behavior of a robot car that is sensible to light changes.

You can download the plugin along with a brief documentation from our webpage.

### 20.1.2.2 Cindy3D

Another project is the use of the JOGL 3D interface as a 3D output device for Cinderella. A Plugin called *Cindy3D* designed by Matthias Reitinger, Jürgen Richter-Gebert and Jan Sommer will be specialized to this task. Using this plugin it will be possible to use 3D drawing functions directly from CindyScript. There will be functions for drawing points, lines, circles, polygons and three-dimensional meshes. The following three pictures which were generated via CindyScript give a rough impression of the possibilities of the plugin.

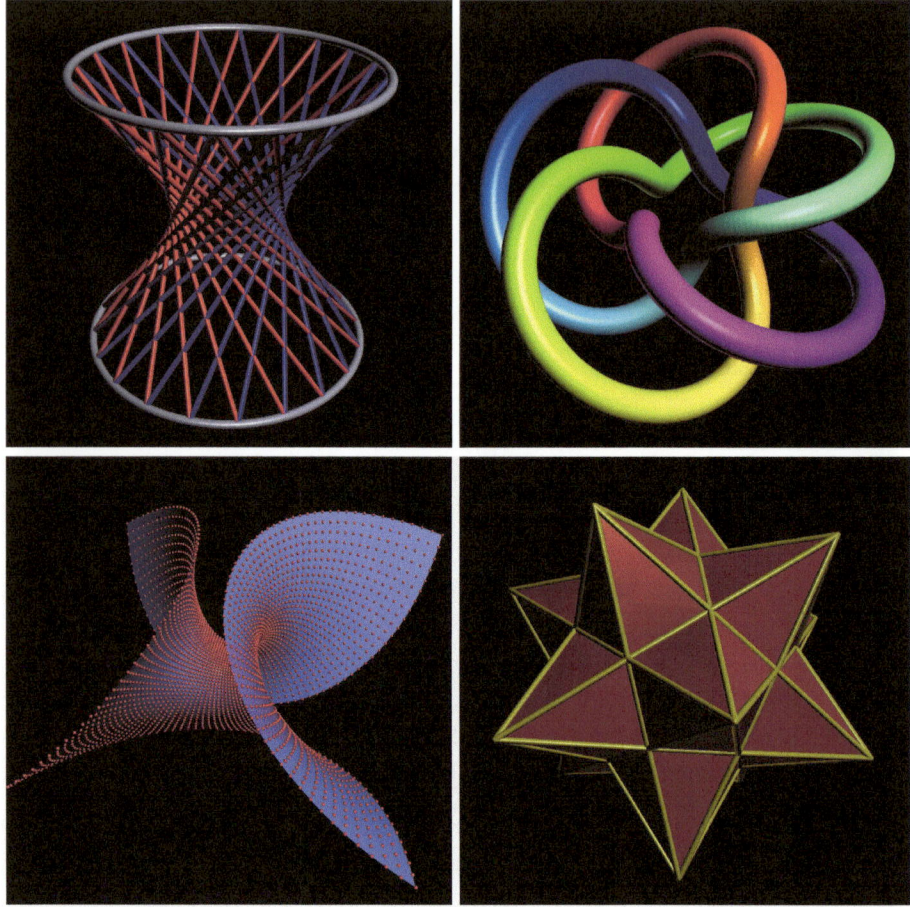

## 20.2 JavaScript

Whenever you use Cinderella as an applet in an internet browser you can link Cinderella and your HTML page via JavaScript. The communication is bidirectional. The Cinderella construction can notify the HTML page of events that occur, and the construction can react to events on the HTML page like a button being pressed.

There are many possible scenarios in which one would like to use the JavaScript extensions. Here are a few of them.

- *Control by HTML buttons:* In many demonstration applets it may be good style to trigger a demonstration by pressing a button in the explanatory text. Such a button could move a construction to a special configuration, for example.

- *Explain parts of the construction:* The mouseover functionality of JavaScript may be used to link certain explanations in the HTML text to events in the applet.

- *Notification:* It may be useful to notify your HTML text of events happening in Cinderella. Thus you may display alert windows or change the text on your HTML page in order to explain something.

### 20.2.1 Calling Cinderella from JavaScript

Basically, you can perform CindyScript commands from JavaScript. By these operations it is for instance possible to

- call functions in CindyScript,
- change the values of variables in CindyScript,
- move elements of your construction, or
- change the appearance of elements.

To that end it is important that your Cinderella applet has an associated name that is used as handle for JavaScript commands. The name is given in the applet tag of your HTML code using the name parameter, for instance with `name="CindyJSDemo"`. Using this handle, JavaScript can call an arbitrary CindyScript statement via the function `doCindyScript(<statement>)`. The statement is given as a string. This string will be parsed and executed whenever the function `doCindyScript` is called. The following piece of code exemplifies this procedure. It shows how to provide four buttons in the HTML text that manage to place a point $D$ in various geometrically interesting positions with respect to $A$, $B$ and $C$.

```
<script language="JavaScript" type="text/javascript">
   function doScript(c) { document.CindyJSDemo.doCindyScript(c);};

</script>


<input type="button" value="Middle of AB"
       onclick="doScript('D.xy=(A+B)/2');" />
<input type="button" value="Middle of AC"
       onclick="doScript('D.xy=(A+C)/2');" />
<input type="button" value="Middle of BC"
       onclick="doScript('D.xy=(B+C)/2');" />
<input type="button" value="Middle of ABC"
       onclick="doScript('D.xy=(A+B+C)/3');" />
<br>

<applet    name="CindyJSDemo"
           code    = "de.cinderella.CindyApplet"
           archive = "cindyrun2.jar"
           width   = 680
           height  = 336>

... applet specifications ...

</applet>
```

In the Cinderella construction only the four points $A,...,D$ are needed to use this JavaScript code.

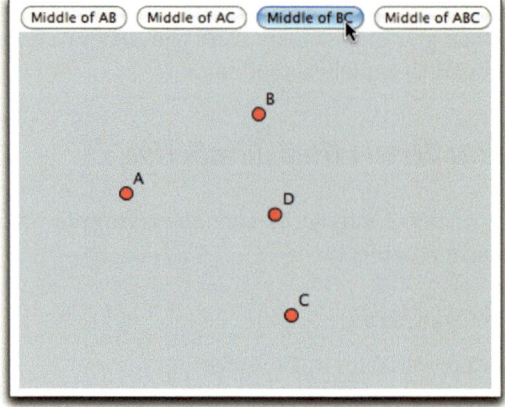

Since the CindyScript calls may be arbitrarily complex this technique may be used for many different purposes. By using a `repaint` statement within the call one can even trigger smooth transitions of a construction from one position to a new one. Unfortunately, in JavaScript, a string is not allowed to contain a usual `newline` character. So it may be the case that complicated script calls look a bit cluttered. One can bypass this problem by subdividing a script into smaller strings. The following piece of code shows the code for a button that smoothly moves the points *A*, *B* and *C* to the positions `[0,0]`, `[4,0]` and `[2,3]`.

```
<input type="button" value="Smooth move of ABC" onclick="doScript(
'oldpos=[A.xy,B.xy,C.xy];'+
'newpos=[[0,0],[4,0],[2,3]];'+
'n=40;'+
'repeat(n,i,'+
'  l=i/n;'+
'  pos=l*newpos+(1-l)*oldpos;'+
'  A.xy=pos_1;'+
'  B.xy=pos_2;'+
'  C.xy=pos_3;'+
'  repaint();'+
'  wait(10);'+
');'
);" />
```

**javascript**

### Calling JavaScript from Cinderella: `javascript(<string>)`

Conversely, it is also possible to call JavaScript from CindyScript. This can simply be done by using the CindyScript function `javascript(...)`. The following piece of code shows how to raise a JavaScript alert whenever point *A* gets to close to the origin (a rather useless piece of code, by the way).

```
if(|A,[0,0]|<1,
  javascript("alert('A is too close to the origin')");
);
```

## 20.3 Interactive Exercises

### 20.3.1 A Little Cinderella History

One of the main features of the older versions of Cinderella (up to 1.4) was the ability to easily create interactive student exercises in elementary geometry. A typical such exercise may for instance ask for *the construction of the midpoint* of two given points. The student was given an interactive web page in which just the two points (say *A* and *B*) are given along with a certain set of construction tools to solve the exercise. While he solved the exercise, the automatic prover in Cinderella monitored his activities and provided useful hints and comments. Finally, when the student had successfully mastered the construction task he was prompted a success message.

Unfortunately, this functionality was no longer available as of version 2.0. There were several reasons why we discontinued this feature. One of them was our goal to to merge student exercises and scripting facilities to support more advanced and flexible exercise formats. Another reason was that only a small fraction of users actually used the possibilities of creating exercises.

With the current version we again provide a possibility to create interactive exercises by including the operator `createtool(...)` (p. 392). This and the constant monitoring of the construction by the integrated proving engine is the key to interactive exercises with automatic checking of students' answers. In contrast to the old version of Cinderella there is no longer an exercise editor, but the exercise must be produced by writing a suitable script. Below we show that this not only provides the old functionality for exercises, but also adds the power and versatility of CindyScript.

In this section we will describe the basic techniques that are necessary to create such exercises, along with some technical explanations. We will do this for a very simple example and start from scratch. On our website at http://cinderella.de we provide templates that may serve as starting point for several typical types of exercises. They will make the generation of exercises quite easy.

### 20.3.2 Exporting Tools

Our task is to generate an exercise that asks for the construction of the *midpoint of two given points A and B*. The student should be asked to solve this by using lines, points and circles. The first task is to create a toolbar with the essential tools for the exercise. For this, we use the `createtool(...)` operator as described earlier (p. 392). We want to provide a small collection of tools essential for the construction. To achieve this, we add the following statement in the *init* block of CindyScript:

```
createtool(["Move","Point","Line","Circle"],2,2);
```

We also add the two points manually that should serve as the start of the exercise. Now our construction looks as follows:

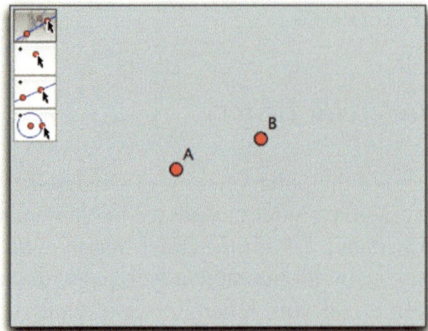

### 20.3.3 Providing a Sample Construction

As a next step we provide a sample construction for the exercise (*construct the midpoint*). This sample construction will later be used as a source to provide hints for the student. The construction is done manually using the construction tools. We also use the Inspector (p. 153) to make the construction look a little nicer. After this the construction may look as follows.

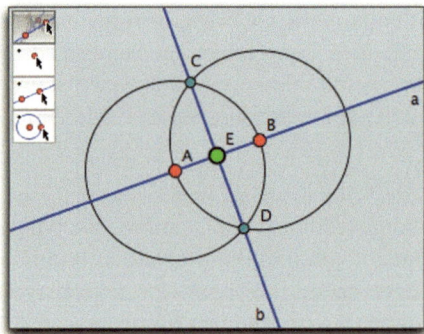

### 20.3.4 Adding the Exercise Text

As a next step we are going to display a text that asks the student to solve the exercise. We will use this text later on also to give some helpful messages while the student is actually solving the exercise. So we will separate the actual message from the statement that displays it. There are several ways to do this, here we will in *init* enter a line

```
message="Construct the Midpoint of A and B";
```

In draw we enter a line like

```
drawtext((-8,0),message);
```

Going back to the construction our example looks like this:

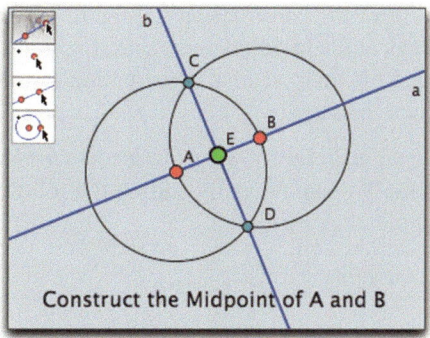

Construct the Midpoint of A and B

### 20.3.5 Hiding the Sample Construction

As mentioned before, the sample construction will only serve as a template that is compared to the student's solution. So we have to hide this part of the construction, except for the starting points. We could either do this manually by using the Inspector (p. 153). Alternatively we can add a line of script to the *init* block that makes all elements but *A* and *B* invisible automatically.

```
apply(allelements()--[A,B],el,el.visible=false);
```

After this the exercise is in perfect shape for the student:

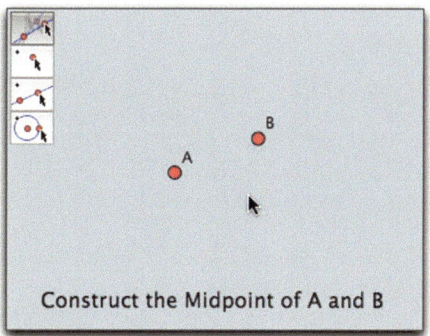

Construct the Midpoint of A and B

### 20.3.6 Providing a Success Message

Now comes the tricky part. We want the student gets a success message when he solved the exercise. For this you have to know the following important fact about Cinderella: Whenever an element is constructed that is already present then Cinderella's proving engine will make sure that the old element is actually re-used and no additional superfluous double element is added. This mechanism even works if the old element was *invisible*. In this case the visibility flag of the old element is simply turned back the `true` state. In our case, if the student constructs the midpoint of *A* and *B* in some correct way, the prover will make sure that the old invisible point

*E* that is already present in the construction will be re-used. As the prover accepts all points that are always at the same position as *E*, regardless of their construction, the student may provide his own creative construction for the midpoint and still be awarded with the success message.

So the only thing we have to do for showing the success message is to check whether point *E* becomes visible. We can do so by adding the following line of code to the *draw* script before the message is printed:

```
if(E.visible,message="YEAH, you got it")
```

If the student manages to construct the midpoint (by any construction!) then automatically point *E* will become visible and the message `"YEAH, you got it"` will be shown.

### *20.3.7 Giving Intermediate Comments*

We can do even better. We can use the sample construction to provide a message whenever the student constructs one of the auxiliary elements of it. For this, we will replace the `if(...)` statement of the last section by a little program that tests which elements of our construction are already visible and adjusts the message accordingly. Here is a code example demonstrating how to do that (we provide the entire code for the *draw* section.)

```
clrscr();
chain=[
 [a,"Good start"],
 [C0,"You need circles"],
 [C1,"You need circles"],
 [b,"Almost done"],
 [E,"YEAH, you got it"]
];

visible=select(allelements(),#.visible);
found=select(chain,el,contains(visible,el_1));

if(found!=[],
  message=found_(-1)_2;
);

drawtext((-8,0),message);
```

In the list `chain` we store pairs of elements and the associated texts that should be shown if the element is constructed. The lines

```
visible=select(allelements(),#.visible);
found=select(chain,el,contains(visi,el_1));
```

collect all elements of `chain` that are already visible in the list `found`. Finally, the following lines of code take the last element of this list (the one that was most recently constructed) and stores the associated comment in `message`:

```
if(found!=[],
  message=found_(-1)_2;
);
```

This message is then shown by:

```
drawtext((-8,0),message);
```

### 20.3.8 What the Student Gets

If the student now solves the exercise in the standard way he will have the following stages of his construction:

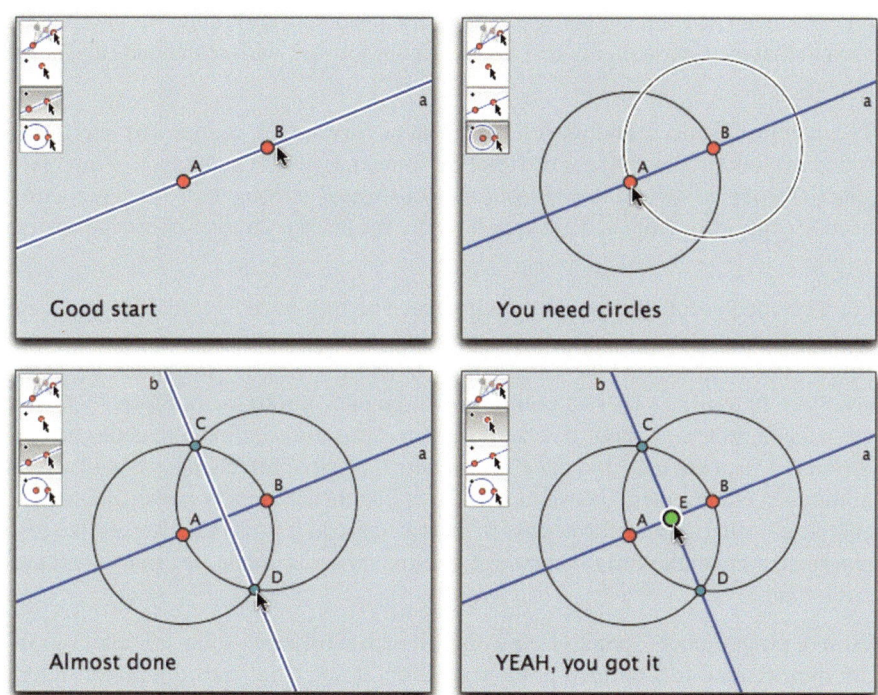

Good start

You need circles

Almost done

YEAH, you got it

### 20.3.9 Exporting

Finally, we will export this exercise to the web as described in the section on HTML Export (p. 419). After this the interactive webpage is easily usable as a general interactive student exercise. There are more elaborate patterns of how to relate the hints to parts of the original construction. On our webpage http://cinderella.de we provide a collection of useful templates, including *hints that help finding the construction*, *timed hint blocking* and *branching hint structure*. They will cover the most common usages of the creation of student exercises.

## 20.4 Multitouch Support

### 20.4.1 Introduction

The computer mouse has been the predominant input device for graphical user interfaces for about 30 years. The interactive component of Cinderella depends on it, as it provides a one-to-one correspondence between the movements of a physical device and a point on the screen. On the other hand, the mouse restricts the gestural expression of a human being to a two-dimensional translation. We have become used to that restriction, but in real life we prefer to work with both hands and several fingers.

Technology advances, and one of the big steps forward are multitouch interfaces as found in modern smartphones or tablet computers. Even the touchpads of laptops are able to recognize several fingers simultaneously, and screens with touch recognition have become affordable and are supported by the latest computer operating systems.

For a geometry tool, multitouch is more than a nice-to-have feature. Mathematically spoken, the interface changes from "translation only" to "multi-transformation capable". For good reasons there is no mode to add a "free line" in Cinderella. Instead you have to use lines by two points (p. 76) or lines through a point (p. 77). If you are using a mouse as input device, only those two tools define a meaningful interaction between the user and the construction. The situation changes drastically in a multitouch environment. Movements of two fingers are already powerful enough to encode a similarity transformation. In such a scenario it is physically possible to use two fingers to get the full interaction for a line: You can rotate and translate it at the same time.

From a programmers' point of view, multitouch introduces some complexity. With the mouse there is a defined pointer position at any time, and the button interaction is always a press-drag-release sequence. This is no longer true for multitouch interaction; obviously there might be any number of current positions (even none, if you don't touch), and the touch-move-release sequences may overlap. As a user, you should not have to care, but if you are going to work with CindyScript, this will be of importance. See [17] for a discussion in more detail.

### 20.4.2 Using Multitouch With Cinderella

Currently, Cinderella supports multitouch events through the community standard TUIO protocol [9], see also http://www.tuio.org. This cross-platform standard has the advantage that there are corresponding drivers for almost every multitouch device. For smartphones and tablets you can find apps on the TUIO home page that will send the multitouch events to your computer.

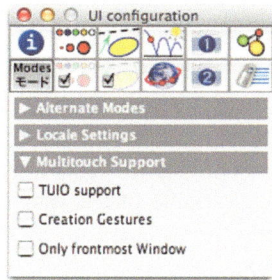

*TUIO settings in the Inspector*

You can enable TUIO support in the mode and language settings of the Inspector (p. 153). Check the TUIO support checkbox and Cinderella will listen on the standard port 3333 for TUIO events in Move mode (p. 71). While TUIO support is on, regular mouse events will be ignored in Move mode to avoid double input if used with standard TUIO drivers on Windows or Mac OS X. Other modes of Cinderella still require mouse operation and do not support TUIO events. If you check *Creation gestures*, you can create points and lines by touching the plane with one or two fingers. As this can lead to a lot of unwanted points and lines this option is unchecked by default and should be considered experimental.

The *Only frontmost window* option should be checked or unchecked depending on the device you use for multitouch gestures. If the device displays the screen contents it usually should be unchecked, as the touches on the screen are mapped to the individual windows of Cinderella automatically. If the device does not have a screen or does not show the Cinderella windows, as it is the case with touchpads on laptop computers or TUIO apps on mobile phones, it might be better to check this option. In that case all events on the touch surface are mapped to the visible view area of the frontmost window of Cinderella.

### 20.4.3 *Programming Multitouch With CindyScript*

Multitouch input for geometry software is still a research area, and besides the obvious behavior in move mode, there are no user interaction standards yet. While we will work on the multitouch integration in other modes for future updates of Cinderella, we also decided to make the multitouch events accessible to CindyScript.

If you enable TUIO support in the Inspector, the mouse event slots *Mouse Down*, *Mouse Up* and *Mouse Drag* will be triggered by TUIO events instead of regular mouse events. The `mouse` function (p. 391) will report the location of the touch event. The `mouse()` location is uniquely defined as every touch event reports about one of the multiple touches only. This implies that you cannot rely on the fact that successive mouse events belong to the same gesture. This is a programmer's nightmare: How can you keep track of the different touches and assign them to the right actions? For your convenience, CindyScript includes the necessary tool, as we describe below.

**mtlocal**       **Declaring touch-local variables: `mtlocal(<var1>,<var2>,...)`**

*Description:* All variables declared to be mtlocal are local variables for a touch sequence. This function is only useful in the *Mouse Down* slot. If a variable is defined as touch local, then for all touch sequences a unique instance of that variable is created. If you use the same variable in the *Mouse Drag* slot or *Mouse Up* slot you will work with the instance of that variable that is assigned to the touch-move-release sequence.

*Example:* We will illustrate the use of the `mtlocal` function with a small example. By placing the following code in the *Draw*, *Mouse Down*, *Mouse Up* and *Mouse Drag* slots you can draw polygons withs the mouse. While the mouse is pressed, the new polygon will be drawn in red, when the mouse is released, the new polygon will be added to a list of polygons that is drawn in blue.

```
// In Draw:
forall(polygons,drawpoly(#));

// In Mouse Down:
polygon=[mouse()];

// In Mouse Drag:
polygon=polygon++[mouse()];
drawpoly(polygon,color->red(1));

// In Mouse Up:
polygons=polygons++[polygon];
```

This code relies on the fact that a polygon will be created first in *Mouse Down*, then points will be added in *Mouse Drag*, and finally the polygon is complete and will be added to the list of polygons in *Mouse Up*.

When the same code runs in a multi touch environment, the (global) variable `polygon` will be used for all touches. Also, if any touch is released, the new polygon is added to the list of polygons, but still other drag events will add points to it. The code is not multitouch-compatible and will badly fail.

By adding a single call to `mtlocal` we can fix the code:

```
// In Draw:
forall(polygons,drawpoly(#));

// In Mouse Down:
mtlocal(polygon); // make sure every touch gets its own polygon
polygon=[mouse()];

// In Mouse Drag:
polygon=polygon++[mouse()];
drawpoly(polygon,color->red(1));

// In Mouse Up:
polygons=polygons++[polygon];
```

Now a new polygon will be created for each touch, and the drag and up events will use the correct polygon variable that is associated to the gesture.

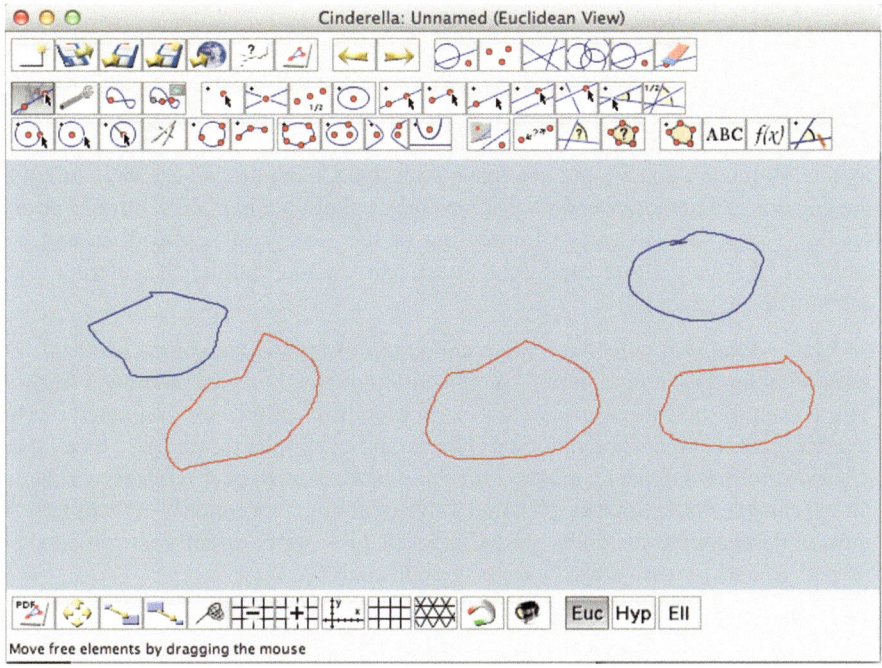

*Drawing three polygons at once using multitouch*

If you try this code example, you will see that the repainting of the red polygons does not work as expected due to the automatic clearing of the screen at every event. We can fix that by explicitly keeping a list of new polygons and drawing them in the draw event only. You find the final code below.

```
// In Draw:

forall(polygons,drawpoly(#));
forall(newpolygons,drawpoly(#,color->red(1)));

// In Mouse Down:
mtlocal(polygon);
polygon=[mouse()];
newpolygons=newpolygons++[polygon];

// In Mouse Up:
newpolygons=newpolygons--[polygon];
polygons=polygons++[polygon];

// In Mouse Drag:
newpolygons=newpolygons--[polygon];
polygon=polygon++[mouse()];
newpolygons=newpolygons++[polygon];
```

Alternatively, you could use layers and the autoclear flag (p. 344) to achieve the same effect without storing all the new polygons.

## 20.5 Mathe-Vital

While reading this manual you may have experienced Cinderella as a powerful tool for doing experiments and constructions in Mathematics, Physics and other related topics. Moreover, you should also have seen that Cinderella is very well suited for the creation of interactive content for web pages about mathematics, physics or even music. If you want to get an impression of the variety of applications and possibilities of Cinderella created content we recommend visiting the internet portal www.mathe-vital.de.

At Mathe-Vital you will find a huge collection of interactive course material. The word *Vital* here is an bareviation for *Visual interactive tools for advanced learning*. The collection not only covers classical topics from university mathematics (like *linear algebra*, *calculus* or *geometry*) but also more popular topics like *The relations of mathematics and music*, *mathematics and plant growth* or *fractals*. Over the last three years the collection has grown to a size of all together roughly 500 applets. For most of these applets the underlying Cinderella files can be directly downloaded and used as a basis for own constructions or code analysis.

The collection in Mathe-Vital is organized in courses, modules and applets. Each course has roughly 12 to 30 modules. Each module contains between 2 and 7 applets. The screenshot of module icons shown below gives a rough impression of the variety of topics and graphical illustration techniques that are covered.

*A collection of Mathe-Vital module icons*

## 20.6 Visage: Visualization of Graph Algorithms

The Visage extension of Cinderella has it origins in a project started in 2005 in the DFG research center MATHEON in Berlin.

We will give only a brief overview over the basic elements of Visage here. You can find more information about Visage (mostly in German language) on its webpage at http://cinderella.de/visage, currently hosted by the Centre for Educational Research in Mathematics Education (CERMAT). We also refer to [7, 6] for more information about how you can use Visage in teaching.

### 20.6.1 Starting Visage

You can access Visage from the Scripting menu. Choosing the Visage menu item will open a window for choosing the basic graph algorithm underlying your explorations.

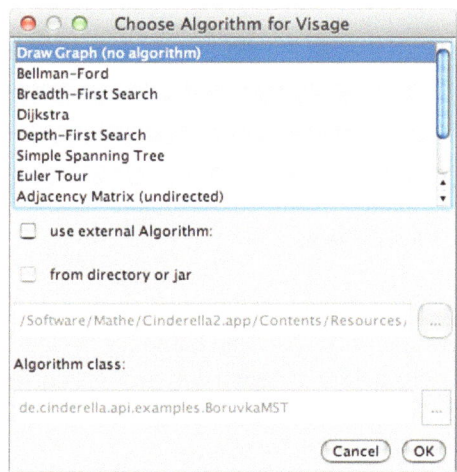

*Choosing the underlying algorithm for Visage*

For now, choose the "Draw Graph" algorithm. The Cinderella toolbar will change to a Visage-specific toolbar that contains tools for drawing vertices and edges.

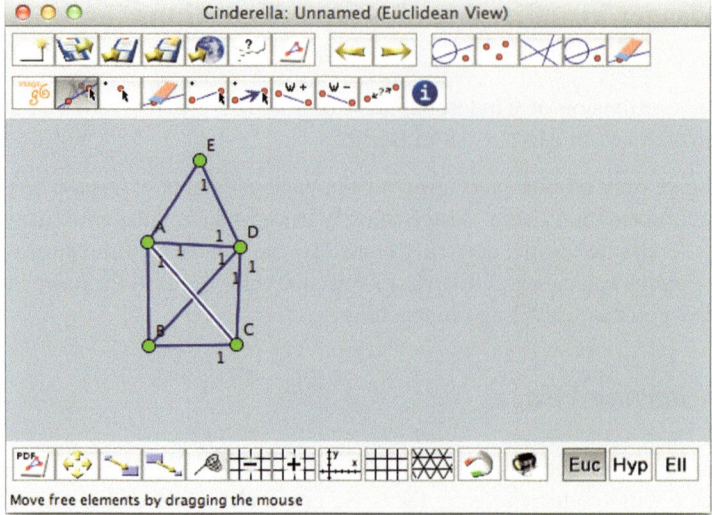

*A famous graph*

You can draw undirected or directed and unweighted or weighted graphs. Usually, the chosen graph algorithms governs the available types. You can always choose a new algorithm (and its corresponding toolbar) using the ![button] button, and, if applicable, the existing graph will be reused.

## 20.6.2 Running Algorithms

Visage comes with several built-in algorithms that can be applied to graphs. By default the algorithms run in step-by-step mode, as Visage is not meant for high-performance calculations but for visualization in educational settings. Pressing the play button ![play] will execute one step each time. The rewind button ![rewind] reinitializes the algorithm and you can start it again.

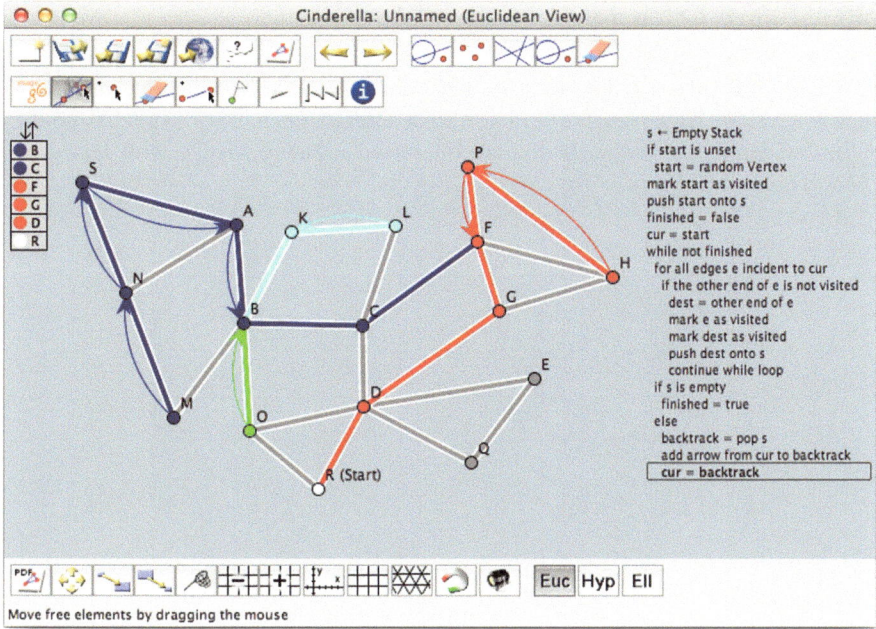

*Visualizing Depth First Search*

### 20.6.3  Interaction With CindyScript

You can write graph algorithms also in CindyScript. On the Visage website you find a small library that makes it easy to access vertices and edges of a graph. This library relies on user attributes of geometric elements. These user attributes are accessible via two CindyScript commands:

**Set a user attribute: `attribute(<geo>,<string1>,<string2>)`**                    `attribute`

*Description:* Sets the user attribute of <geo> identified by <string1> to the value <string2>.

**Read a user attribute: `attribute(<geo>,<string>)`**                              `attribute`

*Description:* Returns the user attribute identified by <string> of the geometric element <geo> .

The Visage extension sets the attribute `"edge"` of undirected edges to `"u"` and of directed edges to `"d"`. The attribute `"vertex"` of vertices is set to "v". Thus, you can identify the vertices and edges of a graph by checking these attributes:

```
alledges():=select(allsegments(),
                  attribute(#,"edge")=="u" % attribute(#,"edge")=="v");
allvertices():=select(allpoints(),attribute(#,"vertex")=="v");
```

If you mark a vertex as a start or end vertex for a graph algorithm it will have the attribute `"specialvertex"` with a value of either `"start"` or `"finish"`. Finally, weighted edges store their weight in the user attribute `"weight"`. For flow algorithms the capacity and flow are stored in the user attributes `"capacity"` and `"flow"`, as expected.

### *20.6.4 Java Extensions*

Actually, it is also possible to write algorithms in Java that interact with the Visage extension. However, the API for such interaction is only available online at http://cinderella.de/visage. If you are interested in using Visage with Java we ask you to contact us at `visage@cinderella.de`.

# Index

# References

1. Bell, E.T.: Men of Mathematics. Touchstone Books, New York (1986 (orig. 1945))
2. Bell, E.T.: The Development of Mathematics. Dover Publishing, New York (1992 (orig. 1945))
3. Coxeter, H.S.M.: The Real Projective Plane, 3rd edn. Springer, New York, Berlin (1992 (orig. 1949))
4. Coxeter, H.S.M.: Projective Geometry, 2nd edn. Springer, New York, Berlin (1994 (orig. 1963))
5. Crapo, H., Richter-Gebert, J.: Automatics proving of geometric theorems. In: N. White (ed.) Invariant Methods in Discrete and Computational Geometry. Kluwer Academic Publishers (1995)
6. Fest, A., Kortenkamp, U.: Teaching graph algorithms with visage. Teaching Mathematics and Computer Science **7**(1), 35–50 (2009). URL http://tmcs.math.klte.hu/Contents/2009-Vol-VII-Issue-I.html
7. Geschke, A., Kortenkamp, U., Lutz-Westphal, B., Materlik, D.: Visage – visualization of algorithms in discrete mathematics. Zentralblatt für Didaktik der Mathematik **37**(5), 395–401 (2005)
8. Greenberg, M.J.: Euclidean and non-Euclidean Geometries. Freeman and Company, New York (1996 (orig. 1974))
9. Kaltenbrunner, M., Bovermann, T., Bencina, R., Costanza, E.: Tuio: A protocol for tabletop tangible user interfaces. In: Proceedings of the 6th International Workshop on Gesture in Human-Computer Interaction and Simulation (GW 2005) (2005). URL http://modin.yuri.at/publications/tuio_gw2005.pdf
10. Klein, F.: Vorlesungen über nicht-euklidische Geometrie. Springer, Berlin (1968 (orig. 1928))
11. Klein, F.: Development of Mathematics in the 19th Century. Math. Sci. Press (1979 (orig. 1928))
12. Knuth, D.E.: The TEXbook. Computers & typesetting. Addison-Wesley (1993)
13. Kortenkamp, U.: Foundations of Dynamic Geometry. Dissertation, ETH Zürich, Institut für Theoretische Informatik, Zürich (1999). URL http://kortenkamps.net/papers/1999/diss.pdf
14. Kortenkamp, U.: The future of mathematical software. In: Proceedings of MTCM 2000. Springer-Verlag (2001). URL http://kortenkamps.net/papers/2001/future.pdf
15. Kortenkamp, U.: Making the move: The next version of Cinderella. In: A.M. Cohen, X.S. Gao, N. Takayama (eds.) Proceedings of the First International Congress of Mathematical Software, pp. 208–216. World Scientific (2002). URL http://www.cs.uleth.ca/~wismath/cccg/papers/ulrich.pdf. A slightly modified version appeared in the proceedings of CCCG 02.
16. Kortenkamp, U.: Combining CAS and DGS – Towards Algorithmic Thinking. In: S. Li, D. Wang, J.Z. Zhang (eds.) Symbolic Computation and Education, pp. 150–173. World Scientific (2007)
17. Kortenkamp, U., Dohrmann, C.: User interface design for Dynamic Geometry software. Acta Didactica Napocensia **3**(2), 59–66 (2010). URL http://dppd.ubbcluj.ro/adn/article_3_2_6.pdf
18. Kortenkamp, U., Fest, A.: From CAS/DGS integration to algorithms in educational math software. In: Proceedings of ATCM 08 (2008)

19. Kortenkamp, U., Materlik, D.: Pen-based input of geometric constructions. In: P. Libbrecht (ed.) Proceedings of MathUI 2004 (2004). URL http://kortenkamps.net/papers/2004/Scribbling-article.pdf

20. Kortenkamp, U., Richter-Gebert, J.: Decision complexity in Dynamic Geometry. In: D. Wang (ed.) Proceedings of ADG 2000, no. 2061 in Lecture Notes in Artificial Intelligence, pp. 167–172. Springer-Verlag, Heidelberg (2001). URL http://kortenkamps.net/papers/2001/36_DecisionComplexity.pdf

21. Kortenkamp, U., Richter-Gebert, J.: Blended experimentation with DGS. In: Proceedings of CADGME 2009 (2009)

22. Kortenkamp, U.H., Richter-Gebert, J.: Geometry and education in the Internet age. In: T. Ottmann, I. Tomek (eds.) Ed-Media & Ed-Telecom 98. Proceedings of the Tenth World Conference on Educational Multimedia and Hypermedia & World Conference on Educational Telecommunications, Freiburg, Germany, June 20-25, 1998. AACE, Charlottesville (1998). URL http://www.cinderella.de/papers/geo-i.pdf.gz

23. Kortenkamp, U.H., Richter-Gebert, J.: Dynamic Geometry II: Applications. In: Abstracts 15th European Workshop Comput. Geom., pp. 109–111. INRIA Sophia-Antipolis (1999). URL http://cinderella.de/papers/antibes-2.pdf

24. Laborde, J.M.: Exploring non-Euclidean geometry in a dynamic geometry environment like Cabri-Géomètre. In: J. King, D. Schattschneider (eds.) Geometry Turned On, pp. 185–191. MAA (1997)

25. Richter-Gebert, J.: Mechanical theorem proving in projective geometry. Annals of Mathematics and Artificial Intelligence **13**, 139–172 (1995)

26. Richter-Gebert, J.: Perspectives on Projective Geometry. Springer-Verlag (2011)

27. Richter-Gebert, J., Kortenkamp, U.: Dynamic aspects in computational geometry. In: A. Montes (ed.) Proceedings of the EACA 2000, pp. 51–61. Barcelona (2000)

28. Richter-Gebert, J., Kortenkamp, U.: A dynamic setup for elementary geometry. In: Proceedings of MTCM 2000. Springer-Verlag (2001). URL http://kortenkamps.net/papers/2001/35_DynamicSetup.pdf

29. Richter-Gebert, J., Kortenkamp, U.: The power of scripting: DGS meets programming. Acta Didactica Napocensia **3**(2), 67–78 (2010). URL http://dppd.ubbcluj.ro/adn/article_3_2_7.pdf

30. Richter-Gebert, J., Kortenkamp, U.H.: Dynamic Geometry I: The problem of continuity. In: Abstracts 15th European Workshop Comput. Geom., pp. 51–53. INRIA Sophia-Antipolis (1999). URL http://www.cinderella.de/papers/antibes-1.pdf

31. Richter-Gebert, J., Kortenkamp, U.H.: Complexity issues in Dynamic Geometry. In: F. Cucker, J.M. Rojas (eds.) Foundations of Computational Mathematics (Proceedings of the Smale Fest 2000). World Scientific (2002). Also available as technical report TRB-2000/22, Freie Universität Berlin

32. Richter-Gebert, J., Orendt, T.: Geometriekalküle. Springer-Verlag (2009)

33. Struik, D.J.: A Concise History of Mathematics. Dover Publishing (1987)

34. Yaglom, I.M.: Felix Klein and Sophus Lie – Evolution of the Idea of Symmetry in the Nineteenths Century. Birkhäuser, Boston, Basel (1988)